Children, Death and Burial

Archaeological Discourses

edited by
Eileen Murphy
and
Mélie Le Roy

OXBOW | books
Oxford & Philadelphia

Childhood in the Past Monograph Series

Volume 1: Childhood and Violence in the Western Tradition
           edited by Laurence Brockliss and Heather Montgomery
Volume 2: The Dark Side of childhood in Late Antiquity and the Middle Ages
           edited by Katariina Mustakallio and Christian Laes
Volume 3: Medieval Childhood: Archaeological Approaches
           edited by D. M. Hadley and K. A. Hemer
Volume 4: Children, Spaces and Identity
           edited by Margarita Sánchez Romero, Eva Alarcón García and Gonzalo Aranda Jiménez
Volume 5: Children, Death and Burial: Archaeological Discourses
           edited by Eileen Murphy and Mélie Le Roy

The monograph series was established to allow scholars from all disciplines a forum for presenting new, groundbreaking or challenging research into themed aspects of childhood in the past. The Society is happy to consider proposals for future monographs.

Proposals should be submitted to the General Editor of the Monograph Series. Details for submission may be found on the Society's webpage at https://sscip.wordpress.com/.

Dr Sally Crawford FSA
General Editor, SSCIP Monograph Series
The Institute of Archaeology
36 Beaumont Street
Oxford OX1 2PG
United Kingdom
sally.crawford@arch.ox.ac.uk

For Gabriel (EM)

For Aydan and Tilden (MLR)

Published in the United Kingdom in 2017 by
OXBOW BOOKS
The Old Music Hall, 106–108 Cowley Road, Oxford OX4 1JE

and in the United States by
OXBOW BOOKS
1950 Lawrence Road, Havertown, PA 19083, USA

© Oxbow Books and the individual contributors 2017

Paperback Edition: ISBN 978-1-78570-712-4
Digital Edition: ISBN 978-1-78570-713-1 (epub)

A CIP record for this book is available from the British Library and the Library of Congress

All rights reserved. No part of this book may be reproduced or transmitted in any form or by any means, electronic or mechanical including photocopying, recording or by any information storage and retrieval system, without permission from the publisher in writing.

Printed in Malta by Gutenberg Press Ltd
Typeset in India by Lapiz Digital Services, Chennai

For a complete list of Oxbow titles, please contact:

UNITED KINGDOM	UNITED STATES OF AMERICA
Oxbow Books	Oxbow Books
Telephone (01865) 241249, Fax (01865) 794449	Telephone (800) 791-9354, Fax (610) 853-9146
Email: oxbow@oxbowbooks.com	Email: queries@casemateacademic.com
www.oxbowbooks.com	www.casemateacademic.com/oxbow

Oxbow Books is part of the Casemate Group

Front cover: Painting by Libby Mulqueeny, based on a photograph taken by Belinda Tibbetts of Sk.4406, a 6–12 month old infant, from Çatalhöyük (Fig. 3.3).

Contents

List of Contributors vii

1. Introduction: Archaeological Children, Death and Burial 1
   Eileen Murphy and Mélie Le Roy

2. How Were Infants Considered at Death during the Neolithic Period
   in France? 19
   Mélie Le Roy

3. Perinatal Death and Cultural Buffering in a Neolithic Community
   at Çatalhöyük 35
   Belinda Tibbetts

4. Burying Children and Infants at Kadruka 23: New Insights into Juvenile
   Identity and Disposal of the Dead in the Nubian Neolithic 43
   Emma Maines, Pascal Sellier, Philippe Chambon and Olivier Langlois

5. Children's Burials in the Eneolithic Cemetery of Sultana-Malu Roşu,
   Romania 57
   Catalin Lazar, Ionela Craciunescu, Gabriel Vasile and Mihai Florea

6. Late Chalcolithic Skeletal Remains and Associated Mortuary Practices
   from Çamlıbel Tarlası in Central Anatolia 77
   Jayne-Leigh Thomas

7. Processed Babies: Early Bronze Age Infant Burials from Bulgarian Thrace 91
   Kathleen McSweeney and Krum Bacvarov

8. 'Missing infants': Giving Life to Aspects of Childhood in Mycenaean
   Greece via Intramural Burials 107
   Katerina Kostanti

9. Bronze Age Child Burials in the Southern Trans-Urals (21st–15th
   Centuries cal. BC) 125
   Natalia Berseneva

10. Juvenile Burial and Age as a Social Category in Funerary Contexts of
    Pre- and Protopalatial Crete 147
    Nathalja Calliauw

11. Geto-Dacian Child Burials in the Second Iron Age 163
    Valeriu Sîrbu and Diana-Crina Dăvîncă

12. Out of the Cradle and into the Grave: The Children of Anglo-Saxon Great
    Chesterford, Essex, England 179
    Christine Cave and Marc Oxenham

13. Emotional Act, Superstition or Ritual? – Evidence from Child Burials
    in the Medieval period. A Case Study from St Clemens Churchyard,
    Copenhagen, Denmark 197
    Jane Jark Jensen

14. Interpreting Cultural and Biological Markers of Stress and Status
    in Medieval Subadults from England 211
    Heidi Dawson-Hobbis

15. Atypical Burial Practice and Juvenile Age-at-death in Later Medieval
    Gaelic Ireland: The Evidence from Ballyhanna, Co. Donegal 227
    Eileen M. Murphy

16. Interring the 'Deserving' Child: The Archaeology of the Deaths and
    Burials of Children at the Kilkenny Workhouse during the Great
    Famine in Ireland, 1845–52 249
    Jonny Geber

Index 263

List of Contributors

KRUM BACVAROV is an Associate Professor of Prehistoric Archaeology, and the Head of the Prehistory Department at the National Institute of Archaeology and Museum of the Bulgarian Academy of Sciences. He has directed multiple excavations and surveys at prehistoric sites in Bulgaria, as well as undertaken research projects on the later prehistory of Southeastern Europe. He has authored, co-authored or edited several monographs and numerous papers. He is the editor of the 2008 volume *Babies Reborn: Infant/Child Burials in Pre- and Proto-history* (BAR International Series 1832).

NATALIA BERSENEVA is a Senior Research Fellow of the Department of Archaeology and Ethnography at the Institute of History and Archaeology (Ural Division, Russian Academy of Sciences, Ekaterinburg) and a researcher in the Scientific-Educational Center of Eurasian Investigations (South-Ural State University, Chelyabinsk). She has published mostly on gender archaeology and the archaeology of childhood. Her book *Social Archaeology: Age, Gender and Status in the Sargat burials* (2011) is almost the first Russian book to have been written about gender and age archaeology.

CHRISTINE CAVE is a doctoral student at the Australian National University, researching the various implications of living to a grand old age in Early Anglo-Saxon England. Recent publications include Cave and Oxenham's (2016) 'Identification of the archaeological 'invisible elderly': an approach illustrated with an Anglo-Saxon example' published in the *International Journal of Osteoarchaeology*.

NATHALJA CALLIAUW is a doctoral student at the Department of Archaeology of the University of Leuven, Belgium, funded by the Research Foundation of Flanders. She first studied juveniles in Bronze Age Crete for her Master's thesis and further pursued the topic of age as an element of social identity in the run-up to her doctoral research. In the latter, she explores how multiple social identities were materialised in Bronze Age Crete and how we may approach their dynamic character archaeologically. She was a member of the Sissi Archaeological Project in Crete from 2010 until 2014.

PHILIPPE CHAMBON is Director of Research in the Centre National de la Recherche Scientifique, Laboratory of Eco-Anthropologie et Ethnobiologie (Musée de l'Homme, Paris, France). He investigates mortuary practices from the Neolithic period, mainly in Europe but also in Sudan. He has published several books including, *Les Morts dans les Sépultures Collectives Néolithiques en France* (2003) and *Les Cistes de Chamblandes et la Place des Coffres dans les Pratiques Funéraires du Néolithique Moyen Occidental* (with P. Moinat in 2007).

IONELA CRACIUNESCU is a doctoral student in the National History Museum of Romania in Bucharest who studies GIS applications in archaeology. Her research interests include the spatial analysis of cemeteries and site distributions in Neolithic and Eneolithic Southeastern Europe.

DIANA-CRINA DĂVÎNCĂ graduated from the Faculty of History in Sibiu and undertook her doctoral studies at the 'Vasile Pârvan' Institute of Archaeology in Bucharest. She is the author of a book published in 2015 entitled *Mortuary Beliefs and Practices Regarding Northern Thracian Children (XIth c. BC – Ist c. AD)* (in Romanian).

HEIDI DAWSON-HOBBIS is a Research Associate at the University of Bristol, currently collaborating on a project to re-associate a collection of high status medieval skeletal remains. Previously a Lecturer in Forensic Archaeology and Anthropology at

Kingston University London, from June 2017 she will take up a lectureship in Biological Anthropology at the University of Winchester. Her PhD research focused on the associations between health and burial status of medieval children which formed the basis for her book published in 2014 and entitled *Unearthing Late Medieval Children: Health, Status and Burial Practice in Southern England* (BAR British Series 593).

MIHAI FLOREA is a doctoral student in the National History Museum of Romania in Bucharest who studies topographical applications in archaeology. His research interests include field survey and the mapping of archaeological sites in Romania.

JONNY GEBER is a Lecturer in Biological Anthropology at the University of Otago in New Zealand. He has a particular research interest in social bioarchaeology, with a focus on the experience of poverty and social marginalisation in the nineteenth century. He is the author of *Victims of Ireland's Great Famine: The Bioarchaeology of Mass Burials at Kilkenny Union Workhouse* (University Press of Florida, 2015).

JANE JARK JENSEN has an MA in Medieval Archaeology and European Ethnology and is a curator with excavation responsibilities at the Museum of Copenhagen. The main focus of her research in recent years has been the large-scale excavations undertaken in Copenhagen in advance of a new Metro Cityring, and in particular at the site of Kongens Nytorv. She is particularly interested in the medieval burial practices of Copenhagen and the Oresund region.

KATERINA KOSTANTI is Curator of the Collection of Prehistoric, Egyptian, Cypriot and Near Eastern Antiquities at the National Archaeological Museum of Athens (Greek Ministry of Culture and Sports). Her research interests focus on the study of children and childhood in archaeology, gender studies, Mycenaean burial customs, social archaeology, Eastern Mediterranean cultural interrelations in the Bronze Age and Museum studies.

EMMA MAINES is currently pursuing her doctorate at the University of Paris 1 – La Sorbonne. Her primary fields of study include funerary archaeology and bioanthropology. Her dissertation topic concentrates on questions of population and funerary practice in the Sudanese Neolithic.

OLIVIER LANGLOIS is an archaeologist and CNRS researcher at the CEPAM (UMR 7264), Université Nice-Sophia Antipolis in Nice. His research mostly concerns the recent history of south Chadian Basin societies (northern Cameroon and south-western Chad). Since 2014 he has been working in northern Sudan, where he co-coordinates the Kadruka project, an archaeological mission of the French Section of the Sudanese Department of Antiquities sustained by the Qatar-Sudan Archaeological Project that relates to the Neolithic societies of Upper Nubia.

CATALIN LAZAR is an archaeologist based at the National History Museum of Romania. His research interests include the survey and excavation of Neolithic and Eneolithic sites in Romania, funerary behaviours, palaeodemography and mortuary practices in Southeastern Europe.

MÉLIE LE ROY currently teaches archaeology (ATER) at Montpellier University. She completed her doctorate in biological anthropology at Bordeaux University. Her research concerned the study of the skeletal remains of children and the social consideration of this part of the population through funerary practices, based on the use in GIS in the analysis of funerary settlements. Her recent publications include Le Roy *et al.* (2016) 'Distinct ancestries for similar funerary practices? A GIS analysis that compares funerary with biological and aDNA data from the Middle Neolithic necropolis Gurgy 'Les Noisats' (Yonne, France)' published in the *Journal of Archaeological Science*.

KATHLEEN MCSWEENEY is a Senior Lecturer in Human Osteoarchaeology at the University of Edinburgh. Her research areas focus on the reconstruction of life from the skeleton and the interpretation of burial practices. In the past she has worked as a freelance bone specialist, working on Scottish inhumations and cremations from the Neolithic to the post-medieval periods.

Her current research projects include: Early Bronze Age infant jar burials from Bulgarian Thrace; Bulgarian Chalcolithic skeletal remains; Early Bronze Age populations from the Arabian Peninsula; Skeletal Remains from Bronze Age tumuli in Transylvania, Romania; Late Mesolithic populations from Romania; Ancient Greek, Roman, Byzantine, medieval and post-medieval remains from the World Heritage site of Nessebar, Bulgaria.

EILEEN MURPHY is a Senior Lecturer in Archaeology in the School of Natural and Built Environment, Queen's University Belfast. Her research focuses on human skeletal populations recovered from prehistoric Russia and from all periods in Ireland. She is particularly interested in the use of osteoarchaeological information to help further our understanding of the daily lives and experiences of the people who lived in the past, as well as mortuary practices. She has published extensively and is the co-editor of the 2015 book *The Science of a Lost Medieval Gaelic Graveyard – The Ballyhanna Research Project* and editor of the international journal *Childhood in the Past*.

MARC OXENHAM reads archaeology and biological anthropology at the Australian National University. Recent publications include Oxenham and Willis (2016) 'Towards a bioarchaeology of care of children' (in Tilley and Schrenk (eds.). *New Developments in the Bioarchaeology of Care: Further Case Studies and Extended*) and Oxenham and Buckley (eds.) (2016) *The Routledge Handbook of Bioarchaeology in Southeast Asia and the Pacific islands*.

PASCAL SELLIER is a Senior Researcher at CNRS (French National Centre for Scientific Research) and a member of the research unit of the Musée de l'Homme UMR 7206 'Éco-Anthropologie et Ethnobiologie'. He is an archaeoanthropologist and bioarchaeologist who also teaches funerary archaeology in Université Paris 1-Panthéon-Sorbonne. After a threefold university education in medicine, archaeology and physical anthropology, he has published on mortuary practices, palaeodemography, palaeopathology and biocultural interrelations, and headed field missions, mostly on prehistoric burial grounds, in Pakistan, France, the Marquesas Archipelago and presently in Sudan.

VALERIU SÎRBU is a Senior Researcher at the Museum Brăila 'Carol I' and at the 'Vasile Pârvan' Institute of Archaeology in Bucharest. He has led 15 archaeological expeditions, organised 24 international scientific events and presented papers at numerous scientific conferences. He was awarded The Romanian Academy Award 'Nicolae Bălcescu' and was the President of the 30th Commission of UISPP. He has published extensively and is the author and co-author of 23 books and the editor/co-editor of 22 volumes.

JAYNE-LEIGH THOMAS is the Director of the Office of the Native American Graves and Repatriation Act at Indiana University in Bloomington, Indiana, United States. Her job entails returning ancestral human remains, funerary objects, and cultural items to appropriate tribal communities in the United States. Her research interests focus on human osteology, cremation studies, mortuary practices, repatriation and NAGPRA. She is the lead author on a 2015 *International Journal of Osteoarchaeology* article entitled: 'Violence and trophy taking: a case study of head and neck trauma in two individuals from the Gant Site (3MS11)'.

BELINDA TIBBETTS received her BSc in Biological Sciences and her BA in Classics and Archaeology from the University of Queensland in Australia. She received an MSc in Human Osteology at the University of Exeter where she is currently undertaking doctoral research. Her research interests centre on infant palaeopathology and the connections between skeletal pathology in very young individuals and maternal health in past populations. Her other interests include the archaeological evidence for maternal care and cultural buffering as a community response to maternal and perinatal mortality.

GABRIEL VASILE is a doctoral student at the 'Vasile Pârvan' Institute of Archaeology in Bucharest who studies physical anthropology. His research interests include biometry, palaeopathology and palaeodemography.

Chapter 1

Introduction: Archaeological Children, Death and Burial

Eileen Murphy[1] *and Mélie Le Roy*[2]

Introduction

Burial is a 'deeply significant act imbued with meaning' (Parker Pearson 2003, 5). The treatment of human dead, past and present, invariably involves the agency of the living. Indeed it would seem that cultural intervention is required to reposition the dead within their society – death is not only a biological reality but also a complex cultural event (Hertz 1905, 48–9). The provision of a last resting place for the body is generally a carefully considered process that would have taken time to execute. Different societies around the world have treated the corpse in a myriad of different ways, which involve various manipulations of the body itself; the use of a variety of forms of burial repository for the remains; the arrangement of the body in a particular manner within its grave; the provision of a range of often carefully selected grave goods; in addition to the spatial differentiation of burial for certain members of society (see e.g. Parker Pearson 2003, 5–15).

The interpretation of archaeological burials has shifted over the years depending on the theoretical perspective of the researcher. Traditional archaeological approaches viewed burials as an expression of religious belief. The processual approaches of the 1970s considered burial to be a reliable reflection of social organisation across cultures, whereas subsequent post-processual approaches focused more on the cultural context and ideological premises of funerary remains which are viewed as particular occurrences that do not have directly comparable universal patterns (Lull 2000, 576–8). The latter approaches draw upon the sociological theory of structuration which considers social structure as active and constantly changing as a result of the actions and activities of people (McHugh 1999, 1). More recent researchers have advocated the use of a more present-past-orientated archaeology which uses present-day examples, and theoretical approaches from other social sciences, as a way of understanding potential past responses and the meanings that may have lain behind aspects of mortuary practice (Taylor 2014, 178, 183).

Writing in her pioneering paper on the archaeology of childhood, Grete Lillehammer (1989, 89) observed that 'the child's world has been left out of archaeological research'. She recognised that research on children and childhood, particularly through the study of burial evidence and toys, had the potential to yield important insights in relation to the agency of children and their relationship with the adults of society (Lillehammer 1989, 102–3). Lucy (1994, 24–5) made the point that, even though we may be dealing with the body of a child in a grave, those who conducted the burial would most probably have been adults. The remains of dead children are manipulated within an adult world and the evidence for funerary processes derived from their burials therefore provides insights in relation to how adults came to terms with such premature deaths, although it should be remembered that other children may have had some agency in the funerary ceremonies associated with their dead siblings and friends (Murphy 2011, 68–9).

Children featured in many of the pioneering studies of funerary archaeology whose theoretical and methodological developments provided a foundation for later studies but the entire population was of interest in these studies and the emphasis was on the adult world (e.g. Saxe 1970; Binford 1971; Tainter 1975; O'Shea 1984; Hodder 1985). Throughout the 1980s, increasing interest in the reconstruction of social, and predominantly gender, identity saw a degree of consideration of burial evidence in relation to children (e.g. Pader 1982). Interest in the funerary archaeology of children gained momentum throughout the 1990s with the child-centred research of scholars including Sally Crawford (1993; 1999) and Sam Lucy (1994), both of whom worked on the inhumation cemeteries of Anglo-Saxon England, leading the charge. Eleanor Scott's (1999) volume, *The Archaeology of Infancy and Infant Death*, was the first study to focus specifically on children, namely infants, and funerary archaeology. Her research provided an overview of the presence of children in the archaeological record in addition to reviewing the evidence for infants in prehistoric and historic burial contexts.

Ten years ago, Mary Lewis (2007, 1) stated that the 'children who were once invisible in the archaeological record are slowly coming into view. The primary data for the archaeology of childhood are the children themselves'. Bioarchaeological studies of juveniles have mushroomed over the past decade and this has seen a refinement in the methods used for assessing age-at-death, physiological stress, disease status and trauma, amongst others. Unfortunately, the reliable determination of the sex of juveniles on the basis of osteological methods continues to elude, although advances in aDNA technology will hopefully eventually result in this being an affordable and accessible method for sex determination (Mays 2013). Bioarchaeological advances go hand-in-hand with developments in funerary archaeology since they enable more nuanced interpretations to be made concerning the world of the living child and the attitudes that were shown towards them in death.

Figure 1.1. Photograph of the speakers in the session 'Archaeological Approaches to the Burial of Children', held at the twenty-first annual conference of the European Association of Archaeologists in Glasgow on Thursday 3rd September 2015.

Alongside the growth in juvenile bioarchaeological research has been a general increase in studies which focus on the archaeology of children from an increasingly diverse range of perspectives. Funerary archaeology has, however, remained a popular tool by which to explore children in the past and such studies have been well represented within the volumes of the journal *Childhood in the Past*, as well as in recent edited collections of essays on past childhood (e.g. Lally and Moore 2011; Hadley and Hemer 2014; Sánchez Romero *et al.* 2015). There has been a plethora of studies undertaken on topics such as atypical mortuary practices for children both within common burial grounds (e.g. Crawford 2007; Craig-Atkins 2014) and in separate burial spaces (Moore 2009; Donnelly and Murphy 2017). Scholars from around the world have undertaken investigations of different chronological and temporal scope in a quest to identify the different stages of the life course and the social role of children (e.g. Lebegyev 2009; Fahlander 2012; Bickle and Fibiger 2014; Le Roy 2015), the emotional aspects of child death (Murphy 2011) and the identification of migration through the burial record of children (e.g. Hadley and Hemer 2011; Bengtson and O'Gorman 2016).

Heidi Dawson's (2014) volume, *Unearthing Late Medieval Children: Health, Status and Burial Practices in Southern England*, demonstrates the value of integrating detailed juvenile bioarchaeological analysis, with a study of mortuary practices, for the purposes of identifying potential differences in social status. The above ground evidence has not been ignored and studies have also focused on commemorative funerary practices invested in children (e.g. McKerr et al. 2009; Mander 2012; Baxter 2013).

The most recent, widely accessible, edited volume to focus solely on mortuary archaeology and children, however, is Krum Bacvarov's (2008), *Babies Reborn: Infant/Child Burials in Pre- and Protohistory*. As such, it seems timely for the production of a new collection of papers on this theme that includes research which draws upon theoretical and methodological advances that have been made over the past decade. The volume originated in a day-long session entitled, 'Archaeological Approaches to the Burial of Children', held at the 21st annual conference of the European Association of Archaeologists in Glasgow in 2015 (Fig. 1.1). The 16 papers included in the book cover a wide geographic area but there is a clear concentration dealing with Europe (Fig. 1.2). This is a not unexpected situation, however, given the nature of the conference for which they were initially gathered. They are also of broad temporal scope extending from the Neolithic through to the 19th century AD. We consider this breadth to be a strength of the book, however, since it provides snap shots of the different burial practices that occurred, thereby facilitating comparisons to be made on a large geographical and chronological scale.

The Place of Juveniles Across the Life Course

A number of recurrent themes and issues are evident in the papers and we will provide a brief overview of these before providing an introduction to each paper. Philippe Ariès (1962) is generally accredited as being the first scholar to give serious academic attention to the lives of the children who lived in the past. He proposed that 'childhood' was a culturally determined concept, an idea that has since been developed, with sociologists James and Prout (2015, 3) more recently stating that childhood is 'to be understood as a social construction. That is, the institution of childhood provides an interpretive frame for understanding the early years of human life. In these terms, it is biological immaturity rather than childhood which is a universal and natural feature of human groups, for ways of understanding this period of human life ... vary cross-culturally although they do form a specific structural and cultural component of all known societies'. What then can be learned about past societies through their treatment of those who died when biologically immature?

Is Everybody Included?

The relative paucity of infants and young children in communal burial grounds is a common occurrence in funerary archaeology (see e.g. Scott 1999 for discussion) and a number of prehistoric papers in the current volume have identified a similar trend (Le Roy, Kostanti, Calliauw, Sîrbu and Dăvîncă). The authors of these papers

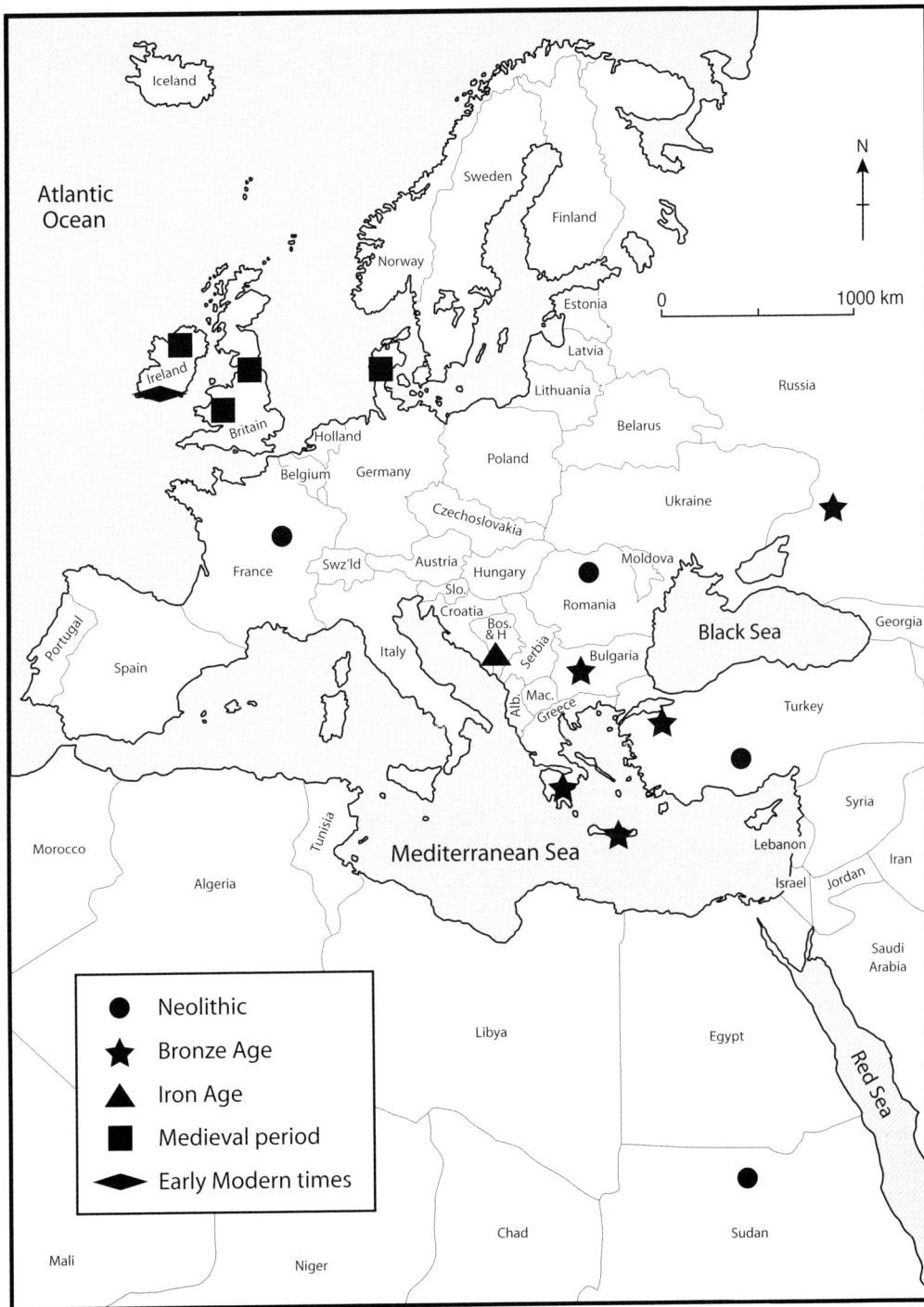

Figure 1.2. Map summarising the chronological and temporal span of the papers included in the volume (prepared by Mélie Le Roy and Libby Mulqueeny).

demonstrate that the under-representation may be a genuine reflection of cultural choice, rather than a product of taphonomic processes. The youngest members of society appear to have been afforded a differential funeral treatment compared to the remainder of the population. Turning to anthropology, Le Roy suggests the funerary practices applied to infants may be invisible to archaeology, citing the example of the Dayaks of the Indonesian peninsula and the Papuans of New Guinea, who deposit the infant corpse inside a dead tree trunk or hang it from the branches of a tree (Hertz 1905, 132–6). The implication is that the death of the very young held different significance to society. In other prehistoric populations juveniles in general are well represented within communal cemeteries (Maines et al., Berseneva), while at the Eneolithic cemetery of Sultana-Malu Roşu, Romania, particular locations within the burial ground appear to have been reserved for their burial (Lazar et al.). Juveniles are relatively well represented in the historic burial grounds included in the volume (Cave and Oxenham, Jark Jensen, Dawson-Hobbis, Murphy, Geber). The Anglo-Saxon, Great Chesterford burial ground is particularly interesting in this respect since the under-representation of children is a common occurrence in cemeteries of this period (Cave and Oxenham). It serves to remind us of the importance of considering local variation when dealing with a particular cultural group.

All members of society, including young infants, were buried within a settlement context at the Neolithic site of Çatalhöyük, Central Anatolia (Tibbetts), but the domestic arena appears to have been largely reserved for the burial of infants and young children in a substantial number of studies included in the volume (Le Roy, Thomas, Kostanti, Sîrbu and Dăvîncă). A number of previous authors have interpreted such burials as foundation deposits (e.g. Moses 2008); a representation of the fact that infants were not full members of society (Wells 1990, 139); related to sympathetic magic to attract fertility to a household (Morris 1987, 63–5; Golden 1990, 85); a reflection of the fact that infant death only had an impact at household rather than community level (Lebegyev 2009, 28), while Ian Hodder (1990, 29) suggested that these very young children were buried within domestic spaces that had been devoted to child-rearing activities. Kostanti (this volume) further observes that this coexistence of the living and the dead can be interpreted in two opposing manners – it may be an expression of absolute oblivion in which the burial is inaccessible and obliterated by the activities of daily life or, conversely, it may represent a profound form of commemoration in which the dead infant continues to share the living space and experiences of its family.

The Social Child

The archaeothanatological approach developed in France during the 1980s (Duday 2009) enables the reconstruction of the behaviour of past populations through detailed studies of burials. The approach focuses on the careful recording of the positioning of the skeleton, and its association with grave goods and furnishings, to provide information about the management and treatment of the body. The identification of repeated patterns can provide information about funerary practices at a cultural level. This approach was followed in the majority of papers included in the volume, many of which identified differential body positions and/or grave goods that appear to have been related to age-at-death.

On the basis of the evidence derived from the spatial analysis (discussed above), and the evidence for funerary rituals, several stages in the life course of a child are observed. Very young children seem to have been viewed as a separate social group in the vast majority of studies. Kostanti (this volume) interprets the inclusion of children older than two years in communal burial grounds as an indication that they had formally entered the Mycenaean community through rites of integration and passage. Berseneva (this volume), however, observes that all children, including neonates were afforded very similar burial practices to adults across the three Bronze Age cultures of the Southern Trans-Urals included in her study. Nevertheless, on the basis of differences in the associated grave goods she was able to determine that childhood ended at around 14 years of age in the Sintashta culture, when their burials began to include similar tools to those of adult burials. She was also able to identify that gendered grave goods were associated with children from the age of three years in the Sintashta population, while potential girls, as young as 9–12 months, were associated with female items in the Alakul' group.

In many of the studies infants and young children were buried inside jars (Thomas, McSweeney and Bacvarov, Kostanti, Calliauw) or boxes (Calliauw). The use of such burial containers would have afforded the infant body an additional degree of protection which may have been considered appropriate for the most fragile members of a population. Some authors have suggested that the ceramic jar was considered to represent the female body and womb and was associated with a hope that the dead infant would be reborn (Goodison 1989, 40; McGeorge 2011, 12). While jars were not used at Neolithic Çatalhöyük, Tibbetts (this volume) notes that infants were inhumed quickly after death in graves that were not reopened. This is in contrast to the situation for older members of society whose remains were not buried immediately, and whose graves could be re-opened to facilitate the removal of parts of the skeleton. Perhaps the idea of the rebirth of the infant meant that its body should not be disturbed? Interestingly, the work of McSweeney and Bacvarov (this volume) has demonstrated that the bodies of perinatal infants buried in jars in Early Bronze Age Bulgarian Thrace were sometimes buried in a partially decomposed state and were subject to a variety of post-mortem manipulations.

The two papers on later medieval burial grounds in Denmark and Ireland (Jark Jensen, Murphy) demonstrate a number of notable parallels in which the burials of children were often associated with grave goods with potential amuletic significance and their bodies were laid out in a variety of positions that did not conform to Church regulations. Burial on the side was a particularly favoured position, particularly for the very young, and both authors considered it to be reminiscent of a sleeping position. They both concluded that the atypical characteristics of the juvenile burials in their respective cemeteries were most likely to be an indication of tenderness on the parts of the families of the dead children who must have played a major role in their interment. Care in the burial of children was also evident in the early modern Kilkenny Union Workhouse mass grave in Ireland which demonstrates that even in an extreme mortality situation, in this case brought on by famine, efforts will be made to afford appropriate burial rites to all members of society (Geber).

All of these examples serve to demonstrate different societal responses in the face of the death of a child that are dependent on the time period and the geographical area. They provide insights as to how different societies viewed the immature members of their communities both during life and after death. There are undoubtedly difficulties in interpretation, however, and in some societies the burials of juveniles are associated with practically the same funerary rituals as for adults. It is clear that age is often not the only element of social organisation that can be identified in burials and other social roles or identities can be materialised in death, including kinship, gender and social status.

Embodying Identity

In this section, we will offer some suggestions regarding the diversity of juvenile identities – 'the domestic child', 'the vulnerable child', 'the high-status child', 'the cherished child', 'the potential child', 'the ritual child' and the 'political child' – that may be encountered throughout the papers of this volume. It needs to be appreciated, however, that these interpretations are neither definitive nor mutually exclusive; in some cases, contrasting or multiple messages may be read in the mortuary record.

The 'domestic child' can be seen through the association of child burials with places of domicile (Le Roy, Tibbetts, Thomas, Kostanti, Sîrbu and Dăvîncă). The potential association of children with apparently mundane everyday items in historic period cemeteries, in addition to their occasional association with pillow stones and their burial on the side may be seen as a further connection to the comfort and security of home (Cave and Oxenham, Jark Jensen, Murphy). The 'vulnerable child' is also evidenced in many of the studies in which infants were afforded further protection through their burial in containers (Thomas, McSweeney and Bacvarov, Kostanti, Calliauw). Certain children in later medieval cemeteries appear to have been deliberately given grave goods, with potentially apotropaic properties, presumably to help them in the afterlife (Cave and Oxenham, Jark Jensen, Murphy). The deliberate burial of children with other children, or indeed adults, may also feed into this notion of vulnerability (Lazar, Berseneva, Cave and Oxenham, Jark Jensen).

A number of studies also yielded burials of children who had been particularly well provisioned for the afterlife and who perhaps signify the 'high status child'. These include the famous example of an infant buried in Grave III of Grave Circle A at Mycenae whose body appears to have been entirely covered with gold foil (Kostanti) and perhaps a child exceptionally buried with almost 500 beads made from marble, malachite, Spondylus shells and snail shells in the Eneolithic Sultana-Malu Roşu

1. Introduction: Archaeological Children, Death and Burial   9

cemetery in Romania (Lazar *et al.*). In addition, an examination of burial location enabled Dawson-Hobbis (this volume) to identify children of higher status in later medieval English cemeteries. The 'cherished child' may be visible in cases where the quality and quantity of grave goods was similar to those of adults (Berseneva), or indeed at sites such as Neolithic Kadruka 23, Sudan, where the graves of younger children were better provisioned than those of adolescents and adults (Maines *et al.*). Even the simple investment of care in the burial of children during the Great Irish Famine, a time of mass death and extreme deprivation for thousands of people, is suggestive that these lives were respected and valued (Geber).

In some cases, children were provided with seemingly inappropriate goods better suited to the adult world that might be interpreted as representing the 'potential child'. Examples of these include two Middle Neolithic infants from France who were buried with objects that form the tool kit of experienced hunters (Le Roy). The Sintashta culture Halvay III burial of a 4–5-year-old child is another potential example. The child was buried in an exceptionally large pit with a massive wooden ceiling supported by posts and had been provisioned with six ceramic vessels, two sets of arrowheads, including some made from bronze, two bronze spearheads, a bronze battle-axe, a bronze knife and stone tools. The richness of the burial was considered to have been highly unusual, and it would have been rare for even adult warrior graves to contain this much wealth (Berseneva). Other examples can be found in Geto-Dacian Child Burials of the Second Iron Age in Romania. Grave 2, Tumulus 44, from the necropolis of Telița in Celic Dere contained the remains of an 8–9 year old child who had been buried with a curved iron knife, five bronze arrowheads, an amber bead and fragments of a pottery bowl, while a curved iron dagger with a scabbard, an iron arrowhead, a fragment from a silver earring, an iron belt buckle, a large iron clamp, a grinding stone and a fragment of bronze foil were recovered from Grave 32, the burial of a child less than seven years of age, from the Grădina Castelului necropolis in Hunedoara (Sîrbu and Dăvîncă). A particularly striking example derived from the Anglo-Saxon burial ground of Great Chesterford, England, where a 1–2-year-old child (skeleton #99) was buried with a spear, knife, buckle and a bronze ring; the child, both physically and developmentally, would have been unable to wield the spear which was of normal size (Cave and Oxenham).

While burial by its very nature has ritual connotations in some cases we can find evidence of a specialised 'ritual child' through the nature of the burial context in cases where child burials are associated with religious sanctuaries (Kostanti, Sîrbu and Dăvîncă). In addition, evidence for the dismemberment and processing of infant bodies in Early Bronze Age Bulgarian Thrace (McSweeney and Bacvarov), has clear ritual connotations.

The 'political child' may be viewed in Tibbetts' findings from Neolithic Çatalhöyük where it has been demonstrated that individuals buried within the same house, including the perinatal infants, were only minimally biologically linked. As such, it would seem that individuals were not grouped together in burial on the basis of biological relatedness but rather for practical kinship purposes that would help forge inter-group alliances and contribute to population stability. We know that living children were used as political actors in many past societies, such as the medieval Irish who practised a system of fosterage for the purposes of creating alliances (Murphy 2015, 107–8), and it would seem that death was not perceived to be a barrier to such strategies in some past societies.

A Way Forward?

A major difficulty within bioarchaeological studies of juveniles is the variation of terminology and age categorisations that exist, making comparative studies of populations very difficult (Halcrow and Tayles 2008; Falys and Lewis 2011). This difficulty is also of relevance to the study of juvenile burial practices and in the current volume a myriad of age ranges are applied to different age categories; for example, some authors considered infants to be less than one year, others younger than two years and others three years and under. The upper end of the juvenile age could extend from 15 up to 20 years. While all of these approaches have their merits it is clear that the creation of a standard system that applies clear terms to well defined age ranges would facilitate more effective comparative studies.

A further issue with archaeological mortuary studies as a whole, and not just in relation to juvenile burials, is the lack of clarity that can be applied to terms related to body position, such as 'flexed' or 'hocker', which can have a variety of meanings for different scholars. To facilitate comparative analysis, it is important to be specific as to what such terms mean by providing details of the degree of flexion and referring to established standards, such as those of Sprague (2005). Another potential term which can be associated with a lack of clarity is 'intramural' which is generally used when human burials are associated with domestic structures but, as Laneri (2011, 44) and Kostanti (this volume) have discussed, more precise terminology is needed to enable, for example, the precise relationship of the burial to the building to be determined and to facilitate an assessment of its level of visibility.

Bioarchaeologists are in unanimous agreement that individuals with expertise in the study of human skeletal remains should always be involved in the excavation of burial contexts to enable the maximum amount of information to be gained about both the deceased individual and their associated burial environment (Duday 2009, 6; Roberts 2009, 74). The work of McSweeney and Bacvarov (this volume) is a prime example of why this should be the case. Their involvement in the micro-excavation of an intact jar burial (Tell Kran 9) from Early Bronze Age Bulgarian Thrace enabled them to verify that bodies could be partially decomposed when they were placed within a jar. Furthermore, they ascertained that dismemberment and the deliberate removal of bones from infant bodies was a genuine component of the repertoire of body processing activities and not simply an artefact of taphonomy and/or curation practices.

The research of both Le Roy and Lazar et al. (this volume) utilise geospatial technologies in their analyses and they have demonstrated that GIS is a powerful platform upon which to undertake the spatial analyses of past burials. Analytical techniques are advancing all the time and future analyses of burial practices could incorporate new research that has demonstrated the ability to differentiate between live and stillborn infants (Booth et al. 2016). Dawson-Hobbis' paper (this volume) clearly demonstrates the potential value of using mortuary practice as an important contextual aid to help inform the interpretation of physiological stressor markers, a process that can be complicated due to the Osteological Paradox (see Wood et al. 1992). By contextualising and integrating the data collected, both in the field and in the laboratory, it is possible to gain a more nuanced understanding of the experiences of the young in the past. We can move beyond generating purely scientific data to gaining an understanding of key facets of past life, including the emotional attitudes that were shown towards children during life and in death, as well as their place in the social strata and the ritual activities of their societies. This volume assembles a panorama of studies about juvenile burials that enable a greater understanding to be gained about the identity of the younger members of society in the past.

Structure of the Volume

The volume commences with Mélie Le Roy's paper which explores the social status of infants in Neolithic France. Using a combination of osteoarchaeological and archaeological data she reviews the nature of burial practices afforded to infants throughout the Early, Middle and Late Neolithic. Her approach involves an investigation of mortality profiles to determine how well represented infants were in the sites of each period. The position of the body and the nature of grave goods in infant burials, as well as their location, are also considered in relation to the practices afforded to older members of society. She found that infants were generally not well represented and differences in burial practice, most notably in relation to location, were also observed across the different periods. It was concluded that infants had a particular social status within these societies. Remaining with the Neolithic, in the second paper Belinda Tibbetts presents an interpretation of the burials of neonates within the context of community, and the cultural response to infant death that is reflected in the archaeological record at Çatalhöyük in Central Anatolia. Her research demonstrates how infant burials can provide significant insights into their cultural identity in past populations. Based on bioarchaeological analysis of the skeletal remains, the burial environment and burial inclusions, it demonstrates that the youngest individuals at Çatalhöyük were recognised as having a social identity and were provided with culturally sanctioned burial, regardless of their viability at birth.

In their paper, Emma Maines, Pascal Sellier, Philippe Chambon and Olivier Langlois provide an overview of the preliminary results of their study of the Nubian Middle Neolithic necropolis of Kadruka 23, Sudan. Excavation of this relatively undisturbed funerary mound revealed a concentration, and seemingly codified treatment, of very young individuals. The inclusion of a large proportion of juveniles is considered suggestive that the mound represented a normal demographic profile and was not reserved for burial of the elite as has been suggested by other scholars. They consider three major elements – the relative placement of the burials within the strata of the mound; the position of the body of the deceased and the nature of the associated mortuary goods – in an attempt to identify differences in the burial practices afforded to younger children and those from approximately 10 to 19 years of age.

The paper of Catalin Lazar, Ionela Craciunescu, Gabriel Vasile and Mihai Florea focuses on Eneolithic child burials in the Sultana-Malu Roşu cemetery of southeastern Romania, a burial ground they consider to be quite typical for this period in the Balkans. Their study involves an analysis of children's burials in relation to funerary rituals, palaeodemographic data and the spatial location of the burials in the cemetery. They note that differences in the treatment of adults and children have the potential to provide insights in relation to the symbolic significance of children as well as the collective identity of the family or community. They also explore how the nature of the funerary processes afforded to children might provide an indication of the impact that death at a young age had on a community.

Jayne-Leigh Thomas explores juvenile skeletal remains and mortuary practices at the Late Chalcolithic site of Çamlıbel Tarlası in Central Anatolia. She identifies that a notable majority of the individuals interred at the site were juveniles, and age-differentiated burial practice was apparent. Infants and young children were buried in pots, while the remains of older children were interred in simple pit burials. The paucity of adult remains is interpreted as an indication that they were buried extramurally or in unexcavated areas of the region. The burial practices identified at Çamlıbel Tarlası correlate well with evidence derived from contemporary sites in the region. The paper also includes palaeopathological data that provides insight in relation to the health status of the children buried at the site.

In their paper, Kathleen Mc Sweeney and Krum Bacvarov present data derived from the analysis of over 50 infants derived from Early Bronze Age jar and pit burials from various sites in Bulgarian Thrace. Their research reveals previously unidentified mortuary practices which involved the deliberate manipulation of the infant body and it clearly demonstrates the importance of having an osteoarchaeologist involved in the excavation of such burials. They conclude that most of the babies had been in a state of partial decomposition prior to their placement in containers and that some of the bodies had been deliberately dismembered. They also consider reasons as to why it may have been necessary to process the infant dead in this manner prior to their burial.

The paper of Katerina Kostanti focuses on infants in Late Helladic burial grounds in Mycenaean Greece. Her work demonstrates how infants are under-represented in these populations and how those individuals aged less than 24 months of age, who have been identified, were frequently buried within domestic contexts. This finding is in contrast to the situation for older children who tended to be buried in communal burial grounds and she argues that this may be an indication that children aged over two years of age had formally entered the Mycenaean community through rites of integration and passage. Propositions are also advanced to explain why the under-twos may have been included within the domestic sphere.

The Bronze Age Sintashta, Petrovka and Alakul' cultural groups of the steppe lands of the southern Trans-Urals form the focus of the paper by Natalia Berseneva who has observed that juveniles comprise 50–80% of all burials identified to date from these cultures. She provides an overview of the burial practices afforded to children in each culture. Although variations between the groups were apparent she concludes that the place of children within the three social structures was of importance and she observes that the juvenile burial practices in each culture largely mirrored those of the adults. On the basis of the grave goods provided to the juveniles she identified that the initial age of labour and gender socialisation was approximately three to five years in all three cultures but differences in the age at which gender-distinctive clothing became apparent were evident across the three groups.

Nathalja Calliauw observes that juvenile burials in Bronze Age Crete have been largely ignored due to focus on socio-political organisation and a lack of appreciation of the information that can be derived from the analysis of human remains. Bearing these limitations in mind she reviews the characteristics of juvenile burials in Pre- and Protopalatial Crete in relation to those of adult burials. She discusses the substantial variation apparent in the Bronze Age Cretan funerary landscape and observes how age-based spatial segregation is apparent in some sites but not present in others. She considers that the deposition of infant bodies within containers in recognised communal burial contexts during the Early Minoan II is suggestive they were an active part of the group and burial community. She notes how the situation changes during the Early Minoan III period, however, when the emphasis appears to shift from the differentiation of infants to one which saw the exclusion of juveniles from burial spaces. She suggests the earlier pattern may be reflective of kinship, whereas the latter situation is suggestive of an age-based social organisation.

Valeriu Sîrbu and Diana-Crina Dăvîncă undertake a review of Geto-Dacian child burials from the fifth century BC to the first century AD. They compare and contrast the nature of the remains recovered from formal burial grounds with those derived from non-funerary contexts, including fortresses, settlements, cult sites and isolated pits. In an attempt to ascertain the nature of the mortuary rituals that may have taken place after death they examine the position and condition of the skeleton as well as the nature of associated grave goods. Their research is hindered, however, as a result of the paucity of osteoarchaeological analyses to have been undertaken on the skeletal remains to date. Juvenile remains appear to be generally under-represented compared to those of adults and the remains derived from both funerary and non-funerary contexts comprise a mixture of complete and partial skeletons as well as isolated bones, thereby suggesting that a variety of mortuary rituals were applied to the bodies of children.

In their paper, Christine Cave and Mark Oxenham aim to provide insights concerning the lived experience of Anglo-Saxon children and, in particular, the infants. They focus on the cemetery of Great Chesterford in England which appears to have been the final resting place of the entire community. They observe that the examination of a child's grave enables inferences to be made about the attitudes shown by adults to the dead child, community concepts of children and childhood, as well as providing a glimpse, albeit through the distorted lens of the grave, of the life of that child. They found that, although some children were buried with exceptional grave accoutrements, they were generally supplied with fewer, and less valuable items than adults. They make the important point that young children, especially infants, have little personal material culture since they are fed by their mothers, and have no need of showy adornments, and warn that the relative absence of grave goods is not necessary a sign that their deaths were without meaning or that the child was not missed.

Jane Jark Jensen's paper presents an overview of the information derived from juvenile burials in the later medieval churchyard of St Clemens in Copenhagen, Denmark, which was used for burial by the lower classes of society. She observed a number of differences in relation to the burial practices afforded to children when compared to adults. Children were more likely to be interred in atypical positions, some of which are suggestive of tenderness, as well as be buried together with the remains of other children in multiple graves. Some of their burials include stones, or other deliberately placed items, that may have had apotropaic purposes. These features, in addition to the inclusion of the remains of unbaptised infants within the burial ground, led her to suggest that particular affection was being shown towards the young. She concludes that the medieval Copenhageners practised a degree of individuality in relation to the funerary practices afforded to children.

The presence of childhood stress markers (dental enamel hypoplasia, cribra orbitalia and periostitis) in relation to social status is explored in Heidi Dawson-Hobbis' paper. She considers the evidence from three medieval (AD 1086–1540) English priory burial grounds – St Peter and Paul, Taunton; St Oswald, Gloucester and St Gregory, Canterbury. She ascribes levels of social status to different geographical areas of each burial ground, with burials within church buildings and associated with coffins, for example, representing individuals of higher status. She suggests that it may be necessary to interpret the prevalence of stress indicators evident in adults and children separately. Her study suggests that adults without stress lesions may have avoided such insults in childhood, possibly due to their higher status. Conversely, children with stress indicators present may have survived the initial physiological stress even though they still died young. The children whose remains displayed no evidence for physiological stress may be the non-survivors of any initial stress insult. Contrary to previous research, she concludes that the remains of advantaged medieval children may be more likely to display stress indicators.

In her paper, Eileen Murphy examines the characteristics of the juvenile burials excavated at the later medieval burial ground of Ballyhanna, Co. Donegal, Ireland. She observes that, while the majority of juvenile burials conform to typical Christian burial practices, subtle variations in relation to orientation, position of the body and inclusion of grave goods were evident. Each of these aspects was examined in a systematic manner in relation to age-at-death to see if children of different ages were more likely to be afforded atypical burial practices. The implications of the findings in relation to the agency of the families of the dead children, as well as to the nature of the management of the burial ground by church authorities, were considered. She observes that positioning of the body in a natural sleeping position seems to have been considered particularly appropriate for babies and young children, while the inclusion of various objects and furnishings with links to the domestic sphere, and a potentially protective function, was considered to be a further sign of tenderness. She concludes that the occurrence of such atypical features is a reflection of individuality in relation to the funerary practices of these later medieval people.

The final paper of the volume moves into the 19th century when Jonny Geber explores the particularly devastating impact that the Great Famine had on the children living in Ireland at that time. Hundreds of thousands of children became institutionalised in union workhouses, such as the Kilkenny Union Workhouse, where excavation of an intramural Famine-period burial ground has provided major insights in relation to the plight of children. He discusses how over half of the individuals buried here were aged less than 15 years when they died. He presents an overview of the archaeological evidence which reveals that burials were undertaken in an organised and structured manner, and that children were treated equally to adults in death. Despite the crisis, with its severe economic difficulties, workhouse officials did their best to ensure the dead were treated with respect and buried with care. Geber considers that the Kilkenny workhouse mass burial ground, with its high proportion of children, typifies the reality of how complete families were destroyed and social bonds severed during the Famine. He poignantly notes that many of the children would have been orphans who had to endure the Famine and enter the workhouse on their own. Some 15 years after the Famine had ended the workhouse guardians covered the burial ground with a thick layer of soil and he suggests this may be indicative of a deliberate intention to obliterate the painful Famine years. Eventually the burial ground became lost to local memory.

The papers included within the pages of this volume have employed a range of methodological approaches but all have a similar objective in mind, namely to understand how children were treated in death by different cultures in the past; to gain insights concerning the roles of children of different ages in their respective societies and to find evidence of the nature of past adult-child relationships and interactions. We hope this volume will provide a positive contribution to this fascinating field of research as it continues to advance.

Acknowledgements

We are grateful to the Organising Committee of the 2015 annual conference of the European Association of Archaeologists for permitting us to organise our session on juvenile burials as part of the conference. Thanks are also due to Dr Anne-Marie Tillier who helped chair the session. We are grateful to Dr Julie Gardiner for her interest in the session and for her encouragement to publish this volume.

We are appreciative to Dr Mark Gardiner, Archaeology and Palaeoecology, School of Natural and Built Environment, Queen's University Belfast, who supported our application for funding for production of the volume's index, and to Ms Libby Mulqueeny, of the aforementioned organisation, for her assistance with illustrations. We are particularly grateful to her for providing our beautiful front cover image which she based on a photograph taken by Belinda Tibbets (Chapter 3) of an infant burial from Çatalhöyük. Finally, we are very grateful to the reviewers who have helped both us and the authors strengthen the volume.

Notes

1. Archaeology and Palaeoecology, School of Natural and Built Environment, Queen's University Belfast, Belfast BT7 1NN, Northern Ireland. Email: eileen.murphy@qub.ac.uk.
2. UMR 5199 – PACEA, Université de Bordeaux, Bat. B8 Allée Geoffroy Saint-Hilaire, CS 50023, F – 33615 Pessac CEDEX, France. Email: melieleroy@hotmail.fr.

References

Ariès, P. 1962. *Centuries of Childhood: A Social History of Family Life* (transl. Baldick, R.). London: Penguin.
Bacvarov, K. (ed.) 2008. *Babies Reborn: Infant/Child Burials in Pre- and Proto-history* (BAR International Series 1832). Oxford: Archaeopress.
Baxter, J. E. 2013. Status, sentimentality and structuration: an examination of 'intellectual spaces' for children in the study of America's historic cemeteries. *Childhood in the Past* 6, 106–122.
Bengtson, J. D. and O'Gorman, J. A. 2016. Children, migration and mortuary representation in the late prehistoric Central Illinois River Valley. *Childhood in the Past* 9, 19–43.
Bickle, P. and Fibiger, L. 2014. Ageing, childhood, and social identity in the Early Neolithic of Central Europe. *European Journal of Archaeology* 17, 208–228.
Binford, L. R. 1971. Mortuary practices: their study and their potential, in Brown, J. A. (ed.), *Approaches to the Social Dimension of Mortuary Practices* (Memoirs for the Society for American Archaeology 25), 6–29. New York: Society for American Archeology.
Booth, T., Redfern, R. C. and Gowland, R. L. 2016. Immaculate conceptions: Micro-CT analysis of diagenesis in Romano-British infant skeletons. *Journal of Archaeological Science* 74, 124–134.
Craig-Atkins, E. 2014. Eavesdropping on short lives: eaves-drip burial and the differential treatment of children one year and under in early Christian cemeteries, in Hadley, D. M. and Hemer, K. A. (eds.), *Medieval Childhood: Archaeological Approaches* (SSCIP Monograph 3), 95–113. Oxford: Oxbow Books.
Crawford, S. 1993. Children, death and the afterlife in Anglo-Saxon England, pp. 83–91 in Filmer-Sankey, W. (ed.), *Anglo-Saxon Studies in Archaeology and History* 6. Oxford: Oxford University Committee for Archaeology.
Crawford, S. 1999. *Childhood in Anglo-Saxon England*. Stroud: Sutton.
Crawford, S. 2007. Companions, co-incidences or chattels? Children and their role in early Anglo-Saxon multiple burials, in Crawford, S. and Shepherd, G. (eds.), *Children, Childhood and Society*.

IAA Interdisciplinary Series Vol. I: Studies in Archaeology, History, Literature and Art (BAR International Series 1696), 83–92. Oxford: Archaeopress.

Dawson, H. 2014. *Unearthing Late Medieval Children: Health, Status and Burial Practice in Southern England* (BAR British Series 593). Oxford: Archaeopress.

Donnelly, C. J. and Murphy, E. M. 2017. Children's burial grounds (*cillíní*) in Ireland: new insights into an Early Modern religious tradition, in Crawford, S., Hadley, D. and Shepherd, G. (eds.), *The Oxford Handbook of the Archaeology of Childhood*. Oxford: Oxford University Press.

Duday, H. 2009. *The Archaeology of the Dead: Lectures in Archaeothanatology*. Oxford: Oxbow Books.

Fahlander, F. 2012. Mesolithic childhoods: changing life-courses of young hunter-fishers in the Stone Age of southern Scandinavia. *Childhood in the Past* 5, 20–34.

Falys, C. G. and Lewis, M. E. 2011. Proposing a way forward: a review of standardisation in the use of age categories and ageing techniques in osteological analysis (2004 to 2009). *International Journal of Osteoarchaeology* 21, 704–716.

Golden, M. 1990. *Children and Childhood in Classical Athens*. Baltimore, MD: John Hopkins University Press.

Goodison, L. 1989. *Death, Women and the Sun: Symbolism of Regeneration in Early Aegean Religion* (Bulletin Supplement 53) London: University of London, Institute of Classical Studies.

Hadley, D. M. and Hemer, K. A. 2011. Microcosms of migration: children and early medieval population movement. *Childhood in the Past* 4, 63–78.

Hadley, D. M. and Hemer, K. A. (eds.) 2014. *Medieval Childhood: Archaeological Approaches* (SSCIP Monograph 3). Oxford: Oxbow Books.

Halcrow, S. E. and Tayles, N. 2008. The bioarchaeological investigation of childhood and social age: Problems and prospects. *Journal of Archaeological Method and Theory* 15, 190–215.

Hertz, R. 1905. Contribution à une étude sur la représentation collective de la mort. *L'Année Sociologique (1896/1897-1924/1925)* 10, 48–137.

Hodder, I. 1985. Postprocessual archaeology. *Advances in Archaeological Method and Theory* 8, 1–26.

Hodder, I. 1990. *The Domestication of Europe. Structure and Contingency in Neolithic Societies*. Oxford: Basil Blackwell.

James, A. and Prout, A. (eds.). 2015. *Constructing and Reconstructing Childhood: Contemporary Issues in the Sociological Study of Childhood*. London: Routledge.

Lally, M. and Moore, A. (eds.). 2011. *(Re)Thinking the Little Ancestor: New Perspectives on the Archaeology of Infancy and Childhood* (BAR International Series 2271). Oxford: Archaeopress.

Laneri, N. 2011. Defining residential graves. The case of Titris Höyük in southeastern Anatolia during the late IIIrd Millennium BC, in Henry, O. (ed.), *Le Mort dans la Ville. Pratiques, Contextes et Impacts des Inhumations Intra-Muros en Anatolie, du Début de l'Age du Bronze à l'Époque Romaine, Actes des 2e Rencontres d'Archéologie, Istanbul 14-15 Novembre 2011*, 43–52. Istanbul: Institut Français d'Études Anatoliennes Georges Dumézil – CNRS USR 3131.

Lebegyev, J. 2009. Phases of childhood in early Mycenaean Greece. *Childhood in the Past* 2, 15–32.

Le Roy, M. 2015. Les Enfants au Néolithique: Du Contexte Funéraire à l'interprétation Socioculturelle en France de 5700 à 2100 ans av. J.-C. Unpublished Ph.D. thesis, University of Bordeaux.

Lewis, M. 2007. *The Bioarchaeology of Children: Perspectives from Biological and Forensic Anthropology*. Cambridge: Cambridge University Press.

Lillehammer, G. 1989. A child is born. The child's world in archaeological perspective. *Norwegian Archaeological Review* 22, 91–105.

Lucy, S. 1994. Children in early medieval cemeteries. *Archaeological Review from Cambridge* 13, 21–34.

Lull, V. 2000. Death and society: a Marxist approach. *Antiquity* 74, 576–580.

Mander, J. 2012. *Portraits of Children on Roman Funerary Monuments*. Cambridge: Cambridge University Press.

Mays, S. 2013. A discussion paper of some recent methodological developments in the osteoarchaeology of childhood. *Childhood in the Past* 6, 4–21.

Moore, A. 2009. Hearth and home: the burial of infants within Romano-British domestic contexts. *Childhood in the Past* 2, 33–54.

McGeorge, P. J. P. 2011. Intramural infant burials in the Aegean Bronze Age, in Henry, O. (ed.), *Le Mort dans la Ville. Pratiques, Contextes et Impacts des Inhumations Intra-Muros en Anatolie, du Début de l'Age du Bronze à l'Époque Romaine, Actes des 2e Rencontres d'Archéologie, Istanbul 14-15 Novembre 2011*, 1–19. Istanbul: Institut Français d'Études Anatoliennes Georges Dumézil – CNRS USR 3131.

McHugh, F. 1999. *Theoretical and Quantitative Approaches to the Study of Mortuary Practices* (BAR International Series 785). Oxford: Archaeopress.

McKerr, L., Murphy E. M. and Donnelly C. J. 2009. "I Am Not Dead, but Do Sleep Here": the representation of children in Early Modern burial grounds in the north of Ireland. *Childhood in the Past* 2, 110–132.

Morris, I. 1987. *Burial and Ancient Society. The Rise of the Greek City-State*. Cambridge: Cambridge University Press.

Moses, S. 2008. Catalhöyük's foundation burials: ritual child sacrifice or convenient deaths? in Bacvarov, K. (ed.), *Babies Reborn: Infant/Child burials in Pre- and Protohistory* (BAR International Series 1832), 45–52. Oxford: Archaeopress.

Murphy, E. M. 2011. Parenting, child loss and the *cillíní* of Post-Medieval Ireland, in Lally, M. and Moore, A. (eds.), *(Re)Thinking the Little Ancestor: New Perspectives on the Archaeology of Infancy and Childhood* (BAR International Series 2271), 63–74. Oxford: Archaeopress.

Murphy, E. M. 2015. Lives cut short – insights from the osteological and palaeopathological analysis of the Ballyhanna juveniles, in McKenzie, C. J., Murphy, E. M. and Donnelly, C. J. (eds.), *The Science of a Lost Medieval Gaelic Graveyard – The Ballyhanna Research Project* (TII Heritage 2), 103–120. Dublin: Transport Infrastructure Ireland.

O'Shea, J. M. 1984. *Mortuary Variability: An Archaeological Investigation* (Studies in Archaeology Series). Orlando, FL: Academic Press.

Pader, E. J. 1982. *Symbolism, Social Relations and the Interpretation of Mortuary Remains* (BAR British Series 130). Oxford: British Archaeological Reports.

Parker Pearson, M. 2003. *The Archaeology of Death and Burial*. Stroud: The History Press.

Roberts, C. A. 2009. *Human Remains in Archaeology: A Handbook* (CBA Practical Handbook 19). York: Council for British Archaeology.

Sánchez Romero, M., Alarcón García, E. and Aranda Jiménez, G. (eds.) 2015. *Children, Spaces and Identity* (SSCIP Monograph 4). Oxford: Oxbow Books.

Saxe, A. A. 1970. Social Dimensions of Mortuary Practice. Unpublished Ph.D. thesis, University of Michigan. Ann Arbor, MI: University Microfilms Inc.

Scott, E. 1999. *The Archaeology of Infancy and Infant Death* (BAR International Series 819). Oxford: Archaeopress.

Sprague, R. 2005. *Burial Terminology: A Guide for Researchers*. Lanham, CA: AltaMira Press.

Tainter, J. A. 1975. Social inference and mortuary practices: an experiment in numerical classification. *World Archaeology* 7, 1–15.

Taylor, N. 2014. Broken mirrors? An archaeological reflection on identity, in Ginn, V., Enlander, R. and Crozier, R. (eds.), *Exploring Prehistoric Identity in Europe*, 175–185. Oxford: Oxbow Books.

Wells, B. 1990. Death at Dendra. On mortuary practices in a Mycenaean community, in Hägg, R. and Nordquist, G. C. (eds.), *Celebrations of Death and Divinity in the Bronze Age Argolid. Proceedings of the 6th International Symposium at the Swedish Institute at Athens, 11-13 June 1988*, 125–140. Stockholm: Paul Åströms Förlag.

Wood, J. W., Milner, G. R., Harpending, H. C. and Weiss, K. M. 1992. The osteological paradox: problems of inferring prehistoric health from skeletal samples. *Current Anthropology* 33, 343–370.

Chapter 2

How Were Infants Considered at Death during the Neolithic Period in France?

Mélie Le Roy[1]

Abstract: In France, during the Neolithic period (5700–2100 BC), several cultural groups have been identified. In the Early Neolithic, burials were simple and gathered in small groups, following by monumentalisation and collectivisation during the Middle Neolithic, before becoming exclusive at the end of the Neolithic. The aim of this research is to focus on infants within these various contexts in terms of funerary practices, age distribution and the location of the burials (either next to domestic structures or in strictly funerary contexts), and to discuss their social status within the society as evidenced through the collected data. The study is based on 237 sites in France and combines biological and archaeological data. The results indicate that infant burials were afforded specific locations and various funerary treatments are also identified on a larger scale. These differences seem to reflect a specific social consideration of infants.

Keywords: infant, Neolithic, France, funerary practices

Introduction

The ultimate goal of an archaeological study is to understand past population behaviours. In relation to funerary practices, more specifically, studies focus on the status of each individual and the place of the dead among the living (Thomas 1975; Baudry 2006). Indeed, the relationship to death and its inclusion in aspects of everyday life is not only expressed during funerals, but also in the funerary practices and the use of the space dedicated to the dead. The nature of the manner in which the deceased is deposited suggests a symbolic consideration between the one who left and those left behind. These gestures may provide information about different social status and aspects of the behaviour of one individual towards another. As such, funerary practices are related to social status, sex and age-at-death (Thomas 1975; Suzuki 2000).

In this study, we consider infants from the Neolithic period in France. In order to avoid too much specific data which could reflect the highly complex nature of the French Neolithic, the chronology is divided into three main periods (according to Tarrête and Le Roux 2008) – the Early Neolithic (5700–4900 BC); Middle Neolithic (4900–3500 BC) and the End of the Neolithic (3500–2100 BC).[2]

During the Neolithic, funerary practices evolved from a single burial tradition in which graves could be grouped and/or isolated to one in which collective burial sites became almost exclusive. This change gradually evolved from the Middle Neolithic but it did not necessarily occur at the same time in the different regions of interest. The relevant architectural structures, which show a wide diversity (Fig. 2.1), also seem to vary throughout the different geographical areas. In this highly complex context one may wonder what places infants have in these Neolithic societies. Indeed, children are rarely, if at all, mentioned in previous studies and their details are often embedded within those of the rest of the population.

Figure 2.1. Some example of burial formats; A) Les Noisats, Gurgy – double burial of children in a simple pit (Photograph: Stéphane Rottier); B) Dolmen les Isserts, Saint-Jean de la Blaquières – megalithic structure which yielded a collective grave (Bec Drelon et al. 2014, fig. 11, 76); C) La Truie Pendue, Passy-Véron – collective burial in a pit (Thiol et al. 2010, fig. 7, 229).

Materials and Methods

In an attempt to fill the gap in the field of archaeological research on Neolithic societies in France an inventory of the skeletal remains of immature individuals (0–19 years) from published and accessible data for this age group has been established. Thus archaeo-anthropological and osteological data, collected within a single database, provides information on the various funerary practices relating to immature individuals. The location of the infant burials (either next to domestic structures or in strictly funerary contexts) then allowed us to define variants and identify potential threshold ages. The results enabled a discussion of the social status of infants based on ethnographic comparisons (Le Roy 2015). Data derived from over 2,000 sources in addition to unpublished data from new research (Bec Drelon *et al.* 2014; Le Roy *et al.* 2014; Le Roy 2015), has resulted in the collection of more or less complete information from 8,124 archaeological sites. Secondly, 1,301 sites containing human remains were identified, from which information about the immature individuals (n=2,790), and especially infants, meaning the children who died before their first year of life (n=237), was isolated. Note that the three time periods are unequally represented (Fig. 2.2).

Several points of view were considered in the analysis. Firstly, the funerary selection inside the burial sites was analysed in order to identify if infants were present or not among the buried population. In an attempt to identify trends, the sites selected were considered to have reliable and usable data. Some 110 sites were analysed to establish the respective mortality profiles, which were then compared with a theoretical mortality pattern to identify any anomalies. Secondly, the location of the graves was taken into account for 237 sites to identify whether infants were more likely to be buried next to domestic structures or in strictly funerary areas. Finally, body positions and associated grave goods were examined for 38 individuals from 20 sites in order to obtain a diachronic view of the funerary treatment of infants.

Results

Missing Infants

Among the 110 funerary sites included in the analysis of mortality profiles, we identified 48 funerary assemblages, which showed a significant lack of young children and especially infants (Table 2.1). Two sites dated from the Early Neolithic presented two distinct contexts. The first is a collective burial in a cave at Aven des Bréguières, Mougins, the oldest collective burial known to date, with 61 individuals interred, including 26 immature individuals that represented all age groups. Among them, 16 infants were present (Soulier 1998; Provost 2013). Only infants are found to be significantly missing when the data is compared to an expected mortality profile, even if some of them were buried there. The second is the open-air necropolis of Les Octrois, Ensisheim, where both infants and children younger than five years at death are significantly under-represented (Jeunesse *et al.* 1993; Jeunesse 1995; 1998).

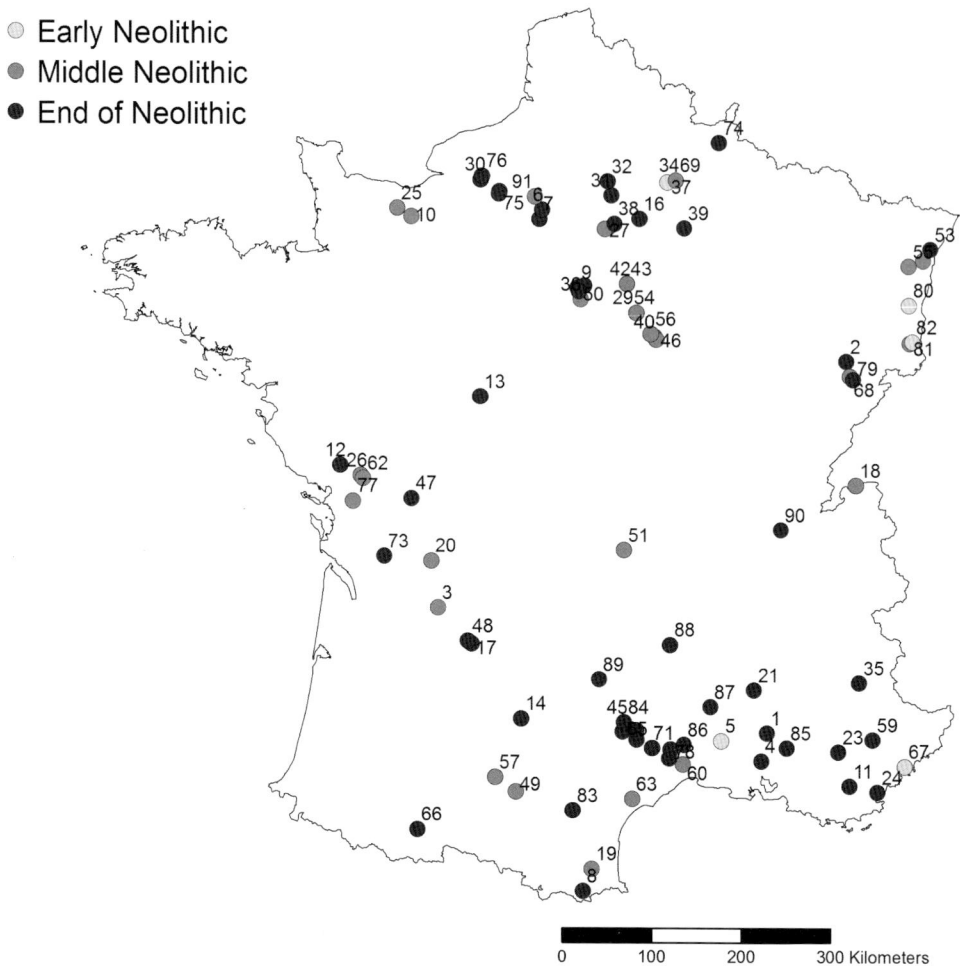

Figure 2.2. Chronological representation of the archaeological sites where infant remains were uncovered. 1 – Abri de Sanguinouse, La Roque sur Pernes; 2 – Aillevans 3, Aillevans; 3 – Barbilloux 2, Saint Aquilin; 4 – Barres, Eyguières; 5 – Baume Bourbon, Cabrières; 6 – Bois du Couturier, Guiry en Vexin; 7 – La Cave aux Fées, Brueil en vexin; 8 – Can Pey, Montferrer; 9 – Chantambre, Buno Bonneveaux; 10 – Derrière les Prés, Ernes; 11 – Dolmen de la Bouissière, Cabasse; 12 – Dolmen de la Pierre, Folletière; 13 – Dolmen de Villaine, Sublaines; 14 – Dolmen 2 du Frau, Cazals; 15 – Dolmen les Places, Nant; 16 – Essomes sur Marne, Essomes sur Marne; 17 – Eybral, Coux et Bigarroque; 18 – Genevray, Thonon; 19 – Grotte de Montou, Corbère les cabanes; 20 – La Grotte du Queroy 1, Chazelles; 21 – Hypogée du Capitaine, Grillon; 22 – La Butte Saint-Cyr, Val de Reuil; 23 – La Grotte Murée 1, Montagnac-Montpezat; 24 – Dolmen de la Haute Suane, Sainte Maxime; 25 – La Hoguette, Fontenay le Marmion; 26 – La Pierre Virante, Xanton Chassenon; 27 – La Porte aux Bergers, Vignely; 28 – La Sablonnière, Passy; 29 – La Truie Pendue, Passy-Veron; 30 – La Vente des Forts, Mauny; 31 – Laris-Goguet, Feigneux; 32 – La Hazoy 1, Compiègne; 33 – Le Paradis 2, Noisy sur Ecole; 3 – Le Vieux Tordoir 2, Berry au Bac; 35 – Le Villard 2, Le Lauzet Ubaye; 36 – Les Fiefs, Orville; 37 – Les Fontinettes 1, Cuiry les Chaudardes; 38 – Les Maillets,

Germigny-l'Evêque; 39 – Les Mournouards 2, Mesnil sur Oger; 40 – Les Noisats, Gurgy; 41 – Les Petits Prés, Léry; 42 – Les Réaudins 13, Balloy; 43 – Les Réaudins 1, Balloy; 44 – Les Terres de la Chapelle, Entzheim; 45 – Les Truels 2, Millau; 46 – Macherin, Monéteau; 47 – Montiou 1, Sainte Soline; 48 – Moulin du Roc, Saint-Chamassy; 49 – Narbons, Montesquieu-de-Lauragais; 50 – Pente de Courcelles, Nanteau sur Essonne; 51 – Pontcharaud 2, Clermont-Ferrand; 52 – La Ferme du Port, Guiry en Vexin; 53 – Reichstett 2, Reichstett; 54 – La Sablière, Passyrichebourg; 55 – Rosenmeer, Rosheim; 56 – Sur les Patureaux, Chichery; 57 – Terrasse Lavimona, Villeneuve tolosane/cugnaux; 58 – Varennes, Val de Rueil; 59 – Villevieille, Demandolx; 60 – Montbeyre la Cadoule, Eteyran; 61 – Dolmen de Devezas, Saint Maurice de Navacelle; 62 – Pierre Levée 8, Nieul sur l Autize; 63 – Le Cres 1, Beziers; 64 – Champ du Quercy 2, La Couvertoirade; 65 – Camp des Armes, Lapanouse de Cernon; 66 – Grotte d'Artigaou, Esparros; 67 – Aven des Breguières, Mougins; 68 – Grotte de la Tuilerie 2, Gondenans Montby; 69 – La Croix Maigret, Berry au Bac; 70 – Cambous, Viols en Laval; 71 – Dolmen de la Prunarède 1, Saint Maurice de Navacelle; 72 – Porte Joie XIV, Val de Reuil; 73 – Grotte 2 de la Trache, Chateaubernard; 74 – La Gandille, Saint Marcel; 75 – Porte Joie 1, Val de Reuil; 76 – La Carrière de Beaulieu, Bardouville; 77 – Champ Chalon I C, Benon; 78 – Dolmen de la Caumette, Notre Dame de Londres; 79 – Dolmen de la Chatre, Santoche; 80 – Rouffach Gallbuehl, Colmar; 81 – Violette, Riedisheim; 82 – Rixheim, Rixheim; 83 – Dolmen des Périères, Villedubert; 84 – Le Monna, Milhau; 85 – La Brémonde, Buoux; 86 – Aven de la Boucle, Corconne; 87 – Aven Ka, Tharaux; 88 – La Coste 2, Saint Haon; 89 – Peyrolebade 3, Espalion; 90 – La Grotte du Gardon 3, Ambérieu en Bugey; 91 – Le Culfroid 1, Boury en Vexin (map: Mélie Le Roy).

Table 2.1. Number of sites where a significant lack of infants was identified relative to the total number of sites considered and details of which age classes are missing.

Time period	Number of sites with a lack of young children	Total number of sites	% of sites with a lack of young children	Lacking age classes (years)
Early Neolithic	2	5	40	0–4
Middle Neolithic	13	31	42	0–4
End of the Neolithic	33	74	45	0–4
Total	48	110	44	

Thirteen sites of Middle Neolithic date show a significant lack of young individuals across the variety of contexts represented. Most of the sites (n=10) comprise open air necropolises (e.g. La Goumoizière, Saint Martin-la-Rivière; Patte 1971; Verjux *et al.* 1998; Chambon 2003), while the others (n=3) are collective burials in megalithic structures (e.g. La Pierre Virante, Xanton Chassenon; Joussaume 1976; Joussaume and Gruet 1977; Chambon 2003).

Finally, the End of the Neolithic includes 33 funerary sites where infants are significantly missing. Most of the burials are collective either in megalithic structure (e.g. Dolmen la Caumette, Notre Dame de Londres; Bec Drelon *et al.* 2014), pits (e.g. Les Réaudins, Balloy; Chambon and Mordant 1996), hypogea (e.g. Essomes sur Marne; Masset 1995; 1997) or in caves (e.g. Clos d'Ayan, Vesc; Beeching *et al.* 1987).

Table 2.2. Number of sites depending on their domestic or funerary nature and the number of sites involving infants' burials. Total number of individuals, immature individuals and infants, depending on the chronology.

Time period	Domestic area with burials	% of sites with infants	Funerary area	% of sites with infants	Number total of sites	% of sites with infants	Number of immature individuals (including infants)	% of infants	Total number of individuals
Early Neolithic	11	27	8	25	19	26	96	8	261
Middle Neolithic	20	35	53	39	73	38	559	9	1431
End of Neolithic	20	25	125	42	145	40	1706	7	5771
Total	51		186		237		2361		7463

Location of the Burials

Burials are located in two main archaeological contexts – a domestic area where graves shared space with domestic structures, and a funerary area reserved only for burials and therefore dedicated to the dead. Considering the individuals buried in these different spaces we can observe a funerary selection dependent on age-at-death (Table 2.2; Fig. 2.3).

The Early Neolithic sites show mostly burials within domestic areas (n=11), among which only three sites yielded infants. These are mostly deposited as single burials, but close to older individuals (immature or adult). For example, an infant who died around birth from the site of Les Fontinettes, Cuiry les Chaudardes, was buried in the same grave as an older individual (Bailloud 1976; Constantin et al. 2003; Dubouloz et al. 2005). Only eight funerary sites have led to the discovery of immature individuals, and only two of these include infants among their buried population. These individuals are again always closely associated with older individuals, especially the infants buried in the collective burial at Aven des Bréguières, Mougins, mentioned above.

A larger number of archaeological sites were identified for the Middle Neolithic. A total of 20 sites, where both domestic and funerary areas are present, contained graves associated with immature individuals, including seven that yielded the remains of infants. Several sites reflect a strong funerary selection, for example, the site of La Croix Maigret, Berry au Bac, which included three individuals (one immature and two adults), and shows the only evidence for the inclusion of perinatal infants among the immature cohort (Dubouloz et al. 2005; Pariat 2007). The other sites showed a wider funerary selection where, in each case, a young individual (zero to four years) is buried in association with older individuals (immature or adult). The site of Terrace Lavimona, Villeneuve Tolosane, for example, contained two individuals aged less

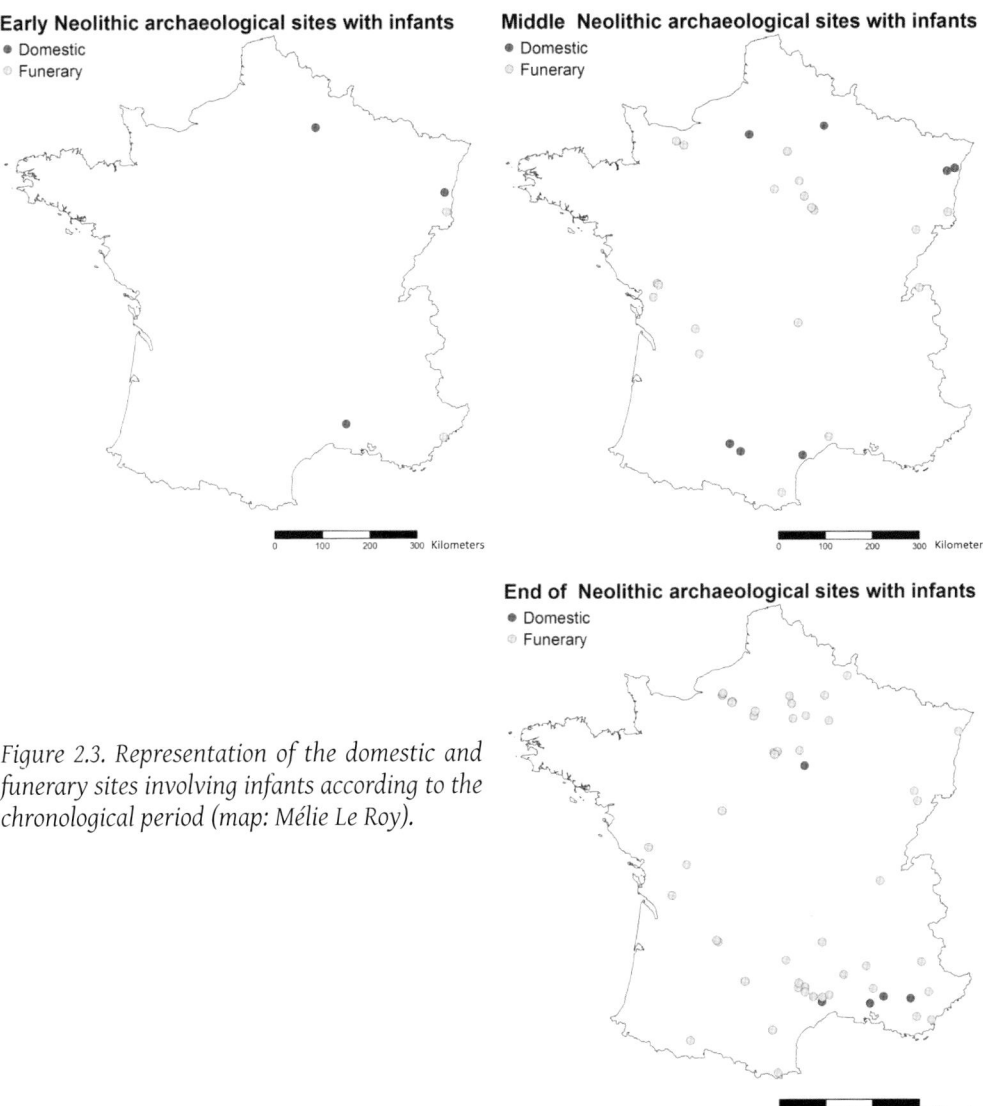

Figure 2.3. Representation of the domestic and funerary sites involving infants according to the chronological period (map: Mélie Le Roy).

than one year, two individuals around nine years old, an individual of 5–14 years, an immature individual of unknown age, and more than 13 adults (Méroc and Simonnet 1979).

A total of 53 Middle Neolithic funerary sites delivered graves containing immature individuals. Some of these only contained one burial, for example, the megalithic structure of Montbeyre La Cadoule, Teyran, was solely dedicated to an infant who died around birth (Laboucarié and Arnal 1989). On the contrary, collective burial sites contained the remains of a lot of individuals including, in certain cases, infants. Among the ten individuals buried in the cave of Montou, Corbère les Cabanes, for example, five immature subjects were identified – a perinatal infant and four individuals less than nine years at death (Beeching and Crubézy 1998). The other funerary contexts are all sites which contain clusters of burials. Although the majority of individuals are buried in individual graves, several cases of multiple burials[3] were also recorded where no age class seems excluded. The funerary site of Les Patureaux, Chichéry, for example, has delivered two double burials involving infants who died around birth. They are respectively associated with an adult woman, and an individual aged over 15 years who was buried in a secondary deposit (Soulier 1998; Thomas *et al.* 2011).

Very few domestic sites (n=20) with associated funerary structures have been identified for the End of the Neolithic, and among these only five included the remains of infants. Three cases of collective burials in habitation contexts were identified. The example of La Truie Pendue, Passy-Véron, included six infants among a total of 68 individuals buried in a pit. A special treatment appears to have been associated with the youngest individuals – they were all gathered around the edges of the structure, whereas older subjects were located all over the surface of the pit (Le Roy *et al.* 2014; Le Roy 2015). The other domestic sites represent isolated burials that contained the remains of infants (e.g. Cambous; Poulain 1978).

In the same period, a total of 125 strictly funerary sites were identified. These sites are represented either by groupings of single, double or multiple burials as well as by collective burials. Nineteen sites represent the first situation, as, for example, the site of Les Petits Prés, Léry, where two individuals who died around birth were buried along with an adult (Billard *et al.* 1998). A lot of sites represent the second case and some 106 collective grave sites were identified. Some sites contain only one age group within their buried population, for example, the site of Réaudins, Balloy, where only one perinatal individual was identified in a grave along with 45 adults. Other sites contained infants associated with older individuals (immature and adult), as was the case at the site of Reichstett where one infant was buried together with three older immature individuals and seven adults (Thévenin *et al.* 1977).

Funerary Practices

Only data for the Middle and the End of the Neolithic were available regarding funerary practices. These comprised some 20 perinatal infants and showed a wide diversity of body positions and associated grave furniture (Fig. 2.4).

Five burial positions were observed during the Middle Neolithic, whereas only the most common position – supine – was observed at the end of the period. We can note that the supine position is exclusive to sites located in the north of France, where it has been identified in the case of 11 infants of both the Middle and End of the Neolithic. It also occurs in both domestic sites (e.g. Rosenmeer, Rosheim; Boës 2003; Leprovost and Queyras 2011) as well as in strictly funerary sites (e.g. Les Réaudins, Balloy; Chambon and Mordant 1996). All of the sites involve burial groupings but the absence of information for collective graves is simply a reflection of the fact that it is difficult to identify individual body positions inside these structures. The other four

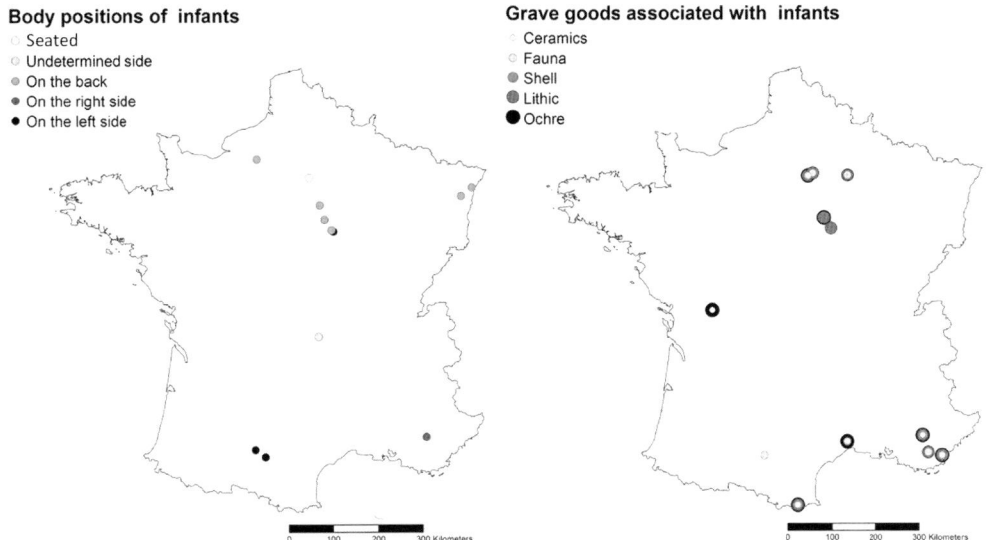

Figure 2.4. Distribution of the various infant body positions from both the Middle Neolithic and the End of the Neolithic; The nature of the grave goods associated with infants from both the Middle Neolithic and the End of the Neolithic (map: Mélie Le Roy).

body positions do not seem to be restricted to any particular area, although very few individuals are involved and the reliability of this trend may be problematic. The case of an infant buried in a seated position at the site of La Porte aux Bergers, Vigney, for example, does not reflect an actual trend but rather an exceptional case since only five other occurrences were registered for the entirety of the Neolithic, none of which involved infants or displayed similarities in relation to the type of site or geographic area (Allard 1999; Chambon and Lanchon 2003; Thomas *et al.* 2011).

Very few immature individuals were buried with grave goods. Indeed, only 20 infants from both the Middle and the End of the Neolithic were buried with ochre, flints, shells, fauna remains or ceramics. None of these grave goods show any particular association with infants over time or geographic area. Only shell elements, transformed into jewellery, occur exclusively at the End of the Neolithic, regardless of whether the infants were buried in domestic or funerary sites or in collective or single burials.

Discussion

The vision of the identity that one can have in archaeology is by definition subjective because we can neither identify with a long-gone culture nor build an identity through interaction with other people and we are therefore influenced by our own culture. The interpretation of the role of individuals within their society is therefore based primarily on understanding the relation that material culture has with cultural identity or, more precisely, how material culture is exploited to produce an identity (Diaz Andreu 2005).

As we have seen from the results there is a significant lack of infants in a variety of sites from the Neolithic. If we exclude all taphonomic causes related to the excavation itself, these under-representations may be a reflection of cultural choice. We demonstrated for the site of Gurgy (Le Roy 2015), that the under-representation of infants indeed illustrated a cultural choice, due to the representation of children aged one to four years which is consistent with theoretical values. Thus, younger individuals were not subject to the same funeral treatment as the rest of the population, and they were excluded from what might be perceived as 'normal' funeral rites. Indeed, this particularity has often been observed in social anthropological studies, for example, among the Dayaks of the Indonesian peninsula and the Papuans of New Guinea, when infants die, they are deposited in a dead tree trunk or hung from the branches of a tree (Hertz 1970). In an archaeological context, this is difficult or impossible to identify. Therefore, it may not be possible to identify the funerary practices that were applied to certain individuals on the basis of their age through the archaeological evidence. This does not necessarily call into question the social consideration of these young children but rather expresses a different meaning to their death to that of the rest of the group.

Thus, it is possible that for some cultures of the Neolithic, when such under-representation is observed, differential funerary treatment could have been given to those members of the population who died before their social integration into the community. Such a conception of the organisation of society can then explain the recurrent under-representation of infants highlighted in this work. The same absence is observed in the rest of Europe, such as at the contemporary German site of Talheim (Whittle 1996; Champion et al. 2009.). This site is interpreted as a pit where an entire group would have been buried after a massacre (34 individuals), including men, women and children. No individuals less than one year of age were discovered within the burial and it has been proposed that these very young children were swept away by the individuals responsible for the massacre and taken to the group of their 'captors' (Taylor and Marshall 1996). This would infer that before a certain age, children are not considered to formally belong to the group into which they were born and can therefore be integrated within another community.

Research has highlighted that the use of space, whether intended as a strictly funerary area or not, can be socially defined by rules inherent to age (Diaz Andreu 2005). In our study of infant graves it was possible to demonstrate a differentiation between sites used for both burials and domestic structures compared to those used for strictly funerary purposes. A first observation was that the use of sites with domestic features, which also accommodate funerary structures, decreased over time. Indeed, most burials identified for the Early Neolithic occur within a habitation context, whereas, at the End of the Neolithic, this trend is reversed. This distancing of burial sites in relation to habitation could be a symbolic action (Baudry 2006). Indeed, graves located inside, or in the immediate vicinity of, a house symbolise the direct involvement of death in everyday life. On the contrary, the use of strict burial sites suggests a symbolic and geographical distancing from the living. Again, Central Europe provides similar observations at the End of the Neolithic and a complete absence of very young children is observed in burials found in a domestic context. This data has been interpreted as a removal of infants from the centres of community power which may mean that their death has a different meaning compared to that of individuals who did have access to these burial places (Chapman 1997).

This study revealed that a specific proportion of immature individuals were preferentially buried in domestic areas. For the Early Neolithic, one can observe the presence of infants, although no age group was specifically chosen to be buried in the context of settlements. This is the case at Les Fontinettes, for example, where two individuals are aged less than five years at death, including one who died in the perinatal period. Other older individuals are also present in these domestic sites, however, and comprised older immature and adult individuals. It must be remembered that most of the graves at that time are associated with domestic structures and that it is not 'abnormal' to find all ages of a population present at these sites. The Middle Neolithic shows continuity in the use of domestic sites to accommodate burials but the number of such sites, relative to those that were strictly funerary in nature, decreases. The age groups present on domestic sites seem subject to selection, however, in contrast to the previous period and only one or two age groups are represented although these are not necessarily consistent throughout the territory. The site La Croix Maigret, for example, contained the remains of individuals who died around birth, and a change in behaviour is therefore noticeable between the Early and Middle Neolithic. During the latter period the remains of immature individuals, including infants, are present in both strictly funerary sites as well as in graves found in domestic contexts. An even greater change between the numbers of funerary and domestic sites which delivered graves is evident at the End of the Neolithic (only 20 sites present burials in a domestic context out of the 145 recorded) and a clear difference in the selection of individuals seems to occur at that time.

At the same time in Central Europe, religious symbolism and ritual behaviours were paralleled with social changes and new ideologies. Children are seen to have an important role providing connections to the supernatural world and infant graves are sometimes discovered as foundation deposits (e.g. Moses 2008). This does not necessarily correspond to sacrifice, but rather may have been an opportunistic exploitation of natural deaths. In addition, Ian Hodder (1990, 29) introduced the idea of child-care in which some domestic spaces were devoted to labour, education and care reserved for children. When children died, they were then buried within the space they would have used during life. The region of the Danube Gorge in Serbia at the beginning of the Neolithic, for example, delivered exceptional sites such as Lepenski Vir and Vlasac (Borić and Stefanović 2004; Stefanović and Borić 2008), where more than 60 perinatal infants have been uncovered in the floors of houses interpreted as nurseries (Stefanović 2006). Such interpretations are not entirely compatible with the results obtained for the French Neolithic sites, with the exception perhaps of a few exceptional cases such as the site at La Croix Maigret where perinatal infants represent the immature cohort (Dubouloz *et al.* 2005; Pariat 2007).

Also, the Neolithic period shows a wide diversity concerning grave goods and no sex related conclusions could be made. Therefore, it is not yet possible to gender a burial through the artefacts. Nevertheless, a study of the previously mentioned site of Gurgy Les Noisats noted the male character of grave goods found with young children (Le Roy 2015). Looking at the specific placement of the burials it was also possible to highlight a recurrent association between young children and adult males, a finding that was interpreted as evidence of a patriarchal society, and which has been confirmed by ancient DNA analysis (Le Roy *et al.* 2016).

Finally, recent work (Thomas *et al.* 2011) has highlighted the strong symbolism that may be apparent in the grave goods associated with a burial. Indeed, in the Cerny Culture of the Middle Neolithic, a significant number of arrowheads were discovered in single burials of male adults or children. These objects have been associated with the practice of archery and hunting and it was suggested that they represent the tool kit of 'experienced hunters'. Two individuals with this type of grave good, however, were too young to have been able to undertake hunting effectively – a two-year-old child from the site of Chichery and a nine-month-old infant from the site of the Sablonnière. The status of hunter would have been awarded posthumously, thereby suggesting a transmission of status from birth, and giving them access to these burial accoutrements alongside adults. Thus it is possible that the exceptional abundance of grave goods apparent in certain infants graves could be related to the status of an important person within the community and that certain associated children were accorded the same importance from birth (van Gennep 1909 [1960]).

Conclusions

The identification of sites showing a significant lack of infants and the fact that they do not belong necessarily to the same burial space as other individuals from society is not shocking. Indeed, these individuals have rarely been integrated in death with the whole population, suggesting that they should have a specific funerary treatment, which illustrates a different social status. Indeed, such individuals would not have been completely integrated within the social group and therefore may not have needed a long and painful departure from it. Following the example of the Dayak tribe discussed above it is possible that infants were afforded a particular treatment in death that is currently archaeologically invisible. Also, the archaeological research undertaken to date suggests that very few infants were provided with grave goods that potentially reflect the heredity nature of social status among these cultural groups. This observation is suggestive that these populations were ranked and that several statuses existed.

Bearing in mind that the idea of childhood in populations of the past may have been different from modern conceptions, however, it seems that this idea is very variable across different chronologies and geographical locations (Ariès 1962; Lucy 2005). In our society, the loss of a child in a family is difficult, if not impossible, to overcome, but others societies do not have the same standards and do not consider the loss of a child in such a desperate point of view, depending on the difficulties encountered in the everyday life or the presence of high infant mortality (e.g. Scheper Hughes 1992), which may have been the case in Neolithic times. This demonstrates that our understanding of the social status of infants, and more generally of childhood, in past societies needs to be enhanced with further studies and the development of new methodological approaches in order to truly gain an understanding of their place within ancient societies.

Acknowledgements

I would like to thank the researchers who allowed me to publish their photography (see Fig. 2.1): Sandrine Thiol, Noisette Bec Drelon and Stéphane Rottier. This research was funded by a ministerial grant from the Research National Agency as a programme of prospects investment ANR-10-LABX-52 (project 'Diversité biologique et culturelle de l'Homme de la fin de la Préhistoire à la Protohistoire'; dir: SR; Université de Bordeaux 1, LaScArBx-ANR; 2012–4) and has also been undertaken thanks to a PhD research grant from the Ministère de l'Enseignement Supérieur et de la Recherche.

Notes

1. UMR 5199 – PACEA, Université de Bordeaux, Bat. B8 Allée Geoffroy Saint Hilaire, CS 50023, F – 33615 Pessac CEDEX, France. Email: melieleroy@hotmail.fr.
2. Traditionally, the end of the period is classified into two sub-periods – the Recent Neolithic and Late Neolithic. In order, once again, to avoid too much particular specificity we will consider both periods as a whole and refer to the period as the 'End of the Neolithic'.
3. Burials containing more than two individuals buried at the same time as determined by Duday (2009).

References

Allard, P. 1999. L'industrie lithique du groupe de Villeneuve-Saint-Germain des sites de Bucy-le-Long (Aisne). *Revue Archéologique de Picardie* 3–4, 53–113.
Ariès, P. 1962. *L'Enfant et la Vie Familiale sous l'Ancien Régime*. Paris: Plon.
Bailloud, G. 1976. Le Néolithique en Picardie. *Revue Archéologique de l'Oise* 7, 10–28.
Baudry, P. 2006. *La Place des Morts. Enjeux and Rites*. Paris: Editions L'Harmattan.
Bec Drelon, N., Le Roy, M. and Recchia Quiniou, J. 2014. Autour de la chambre: nouveaux éléments de réflexion sur les structures tumulaires. Apport des fouilles récentes de cinq dolmens de l'Hérault, in Sénépart, I., Léandri, F., Cauliez, J., Perrin B. and Thirault, E. (eds.), *Chronologie de la Préhistoire Récente dans le Sud de la France: Acquis 1992-2012/Actualité de la Recherche*, 569–582. Porticcio: Archives d'Ecologie Préhistorique.
Beeching, A. and Crubézy, E. 1998. Les sépultures chasséennes de la vallée du Rhône, in Guilaine J. (ed.), *Sépultures d'Occident and Génèses des Mégalithismes (9000-3500 avant notre ère)*, 147–164. Paris: Editions Errance.

Beeching, A., Brochier, J. L., Matteucci, S., Pahin A. C. and Thiercelin, F. 1987. Les sépultures et dépôts d'ossements humains dans le chasséen de la moyenne vallée du Rhône, in Beeching, A. (ed.), *Actes des Rencontres Néolithique de Rhone-Alpes*, vol. 3, 75–83. Lyon: Centre d'Archéol. Préhist. de Valence.

Billard, C., Querre, G. and Salanova, L. 1998. Le phénomène campaniforme dans la basse vallée de la Seine: chronologie and relation habitats-sépultures. *Bulletin de la Société Préhistorique Française* 95, 351–364.

Boës, E. 2003. Comportements funéraires, modifications sociales and mentalités, aux VIe and Ve millénaires avant J.-C. en Alsace, in Chambon, P. and Leclerc, J. (eds.), *Les Pratiques Funéraires Néolithiques avant 3500 av. J.-C. en France and dans les Régions Limitrophes*, 33–43. Saint-Germain-en-Laye: Société Préhistorique Française.

Borić, D. and Stefanović, S. 2004. Birth and death: infant burials from Vlasac and Lepenski Vir. *Antiquity* 78, 526–546.

Chambon, P. 2003. *Les Morts dans les Sépultures Collectives Néolithiques en France: du Cadavre aux Restes Ultimes.* Paris: Gallia Préhistoire.

Chambon, P. and Lanchon, Y. 2003. Les structures sépulcrales de la nécropole de Vigneły (Seine-and-Marne), in Chambon, P. and Leclerc, J. (eds.), *Les Pratiques Funéraires Néolithiques avant 3500 av. J.-C. en France et dans les Régions Limitrophes*, 159–173. Saint-Germain-en-Laye: Société Préhistorique Française.

Chambon, P. and Mordant, D. 1996. Monumentalisme et sépultures collectives à Balloy (Seine-et-Marne). *Bulletin de la Société préhistorique française* 93, 396–402.

Champion, T., Gamble, C., Shennan, S. and Whittle, A. W. 2009. *Prehistoric Europe.* Walnut Creek, CA: Left Coast Press.

Chapman, J. 1997. Changing gender relations in the later prehistory of Eastern Hungary, in Moore, J. and Scott, E. (eds.), *Invisible People and Processes: Writing Gender and Childhood into European Archaeology*, 131–149. London: Leicester University Press.

Constantin, C., Farrugia, J.-P., Bonnardin, S., Guichard, Y. and Sidéra, I. 2003. Les tombes de la vallée de l'Aisne. Présentation, in Chambon, P. and Leclerc, J. (eds.), *Les Pratiques Funéraires Néolithiques avant 3500 av. J.-C. en France et dans les Régions Limitrophes*, 55–63. Saint-Germain-en-Laye: Société Préhistorique Française.

Diaz Andreu, M. 2005. Gender identity, in Diaz Andreu, M., Lucy, S., Babic, S. and Edwards, D. N. (eds.), *The Archaeology of Identity: Approaches to Gender, Age, Status, Ethnicity and Religion*, 13–42. London: Routledge.

Dubouloz, J., Bostyn, F., Chartier, M., Cottiaux, R. and Le Bolloch, M. 2005. La recherche archéologique sur le Néolithique en Picardie. *Revue Archéologique de Picardie* 3-4, 63–98.

Duday, H. 2009. *The Archaeology of the Dead: Lectures in Archaeothanatology.* Oxford: Oxbow Books.

Hertz, R. 1970. *Sociologie Religieuse et Folklore.* Paris: Presses Universitaires de France.

Hodder, I. 1990. *The Domestication of Europe. Structure and Contingency in Neolithic Societies.* Oxford: Basil Blackwell.

Jeunesse, C. 1995. Les groupes régionaux occidentaux du Rubané (Rhin and Bassin parisien) à travers les pratiques funéraires. *Gallia Préhistoire* 37, 115–154.

Jeunesse, C. 1998. Pratiques funéraires and sociétés danubiennes au Néolithique ancient, in Guilaine, J. (ed.), *Sépultures d'Occident and Genèse des Mégalithes (9000-3500 avant notre ère)*, 41–58. Paris: Editions Errance.

Jeunesse, C., Lambach, F., Mathieu, G. and Mauvilly, M. 1993. La nécropole rubanée d'Ensisheim Les Octrois (Haut-Rhin): Conclusions: Ensisheim Les Octrois (Haut-Rhin), une nécropole rubanée de Haute-Alsace. *Cahiers de l'Association pour la Promotion de la Recherche Archéologique en Alsace* 9, 81–88.

Joussaume, R. 1976. Etude architecturale and archéologique. *Gallia Préhistoire* 19, 1–38.

Joussaume, R. and Gruet, M. 1977. Le mégalithe de La Pierre Virante à Xanton-Chassenon (Vendée). *L'Anthropolgie* 81, 5–66.

Laboucarié, S. and Arnal, G. B. 1989. La sépulture chasséenne (L. IV) du gisement de Montbeyre-la-Cadoule, Teyran (Hérault). *Archéologie en Languedoc* 4, 27–33.

Le Roy, M. 2015. Les Enfants au Néolithique: Du Contexte Funéraire à l'interprétation Socioculturelle en France de 5700 à 2100 ans av. J.-C. Unpublished Ph.D. thesis, Université de Bordeaux.

Le Roy, M., Rottier, S., De Becdelièvre, C., Thiol, S., Coutelier, C. and Tillier, A.-M. 2014. Funerary behaviour of Neolithic necropolises and collective graves in France. Evidence from Gurgy 'Les Noisats' (middle Neolithic) and Passy/Véron 'La Truie Pendue' (late Neolithic). *Archäologisches Korrespondenzblatt* 3, 337–351.

Le Roy, M. and Rivollat, M., Mendisco, F., Pemonge, M. H., Coutelier, C., Couture, C., Tillier, A.-M., Rottier, S., Deguilloux, M. F. 2016. Distinct ancestries for similar funerary practices? A GIS analysis comparing funerary, osteological and aDNA data from the Middle Neolithic necropolis Gurgy 'Les Noisats' (Yonne, France). *Journal of Archaeological Science* 73, 45–54.

Leprovost, C. and Queyras, M. 2011. La nécropole d'Entzheim (Bas-Rhin): nouvelles données sur le Néolithique moyen alsacien, in Denaire, A., Jeunesse, C. and Lefranc, P. (eds.), *Nécropoles et Enceintes Danubiennes du Ve millénaire dans le Nord-Est de la France et le Sud-Ouest de l'Allemagne*, 115–126. Strasbourg: Université Marc Bloch.

Lucy, S. 2005. The archaeology of age, in Diaz Andreu, M., Lucy, S., Babic, S. and Edwards, D. N. (eds.), *The Archaeology of Identity*, 43–66. London: Routledge.

Masset, C. 1995. Une demeure d'éternité construite dans du sable: la sépulture collective d'Essômes-sur-Marne (Aisne). *Revue Archéologique de Picardie* 9, 131–133.

Masset, C. 1997. La sépulture collective d'Essômes-sur-Marne (Aisne). *Revue Archéologique de Picardie* 1–2, 5–17.

Méroc, L. and Simonnet, G. 1979. Les sépultures chasséennes de Saint-Michel-du-Touch à Toulouse (Haute-Garonne). *Bulletin de la Société Préhistorique Française* 76, 379–407.

Moses, S. 2008. Catalhöyük's foundation burials: ritual child sacrifice or convenient deaths? in Bacvarov, K. (ed.), *Babies Reborn: Infant/Child burials in Pre- and Protohistory* (BAR International Series 1832), 45–52. Oxford: Archaeopress.

Pariat, J.-G. 2007. *Des Morts sans Tombe? Le Cas des Ossements Humains en Contexte Non Sépulcral en Europe Tempérée entre les 6° et 3° Millénaires av. J.-C* (BAR International Series 1683). Oxford: Archaeopress.

Patte, E. 1971. La grotte sépulcrale de Larris Goget à Feigneux (Oise). *Bulletins and Mémoires de la Société d'Anthropologie de Paris* 12, 381–452.

Poulain, T. 1978. Etude des vestiges osseux de la cabane 11. *Gallia Préhistoire* 21, 183–188.

Provost, S. 2013. La Galerie Sépulcrale des Bréguières (Mougins, Alpes-Maritimes): Paramètres Quantitatifs et Fonctionnement d'une Sépulture Collective entre le VIème et le Vème Millénaire avant J.-C., Unpublished Master's thesis, University of Bordeaux.

Scheper Hughes, N. 1992. *Death without Weeping: The Violence of Everyday Life in Brazil*. Berkeley, CA: University of California Press.

Soulier, P. 1998. *La France des Dolmens et des Sépultures Collectives (4500–2000 avant J.-C.)*. Paris: Editions Errance.

Stefanović, S. 2006. The domestication of human birth. *Documenta Praehistorica* 33, 159–164.

Stefanović, S. and Borić, D. 2008. New-born infant burials underneath house floors at Lepenski Vir: in pursuit of contextual meanings, in Bonsall, C., Boroneant, V. and Radovanovic, I. (eds.), *The Iron Gates in Prehistory: New Perspectives* (BAR International Series 1893), 131–169. Oxford: Archaeopress.

Suzuki, H. 2000. *The Price of Death: The Funeral Industry in Contemporary Japan*. Stanford, CA: Stanford University.

Tarrête, J. and Le Roux, C. T. 2008. *Le Néolithique*. Paris: Editions Picard.

Taylor, T. and Marshall, Y. 1996. *The Prehistory of Sex: Four Million Years of Human Sexual Culture*. London: Fourth Estate.

Thévenin, A. G., Sainty, J. and Poulain, T. 1977. Fosses et sépultures michelsberg, sablière Maetz à Rosheim (Bas-Rhin). *Bulletin de la Société Préhistorique Française* 74, 608–621.

Thiol, S., Chevrier, S., Labeaune, R., Boitard Bidaut, E., Clerget, J., Lecomue, J., Ligouis, B., Goutelard, A., Desbat, A. and Malette, C. 2010. Passy-Véron (89). Rapport Final d'Opération. Unpublished INRAP, SRA Bourgogne report, Dijon.

Thomas, A., Chambon, P. and Murail, P. 2011. Unpacking burial and rank: the role of children in the first monumental cemeteries of Western Europe (4600–4300 BC). *Antiquity* 85, 772–786.

Thomas, L. V. 1975. *Anthropologie de la Mort*. Paris: Payot.

van Gennep, A. 1909 [1960]. *The Rites of Passage* (transl. Vizedom, M. B. and Caffee, G. L.). Chicago, IL: Chicago University Press.

Verjux, C., Simonin, D. and Richard, G. 1998. Des sépultures mésolithiques aux tombes sous dalles du Néolithique moyen I en région Centre and ses marges, in Guilaine, J. (ed.), *Sépultures d'Occident et Genèse des Mégalithes (9000–3500 avant notre ère)*, 61–70. Paris: Editions Errance.

Whittle, A. W. 1996. *Europe in the Neolithic: The Creation of New Worlds*. Cambridge: Cambridge University Press.

Chapter 3

Perinatal Death and Cultural Buffering in a Neolithic Community at Çatalhöyük

Belinda Tibbetts[1]

Abstract: Identifying cultural responses to infant mortality in past populations can reveal beliefs and approaches toward the funerary treatment of juveniles. The burial remains of very young juveniles are an important source of information and hold considerable potential for providing significant insight into their cultural identity in past populations. This paper presents an interpretation of the burial remains of neonates within the context of community, and the cultural response to infant death that is reflected in the archaeological record at Çatalhöyük. In order to move toward a more holistic understanding of the social identity of these neonatal individuals, results from the bioarchaeological analysis of the skeletal remains, burial environment and burial inclusions will be considered in relation to the cultural context of the contemporary living community. It will be demonstrated that within this Neolithic community, the youngest members were recognised as having a social identity and were provided with culturally sanctioned burial, regardless of their viability at birth.

Keywords: perinatal death, cultural buffering, bioarchaeology, Neolithic, Çatalhöyük.

Introduction

The Neolithic site of Çatalhöyük is located on the Konya plain in central Anatolia, and is one of several mounds that remain visible today (Fig. 3.1). The site was first identified and subsequently excavated by Mellaart in 1961 and is one of several sites within the Konya plain that have been the focus of archaeological investigation. The area of the plain in which Çatalhöyük is located is now dominated by intensively irrigated agricultural land. During the Neolithic period the landscape of the plain was a mixture of alluvial land and freshwater marshes (Kuzucuoğlu 2002, figs. 2–7). Çatalhöyük is one of the earliest known Neolithic settlements, and consisted of adjoining dwellings, accessed through the roof (Hodder 2016, 1). The site is particularly

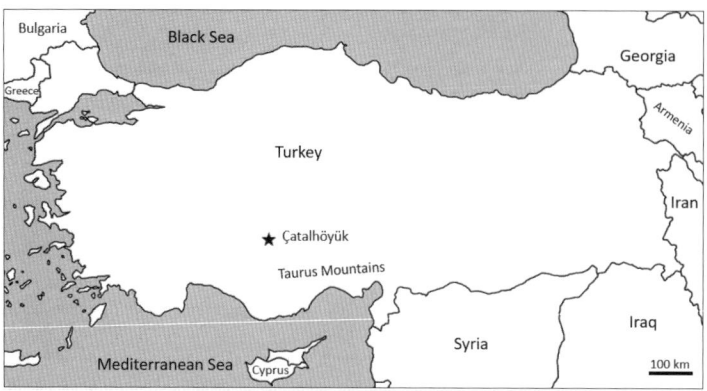

Figure 3.1. Location of Çatalhöyük in central Anatolia, Turkey (prepared by Belinda Tibbetts).

well known for the decorated, plastered interiors of the dwellings and the occupants' practice of intramural burial. The east mound settlement was occupied in the Neolithic between approximately 7100 and 6000 BC (Bayliss *et al.* 2015, 17), with the highest population density occurring between 6700 and 6400 BC. The human remains from this period display increased adult palaeopathology and include the majority of neonatal burials from the site's occupation. The exceptional sequence and preservation at Çatalhöyük, along with artefacts and skeletal remains, provide a wealth of detail for bioarchaeological analysis. By incorporating a broad range of information along with the biological analysis of the human remains it is possible to gain a clearer image of the behaviours that resulted in the burial deposits, allowing us to approach an understanding of community response to mortality within this past population.

Infant Analysis

The remains of individuals who died prior to, or within a short period after, birth represent a record for a very specific period of time in the life of reproductive females within the population. These infant remains have the capacity to inform on maternal health during the pregnancy and in some cases immediately prior to pregnancy. For those that survived into early infancy, the remains reveal aspects of the infant's health in the short period following birth.

Biological analysis of the infant skeletal remains dating to a period spanning the population density peak was undertaken in order to assess their development, growth and pathology. The skeletal analysis incorporated metrics, ossification, growth and fusion, as well as dental development, and utilised methods and recording drawn from standard procedures (Schour and Massler 1941; Fazekas and Kósa 1978; Ubelaker 1989; Scheuer and Black 2004; Schaefer *et al.* 2009; Al Qahtani *et al.* 2010). Each skeletal element was examined for evidence of pathology, and congenital and non-metric traits were noted. This analysis revealed that the infants included individuals from approximately 26 weeks gestation to several months postnatal. Although older infants and children are also present in the burial population, it is more difficult to differentiate between the impacts of maternal health during pregnancy and the external influences during infancy when analysing their skeletal remains. The stages of infant skeletal and dental development at Çatalhöyük correspond to one another for the majority of individuals that died during the perinatal period. However, for some infants that survived into early infancy there is evidence of a delay in skeletal development. Although this observation is yet to be fully investigated, the initial assumption is that environmental stress experienced during periods of high population density would have been sufficient to create such a difference in developmental timing.

Community Response

The majority of individuals buried during the Neolithic occupation at Çatalhöyük were tightly flexed, although neonatal infants were not as tightly flexed as older individuals (Fig. 3.2). Skeletal elements often retain surface staining where binding or strapping was once positioned and occasionally remnants of binding or strapping survive *in situ* within the burial environment. This practice of tightly binding and otherwise constraining the body, although clearly the result of complex behaviours, also appears to be directly associated with convenience of size for burial, and perhaps to a period of prior storage.

The community at Çatalhöyük did not exclude individuals from burial based upon age-at-death. There is considerable variation in other features such as body position, grave goods and inclusions, and burial location, but these do not appear to be dependent upon age or sex. The number and inclusive nature of the burials at Çatalhöyük suggest that the majority of deaths were managed by the resident community and deposited within the settlement. Indications of the taphonomic processes at play within the burial environment are widely encountered in the skeletal remains. The more fragile infant cranial vault elements rarely survive intact and are often highly fragmented, while the post-cranial elements are usually very well preserved. Post-deposition sediment compaction accounts for the majority of skeletal trauma present in infant remains. It is not unusual to have burial disturbance due to rodent burrowing, however there is very little evidence of animal gnawing, which indicates that infant remains were not retained in a more accessible location prior to burial. The practice of re-opening adult burials and removing selected elements, widely practised in this region during the Neolithic (Mellaart 1967, 84; Andrews *et al.* 2005, 274; Molleson *et al.* 2005, 279; Boz and Hager 2013, 433; Haddow *et al.* 2016, 21), was not extended to include young infants, the primary burials of which do not appear to have been deliberately disturbed. As with the many successive events of internal plastering and decoration, the intramural burials were deposited throughout the use of a building space, which in some cases was as long as 100 years (Hodder and Cessford 2004, 22; Matthews 2005, 141–3; Twiss *et al.* 2008, 44). The Neolithic burials at Çatalhöyük appear to be connected with the 'life' of a building. This interpretation is

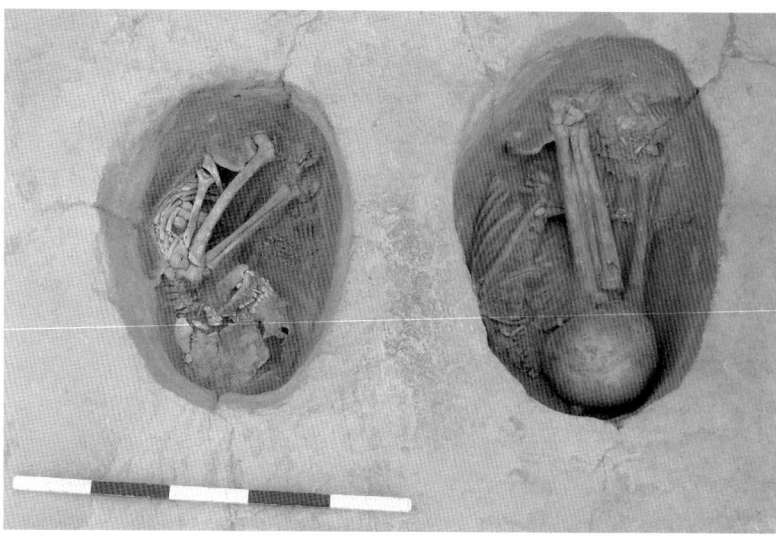

Figure 3.2. Tightly flexed juvenile burials within the east platform of Building 80. Note the burial cuts are specific to the dimensions of the skeletal remains (L-R: Sk.21777, 6 years and Sk.21778, 8 years) (photograph taken by Belinda Tibbetts).

further supported by the association of some infant burials with the lowest levels of a structure's sequence, referred to as foundation deposits (Boz and Hagar 2013, 413), and stratigraphically connected with the establishment of a new building. Infants have most frequently been recovered from a range of intramural locations including under sleeping platforms and in smaller side rooms (Boz and Hager 2013, 419). The majority of these young infants were buried with limited or no associated grave goods. For the cases in which items were recovered with the skeletal remains, these included numerous small shell beads (1mm diameter) and bivalve shells. Many of the infant burials have revealed preserved organic material in close contact with the skeletal remains. These surviving organic materials and phytoliths confirm that infants were typically buried in woven containers, sometimes with lids, or placed within woven mats (Boz and Hager 2013, 421). It is noteworthy that the organic materials contained in the burials of older juveniles and adults were predominantly sedge textiles, while the organic material recovered from the burials of younger infants was derived from a wider range of plant species and included flowering parts (Rosen 2005, 209–10). This suggests that a higher number of young infants were buried during warmer seasons when a wider variety of plant material was available.

The more open body position of infant skeletal remains and the composition of organic material included in infant burials may be explained by the practice of burying infants shortly after their death and the possibility of seasonally higher birth rates. Although the smaller body size and immature development of infants allows for tight flexion, the community evidently did not consider this to be a required body treatment for infants. The tightly flexed body position observed in the burials of older individuals would be more easily achieved following soft tissue reduction through decomposition, suggesting that these remains were being retained for a limited period prior to intramural burial.

The evidence for the more immediate burial of young infants sheds light on the community's perception of these individuals. Although young infants received funerary treatments that resulted in a very similar archaeological expression to those of older individuals, it is apparent that their burial was viewed within a different temporal scale for cultural response. Some burial deposits are stratigraphically contemporary clusters of young infants that are of similar developmental ages and may represent contemporary deaths. While the majority of young infants are in single burials, others were added to existing burial deposits of older individuals. If the normal funerary practice was to retain older individuals for a period prior to burial, it is unusual that following a maternal death, the mother was buried shortly after death, with a more open body position in which the upper limbs were loosely extended beyond the knee joints. This burial may be an indication of the importance given to the rapid burial of infants, born or otherwise, compared with a funerary practice for older individuals that involved the retention of remains for postponed burial.

The burial remains at Çatalhöyük also provide strong support for a social memory of location and form of burial within the Neolithic community. This is seen in the repeated deposition location of subsequent burials with minimal disturbance to surrounding burials, and the repetition of burial form. It may be that there were visible markers or cultural behaviours that resulted in similar burials within buildings.

Discussion

It is clear from analysis of the skeletal remains that the Neolithic community of Çatalhöyük experienced considerable environmental pressures during the period of peak population density. Assuming consistency in funerary practice, an increase in the number of perinates buried during this period suggests increased fecundity and indicates that the population had developed a robust means of coping with the individual stresses these early deaths placed on smaller groups. The impacts of perinatal mortality on individuals within western cultures today are socially recognised in terms of grief and the need for others within the community to provide support both emotionally and logistically. The same impacts are not necessarily applicable to the community of Çatalhöyük, but it is undeniable that the community followed socially recognised and culturally sanctioned behaviours throughout their occupation of the site. These behaviours developed into a relatively consistent burial practice, the end results of which are observed in the burial record.

The social impacts of perinatal mortality in a Neolithic community cannot be determined, but the implications of perinatal mortality for the community at Çatalhöyük can be hypothesised on the basis of our present understanding of their culture. Any loss to the community must have created a need to compensate, whether that would have been through emotional buffering, replacement or substitution, or an imposed delay in planned activities. This relies on the assumption that perinatal mortality, and the loss of this potential human resource, was negatively perceived within the community. It is possible that the community was intentionally managing their resources through selective removal of infants at a time when population pressures were high.

The available evidence reveals that regardless of whether the community was creating or responding to increased perinatal mortality, it was certainly recognising each perinatal infant as an individual who was entitled to a certain manner of treatment following their death. This social acknowledgement of very young individuals, including those that were not viable at birth, demonstrates that the community valued these perinatal individuals not in terms of potential contribution to the living population, but perhaps for kinship or inter-group alliances that would stabilise populations in flux. As such, their value was not altered by their early death, and may even have encouraged a stronger bond between groups within the population. Indeed, Pilloud and Larsen (2011, 527) have demonstrated that individuals buried within the same house were only minimally biologically linked. This finding highlights that individuals were not grouped together in burial on the basis of

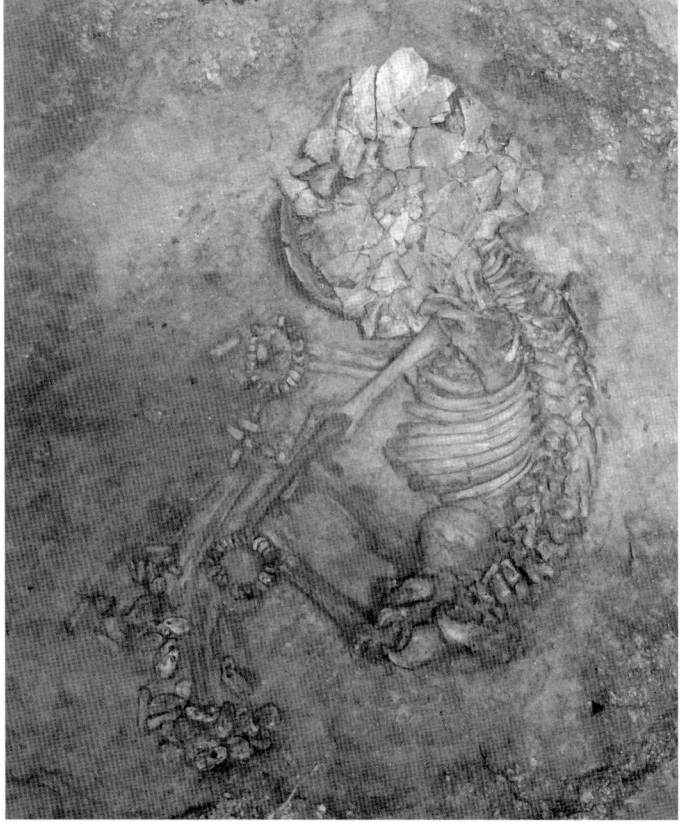

Figure 3.3. Infant burial with items of personal adornment (Sk.4406, 6–12 months, southwest platform of Building 6) (photograph taken by Belinda Tibbetts).

biological relatedness. It is likely that the placement of burials of biologically unrelated individuals within the same building reflects a social organisation based on practical kinship (Pilloud and Larsen 2011, 526; Hillson et al. 2013, 348).

The persistence of social memory and the importance of funerary practice within the community are demonstrated through the repetition of burial location and form (Hodder 2007, 32). This behaviour may also indicate the continuation or re-establishment of connections between groups within the community through burial. Although infants were not necessarily buried with biological relatives, the inclusion of objects in the burials of young infants strongly suggests personal connections with the infant. Items of personal adornment and grave goods are likely to be an integral part of cultural buffering through association (Fig. 3.3). Whether the items included in burials were placed there by biological relatives or gifted to the dead infant by others, the result is a socially visible means of processing the death that assimilates the infant into the community and reaffirms intra-population connections. The evidence from Çatalhöyük strongly suggests that burial location, body treatment and grave goods were all elements of the cultural buffering behaviours of the community in response to infant mortality.

Conclusion

Analysis of the infant remains from the burial population of Neolithic Çatalhöyük provides strong evidence that these youngest members of the community were provided with socially sanctioned funerary treatments and burial. It is clear from the evidence presented above that there was no discrimination based upon age-at-death or viability at birth in the burial of infants. Although a cause of death is not currently accessible for many of the young infants through analysis of their remains, this criterion may not have been an important factor to the community who buried them. Despite their young age, these infants appear to have been recognised as individuals by the living community.

Acknowledgements

The author thanks the Çatalhöyük Project for research access, the Human Remains team for their support, and the editors of this volume for the opportunity to publish.

Note

1. Department of Archaeology, University of Exeter, Exeter, UK. Email: bwt201@exeter.ac.uk.

References

Al Qahtani, S. J., Hector, M. P. and Liversidge, H. M. 2010. Brief communication: The London atlas of human tooth development and eruption. *American Journal of Anthropology* 142, 481–490.
Andrews, P., Molleson, T. and Boz, B. 2005. The human burials at Çatalhöyük, in Hodder, I. (ed.), *Inhabiting Çatalhöyük: Reports from the 1995-9 seasons*, 261–278. Cambridge: McDonald Institute for Archaeological Research.

Bayliss, A., Brock, F., Farid, S., Hodder, I., Southon, J. and Taylor, R. E. 2015. Getting to the bottom of it all: A Bayesian approach to dating the start of Çatalhöyük. *Journal of World Prehistory* 28, 1–26.

Boz, B. and Hager, L. D. 2013. Living above the dead: Intramural burial practices at Çatalhöyük, in Hodder, I. (ed.), *Humans and Landscapes of Çatalhöyük: Reports from the 2000-8 Seasons* (British Institute at Ankara Monograph 47), 413–440. London: British Institute at Ankara.

Fazekas, I. G. and Kósa, F. 1978. *Forensic Fetal Osteology*. Budapest: Akadémiai Kiadó.

Haddow, S. D., Sadvari, J. W., Knüsel, C. J. and Hadad, R. 2016. A tale of two platforms: Commingled remains and the life-course of houses at Neolithic Çatalhöyük, in Osterholtz, A. J. (ed.), *Theoretical Approaches to Analysis and Interpretation of Commingled Human Remains, Bioarchaeology and Social Theory*, 5–29. Cham: Springer International Publishing Switzerland.

Hillson, S. W., Larsen, C. S., Boz, B., Pilloud, M. A., Sadvari, J. W., Agarwal, S. C., Glencross, B., Beauchesne, P., Pearson, J. A., Ruff, C. B., Garofalo, E. M., Hager, L. D. and Haddow, S. D. 2013. The human remains I: Interpreting community structure, health and diet in Neolithic Çatalhöyük, in Hodder, I. (ed.), *Humans and Landscapes of Çatalhöyük: Reports from the 2000-8 Seasons* (British Institute at Ankara Monograph 47), 339–396. London: British Institute at Ankara.

Hodder, I. 2007. Summary of results, in Hodder, I. (ed.), *Excavating Çatalhöyük: South, North and KOPAL Area Reports from the 1995-9 Seasons* (British Institute at Ankara Monograph 37) 25–37. London: British Institute at Ankara.

Hodder, I. 2016. More on history houses at Çatalhöyük: a response to Carleton et al. *Journal of Archaeological Science* 67, 1–6.

Hodder, I., and Cessford, C. 2004. Daily practice and social memory at Çatalhöyük. *American Antiquity* 69, 17–40.

Kuzucuoğlu, C. 2002. Environmental setting and evolution from the 9th to the 5th millennium cal BC in Central Anatolia: an introduction to the study of relations between environmental conditions and the development of human societies, in Thissen, L. and Gerard, F. (eds.), *The Neolithic of Central Anatolia: Internal Developments and External Relations During the 9th-6th millennia cal BC* (Proceedings of the International CANeW Round Table, Istanbul 23-4 November 2001) 33–58. Istanbul: Ege Yayinlari.

Matthews, W. 2005. Life-cycles and life-courses of buildings, in Hodder, I. (ed.), *Çatalhöyük Perspectives: Themes from the 1995-9 Seasons*, 125–149. Cambridge: McDonald Institute for Archaeological Research.

Mellaart, J. 1967. *Çatal Höyük. A Neolithic Town in Anatolia*. London: Thames and Hudson.

Molleson, T., Andrews, P. and Boz, B. 2005. Reconstruction of the Neolithic people of Çatalhöyük, in Hodder, I. (ed.), *Inhabiting Çatalhöyük: Reports from the 1995-9 Seasons*, 279–300. Cambridge: McDonald Institute for Archaeological Research.

Pilloud, M. A. and Larsen, C. S. 2011. 'Official' and 'practical' kin: Inferring social and community structure from dental phenotype at Neolithic Çatalhöyük, Turkey. *American Journal of Physical Anthropology* 145, 519–530.

Rosen, A. M. 2005. Phytolith indicators of plant and land use at Çatalhöyük, in Hodder, I. (ed.), *Inhabiting Çatalhöyük: Reports from the 1995-9 Seasons*, 203–212. Cambridge: McDonald Institute for Archaeological Research.

Schaefer, M., Black, S. and Scheuer, J. L. 2009. *Juvenile Osteology: A Laboratory and Field Manual*. Amsterdam: Elsevier.

Scheuer, L. and Black, S. 2004. *The Juvenile Skeleton*. London: Elsevier and Academic Press.

Schour, I. and Massler, M. 1941. The development of the human dentition. *Journal of the American Dental Association* 28, 1153–1160.

Twiss, K. C., Bogaard, A., Bogdan, D., Carter, T., Charles, M. P., Farid, S., Russell, N., Stevanović, M., Yalman, E. N. and Yeomans, L. 2008. Arson or accident? The burning of a Neolithic house at Çatalhöyük, Turkey. *Journal of Field Archaeology* 33, 41–57.

Ubelaker, D. H. 1989. *Human Skeletal Remains: Excavation, Analysis, Interpretation*. Washington DC: Taraxacum.

Chapter 4

Burying Children and Infants at Kadruka 23: New Insights into Juvenile Identity and Disposal of the Dead in the Nubian Neolithic

Emma Maines,[1] Pascal Sellier,[2] Philippe Chambon[2] and Olivier Langlois[3]

Abstract: Kadruka 23 is a relatively undisturbed funerary mound located in the Northern State of the Sudan, dating to the fifth millennium, also known as the Nubian Middle Neolithic. Begun in 2014, the excavations at the site have already brought to light nearly 40 burials, a majority of which belongs to juvenile subjects. A concentration and seemingly codified treatment of very young individuals was immediately remarked upon. With the knowledge that such practices have been observed elsewhere, this hypothesis about differential treatment based on age, is one we wish to develop in further detail. The demarcation of the very young deceased is underscored by several phenomena. We have identified a preferential placement within the cemetery (at the top of the mound), the association of mortuary goods not found with older individuals (bead waistbands, ceramic and shell spoons, etc.), consistent placement of ceramics (slightly above the burial level), as well as a variety of burial positions (not seen in the significantly more codified disposal of the older deceased). Our aim is to improve comprehension of the identity of the young dead, and their place in society before and after death.

Keywords: Immature, burial, cemetery, mortuary goods, archaeology of death, Neolithic, Nubia, Sudan

Introduction

At what moment does childhood end and adolescence begin? Should adolescence be considered as a merely transitory phase on the more important passage to adulthood? Or can adolescence be interpreted as an actual life stage recognised in its own right? Since the dawn of anthropology as a discipline, and certainly since Arnold van Gennep's (1909) seminal work, anthropologists have been fascinated by these transitional periods and the symbolism attached to life, growth, decline and death. 'The life of an individual in any society is a series of passages from one age to another and from one occupation to another. ... Transitions from group to group and from one social situation to the next are looked on as implicit in the very fact of existence, so that a man's life comes to be made up of a succession of stages with similar ends and beginnings: birth, social puberty, marriage, fatherhood, advancement to a higher class, occupational specialization, and death' (van Gennep 1909 [1960], 2–3).[4] For biologists, and even ethnographers, the analysis of the shift from childhood to adolescence or adolescence to adulthood, while varied in its manifestations, remains relatively simple to identify. When the analysis of this process is transferred to the domain of physical and biological anthropology or archaeology of death and of the deceased, it becomes more complex. The researcher's vantage point is temporally removed from the context being studied, and therefore the boundary between the biological and the social may become blurred. van Gennep differentiates, for example, between physiological puberty and social puberty. In the context of this project, the archaeological record necessitated taking into consideration a potential divide between the societal definitions of life stages as opposed to biological realities.

To speak of religion for prehistoric periods is a difficult and in some ways an impossible task although we can theorise about phenomena adjacent, and sometimes correlated, to notions of religion and the domains of the symbolic and ritual. When we examine prehistory we are confronted with the reality of a distant past, however, devoid of written testimony and therefore of 'proof' for our hypotheses. These problems are the same ones that we have confronted since the era of André Leroi-Gourhan and it is he who stated that: 'In order to comprehend the position of the prehistorian faced with a religious phenomenon, one must simply imagine an intelligent being, arriving from a distant astral system (completely ignorant of the notion that man might be religious), who encounters a non-decorated chalice and a champagne glass, a butcher's knife and a sacrificial blade. By what means would he infer, even vaguely, the meaning of sacrifice?' (Leroi-Gourhan 1971, 21).[5] If we cannot then speak of religious practices, we can still speak of practices, which may touch upon the realms of the aesthetic, the symbolic or simply the functional. The present study is concerned with those practices that are considered 'funerary' that is, practices associated with, or related to, death be it before, during or after the interment. It is of course tempting to associate funerary practices with symbolic intentions, to a set of systems of transcendental or even religious beliefs. The former cannot, however, always be accurately interpreted as a direct testimonial of the latter. Nonetheless, we benefit from a set of techniques that allow us to interpret and to reconstruct some elements of the symbolic in funerary practices. In this case, we speak of the identification, analysis and interpretation of the disposal of the dead interpreted as 'funerary gestures' (Duday *et al.* 1990) or even as 'mortuary chaînes opératoires' (Valentin *et al.* 2014; Sellier 2016). The interpretation of these 'gestures' is acquired by means of taphonomic observations carried out on the remnants of the deceased, the bones and goods within the grave, as well as the burial space itself. The observation of the funerary goods, the positioning of the body as reconstructed through an interpretation of the location of the remaining bones, as well as the nature of the objects allow researchers to reach a certain number of conclusions. Analysis of biological elements, including age and sex estimations, measurements of the deceased and observation of osteological or dental discrete traits can further inform potential interpretations. Each of these methods can bear witness to particular and differential practices based on varying criteria.

This study analyses juvenile burial practice in the context of the Sudanese Neolithic (5500–3100 BC), with a concentration on the Middle Neolithic (5000–4000 BC) (Salvatori and Usai 2007; Sadig 2013). It examines notions of selective criteria employed among the deceased within the burial ground. Specifically, it considers the differences between what will be distinguished as juvenile and subadult individuals. The distinction between these two categories emerged as part of a doctoral research in progress[6] as well as during excavations carried out by a French research team currently working in the Sudan at the Kadruka 23 funerary site.[7] The results are of course preliminary findings, as they represent only two years of collected data thus far, but will continue to be explored as both the doctoral and excavation project go forward.

A Brief Presentation of the Project History and Context

Kadruka 23 is a relatively undisturbed funerary mound (locally called *kôm*) situated in Upper Nubia, in the Northern State of Sudan (Fig. 4.1). Renewed fieldwork began there in January 2014. Today the region surrounding Kadruka 23 can be characterised as arid. The site is situated not far from the third cataract of the Nile, in what is now a desert region removed by several kilometres from the Nile East bank's current position. Substantial irrigation makes this landscape much greener than one would expect for a desert. Access to water is also known to have played an important role during the Neolithic period in this area. Indeed, Kadruka 23 is situated within the flood plain of the Wadi el Khowi, a former distributary of the Nile River, since dried up, but that would have flowed in the Neolithic period (between 7000 and 3000 BC).

This area is a particularly rich area for archaeology and appears to have been an important geographic zone in the Neolithic period. A remarkable density of funerary sites has been surveyed and recorded for this region. It would appear then that this entire zone has great potential and importance for assessing funerary practices, and by extension cultural traditions, during the Neolithic in Northern Sudan. Indeed, David Wengrow and colleagues (2014) suggest that the flow of culture and populations in this region functioned not only on a north–south axis, but also from east to west. They also suggest that the cultural entities present in Northern Sudan, responsible for the funerary mounds to which Kadruka 23 belongs, may have played a pivotal role in this transfer thanks to its central geographical placement

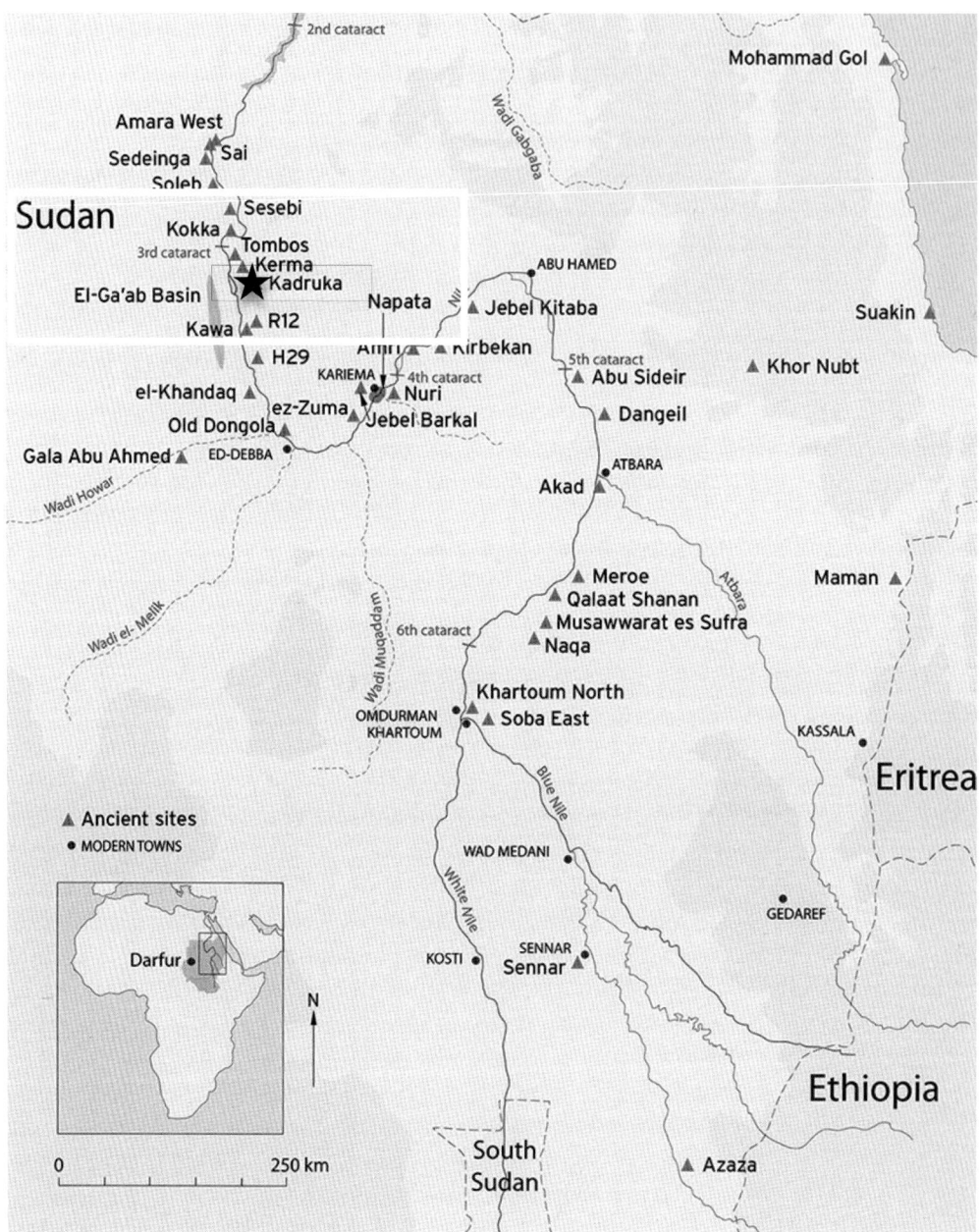

Figure 4.1. Map of the geographical and archaeological context for Kadruka (courtesy of the Sudan Archaeological Research Society).

(Wengrow *et al.* 2014, 102, 107). In this manner, the Sudan may be interpreted as having functioned as a sort of gateway between multiple cultural entities and geographic zones.

To date, the most prevalent hypothesis is that Neolithic populations from the fifth millennium BC were semi-nomadic pastoralists who moved along the banks of the Nile and its flood plains and tributaries. The role of domesticated animals is well established, while a lack of evidence for the organised implementation of agriculture makes it difficult to assess the importance of this practice for these populations (Honegger 2001; Salvatori and Usai 2008; Sadig 2013, 30). This circumstance has potentially interesting implications for societal organisation, as intensifying hierarchisation is often associated with the process of adopting a more sedentary lifestyle. One of the primary hypotheses advanced by Jacques Reinold, the archaeologist formerly in charge of excavations at Kadruka (1986–2009), was that these burial ensembles represent a selection of elite individuals from a larger population, typically featuring one particularly wealthy central grave (Reinold, 1985; 2000, 46; 2001, 2, 6). As we will see, this theory has not been supported by our findings to date.

Interest in Sudan's prehistory, and the archaeological study of it, began in earnest during an excavation undertaken by A. J. Arkell during the 1940s in the Khartoum region (Arkell 1949; Sadig 2013, 23). While the Kadruka concession has been excavated since the 1980s, few publications exist. J. Reinold excavated within the region between 1986 and 2009, discovering over 120 sites, and excavating over 700 burials at six of these sites (Kadruka 1, 2, 13, 18, 21 and 22), but there is as yet no monograph dedicated exclusively to the Kadruka sites in print and only one article concerning the burial mound at Kadruka has been published to date (Reinold 1994). In fact, few publications concerning Neolithic cemeteries exist for the whole of Upper Nubia and only a very small number of these publications apply what might be termed a biocultural approach (Reinold 1985; 1986; 1994; 2001; Simon 1997; Salvatori and Usai 2001; 2002; 2008; Honegger 2004; Wengrow *et al.* 2014). Interest within the scientific community for the Sudanese Neolithic has been firmly established and the terrain is as fertile from a scientific standpoint, as it is precarious from a conservation point of view. Ever increasing climate change and encroaching agriculture have endangered this region's remarkable archaeological record.

In light of the relative wealth of sites and potential data, as well as the relative dearth of studies that have thus far taken all elements of analysis into consideration, the present study has several overarching goals. The project aims to improve understanding from a funerary archaeological perspective. This implies exploring notions of the 'ritual' activity at these sites, as well as their usage by their contemporaries. This necessitates taking into consideration whether ritual and function evolve over time. A further aim is to improve the biological anthropological analysis. Significant developments have been brought to this field of study since the early 2000s and these improvements can be brought to bear on the data existing for the Kadruka sites since it dates to the 1980s and 1990s. The hope is to thereby explore questions of population and relationships, as well as sex, age, health and markers of activity.

The Kadruka 23 Funerary Mound, and its Occupants: A 'Normal' Ancient Mortality

Over the course of the two first field expeditions (2014–5), a total of 48 individuals have been excavated at Kadruka 23. Of those individuals, 25 were non-adults, and 23 were

Table 4.1. Summary of age estimations for juvenile and subadult individuals at Kadruka 23, calculated using both the evaluation of dental development and long bone measurements, which were then correlated to existing knowledge of corresponding age ranges according to the methods cited in text.

Individual	Age Estimation
Reinold 1	perinatal
ST. 19B	perinatal
ST. 32	> 2 weeks - 4 months
ST. 37	> 2 weeks - 4.5 months
ST. 33	> 2 weeks - 5 months
ST. 35	2-3 months
ST. 36	> 6 months
ST. 31	> 2 weeks - 6 months
ST. 25	2 weeks - 6 months
ST. 27	2 weeks - 6 months
ST. 30	2 weeks - 6 months
Reinold 2	6-9 months
ST. 40	2 weeks - 6 months
ST. 60	5 months - 1.5 year
ST. 54	1.5 years - 6.8 years
ST. 52B	1.7 years - 3.8 years
ST. 39	2.2 years - 7.9 years
ST. 38	2.3 years - 6.8 years
ST. 44	3.1 years - 7.9 years
ST. 15B	3.7 years - 6.9 years
ST. 41	3.7 years - 8.7 years
ST. 52A	5-8 years
ST. 19A	4.10 years - 7.9 years
ST. 28	13-14 years
ST. 26	15-16 years

0	0 or 1 to 4	1 to 4	1 to 4 or 5 to 9	10 to 14	15 to 19
13	1	0	2	1	1

TOTAL JUVENILE	TOTAL SUB-ADULT
23	2

adults. In the context of this study, stillborns, newborns, and infants, are terms, which refer to those individuals under the age of one year old, and the term children, refers to those individuals between the ages of one and approximately nine years of age. Stillborns, newborns, infants and children will be referred to throughout as juvenile subjects. Those that will be referred to as 'subadult' are a group of individuals, not yet fully biologically mature, or adolescent, between the ages of approximately 10 and 19 years of age. Once analysis of the data from the terrain began and age estimations started to garner results, it was immediately evident that an important number of juvenile graves contained very young individuals – stillborns, newborns and infants (Table 4.1). This phenomenon is representative of a 'normal', attritional mortality for this kind of ancient population, with high probabilities of death for the youngest comparable to model life tables with low life expectancies at birth (Ledermann 1969; Sellier 1996; Fig. 4.2). This finding was particularly thought-provoking as it aligns itself in direct opposition to Reinold's theory of selection within the burial ground (Reinold, 1982; 2000, 46). This normality would indeed suggest no argument for any kind of selection between age classes in terms of inclusion in the burial mound, and undermines the notion that only an elite subset would have access to a burial within a *kôm*. Finally, another trend began to appear, suggesting a selection between what will be termed 'juvenile' or very young individuals (stillborns, newborns, infants and children), and those that will be referred to as 'subadult' (adolescents) and 'adult' subjects.

Before delving into the analysis of this last trend, the definition of these burial categories must be clarified. A clear difference in the treatment of the individuals, along with their grave goods, appeared that did not correspond to a biological reality. This, in turn, necessitated a reconsideration of the categories used for grouping

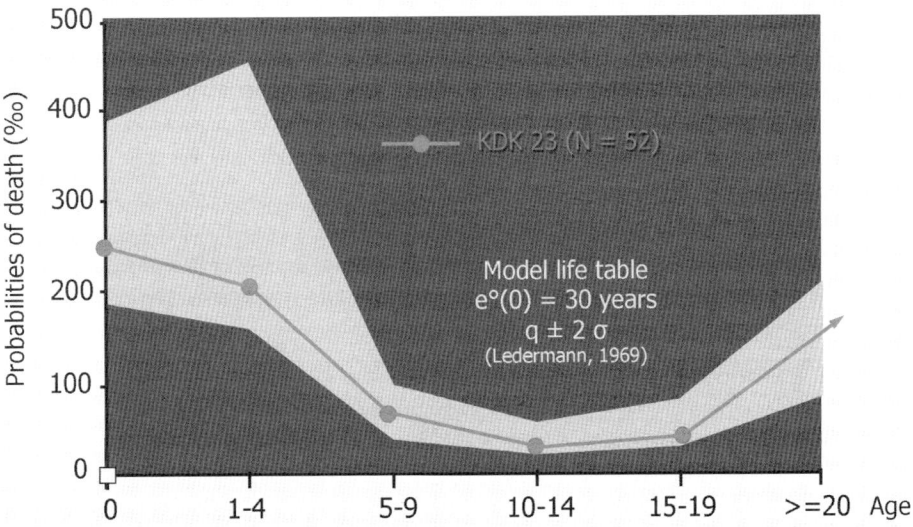

Figure 4.2. Mortality profile for Kadruka 23 demonstrating 'normal', attritional mortality.

individuals. Age estimations for the burials that had been termed 'juvenile', began with the separation of individuals according to dentition (Moorrees *et al.* 1963a; 1963b): thus any individuals presenting mixed deciduous and permanent dentition or dentition in the process of formation (with incomplete calcification of any dental element, except for the third molars) were regarded as belonging to this category. Primary or secondary points of ossification, maturation and fusion of osteological elements, and long bone measurements were also taken into consideration (Scheuer and Black 2004). It became apparent, however, that those individuals who might typically be termed 'adolescent' were in fact treated in this burial context in the same way as fully adult individuals. The term subadult was thus implemented, instead of adolescent, the latter being charged with significance about the cultural understanding of what is otherwise recognised as an intermediate biological state of development (van Gennep 1909 [1960]).

Differential Funerary Practices

Indeed, in the Neolithic burial context, there seems to have been no difference of treatment between these subadult individuals and fully adult ones. There is however a marked difference between those individuals we have referred to as juvenile or, the very young, perinatal, infant or child, and this category of subadult, which refers to an older subset of not-yet-biologically-mature individuals.

Criteria of Selection

With these categories clarified, the marked difference in treatment of these two categories can be explored. Three major elements were taken into consideration when examining a potential difference of treatment specific to subadult and adult or juvenile individuals. The first was the relative placement of the tombs within the strata of the funerary mound. The second was the manner of disposal of the deceased, and the third was the nature of the associated mortuary goods.

Principles of Placement and Disposal of the Dead

Upon analysis of placement within the funerary mound strata, placement of grave goods, burial position, as well as the wealth, variety and nature of mortuary goods, two very distinct types of treatment emerged for juvenile subjects on the one hand and subadult and adult individuals on the other. Juveniles were placed in the upper strata of the funerary mound, with ceramic bowls consistently above the burial level and sometimes even directly atop the deceased. Subadult and adult individuals were consistently interred in the lower strata with no preferential placement of grave goods.

In terms of positioning within the burial, we noticed an extreme variety for juvenile subjects, as opposed to their subadult and adult counterparts. For example, one

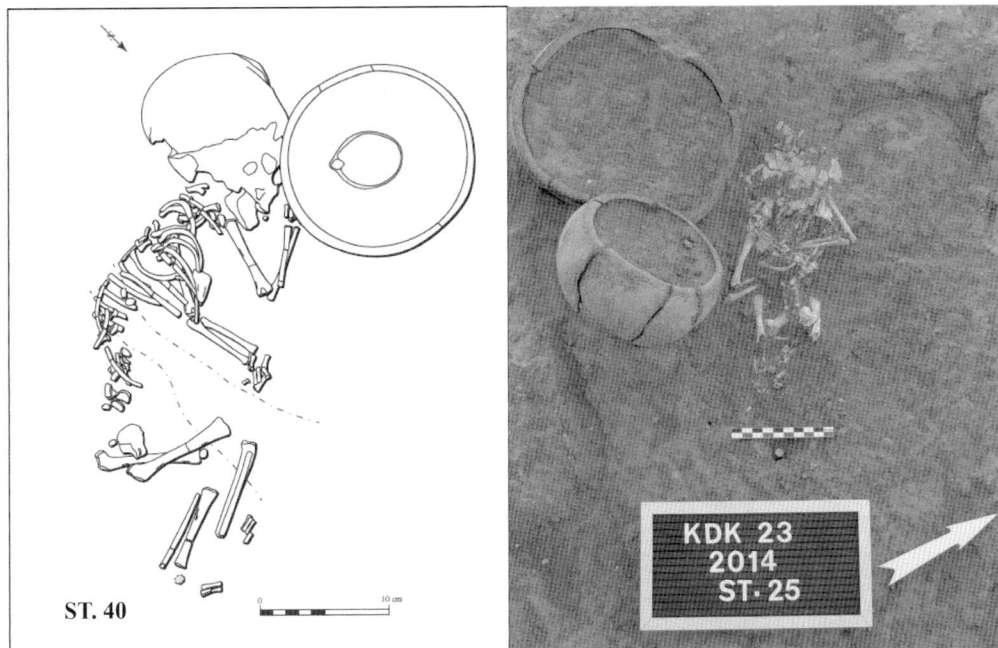

Figure 4.3. Example of two juvenile burials at Kadruka 23 – ST. 40 (on the left), whose age was estimated at between two weeks and six months old, presenting a very loosely contracted position (as opposed to the extreme contraction seen in adult burials), and with the face angled toward the ground. ST. 25 (on the right), presenting a unique positioning, also estimated at between two weeks and six months old, discovered in a crouched, leap-frog-like position, with the front of the head facing entirely downward, toward the ground (photo and drawing prepared by Philippe Chambon).

individual (ST. 40) whose age was estimated at between two weeks and six months old, presented a very loosely contracted position (as opposed to the extreme contraction seen in adult burials). Furthermore, the face of this individual was angled toward the ground. Another juvenile individual (ST. 25), presenting a unique positioning and also estimated at between two weeks and six months old, was discovered in a crouched, leap-frog-like position, and with the front of the head facing entirely downward, toward the ground (Fig. 4.3). The following instances of both subadult and adult burials can be contrasted with these two examples. The first was a subadult (ST. 26), whose age was estimated between 15 and 16 years old, while the other was an adult burial (ST. 45) for whom, unfortunately, an estimation of age was not possible due to poor preservation of the hip bones. Both demonstrated the extremely codified position exhibited by this class of individual. In the case of this age group, extreme contraction of the body was systematic, as was placement on one side or the other. This age group also frequently presented with the placement of one hand under the head, on the side on which the body was laid to rest (Fig. 4.4).

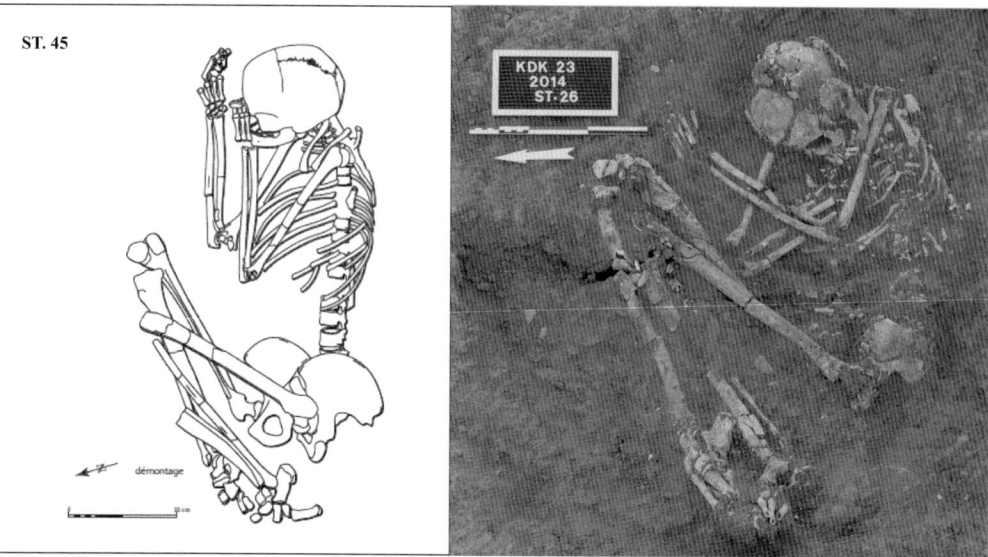

Figure 4.4. Example of an adult and subadult burial at Kadruka 23 – ST. 45 (on the left) is an adult burial for whom, unfortunately, an estimation of age was not possible due to poor preservation of the hip bones. ST. 26 (on the right) is a subadult, whose age was estimated at 15-16 years old. Both demonstrated the extremely codified position exhibited by this class of individual (photo and drawing prepared by Philippe Chambon).

Pattern Recognition

The meticulous plotting of burial orientations within a site and their comparison between the different inhumations is a practice already commonly used in the prehistoric, funerary archaeological context of the Sudan (Reinold 2007; Salvatori and Usai 2008). We proceeded with the collection of this data in order to have a comparable corpus with which to test out the site's unique qualities or lack thereof, as well as hypotheses about seasonal use based on varied body orientation in relation to shifting sun location as has been advanced for other sites (Salvatori and Usai 2008, 132). While there is insufficient data for these observations to be statistically significant, some first impressions can be offered. In general, a majority of adults and subadults were orientated with their heads facing east or south-east. Juvenile burials, however, presented a different schematic, with most of these individuals facing north-west. Furthermore, in the juvenile burials, a majority of individuals were laid to rest on their left side. This is in direct opposition to the majority of adult and subadult graves, wherein most individuals were laid to rest on their right side.

Mortuary Goods and Treatment Specific to Juvenile Burials

Finally, the variety and wealth of grave goods associated with juvenile burials is in direct contrast to the types and quantity of mortuary goods present in the case of adult and subadult graves. Not only have the juvenile burials presented with a greater variety of goods thus far, but they have also presented with a greater concentration of material culture per grave. Furthermore, there were goods found that were specific *only* to juvenile burials. These goods include bead waistbands (made of ostrich egg shell), and spoon-like objects (fabricated from ceramic or with a half shell of a bivalve), frequently found within ceramic bowls also associated with the individual and never found within the context of burials of older individuals. Several instances of multiple ceramic vessels associated with a juvenile burial were also observed. A majority of adult and subadult graves appeared with only one ceramic container in association with the deceased. Adult and subadult graves have thus far presented very few instances of personal adornment (two instances out of a total of 23 to be exact). Several adult and subadult burials, however, were found to include goblet shaped ceramic vessels, not found with younger individuals. This has also been true for what might be termed objects of quotidian life and activity, such as, reaping tools made of lithics mounted in wood or animal bone, polished axe heads or most spectacularly a small bead fabrication workshop, represented by a concentration of beads in progress, some pierced, some not, along with several accompanying fragments of the raw material of production, all found in association with an adult burial. In general, however, there was simply nothing as systematic, nor as plentiful per burial for the adult and subadult graves in comparison with the burials of their younger counterparts.

Future Directions for Further Research

The apparent dichotomy between the treatment of deceased juveniles and subadult and adult individuals is an avenue that we will continue to explore at the Kadruka 23 necropolis. In terms of the site itself, we will look to see whether juvenile burials are found elsewhere in the funerary mound. If so, their density will be examined in order to determine if it is similarly dense or not to what has been observed in the upper strata. Population demographics and questions regarding selection of the deceased within the burial mound will also be examined. In terms of the elaboration of the exact nature of this division, we will seek to define more clearly the moment at which this differential treatment is established. Several boundaries may have separated different periods of the life and death of the immature individuals and these will also be explored. Finally, we must continue to confront how one should understand the difference between biology and society in an archaeological context.

Hopefully comparisons with other sites, wherein selective treatment of the very young in a funerary context has been remarked upon, will also bear fruit. For example, in the Neolithic cemetery of el-Kadada, a selective treatment of juveniles was documented, but manifested itself differently. In this context, juvenile individuals were interred in ceramic containers, and on the margins of the burial site, as opposed to the more centralised adult burials (Reinold 2007, 275).

Neolithic sites of the region demonstrate an incredible diversity in burial practice. It is through pursuing the analysis of these variations that we may hope to draw the most fruitful conclusions about the identity of the cultures using these sites. At Kadruka 23 the hope is to continue the exploration of this separation of juvenile and subadult individuals in order to better elucidate the moment at which this separation occurs. For the present, it seems reasonable to suggest that the population using this cemetery defined subadults culturally (i.e. not biologically) as adults. In other words, perhaps as juveniles passed into puberty they came to be regarded culturally as adults even though developmentally they were not yet biologically fully mature.

Notes

1. Université Paris 1-Panthéon-Sorbonne, Institut d'art et d'archéologie, ED 112, 3, rue Michelet, 75005 Paris, France. Email: elmaines@gmail.com.
2. CNRS, UMR 7206 Éco-Anthropologie et Ethnobiologie, Musée de l'Homme, 17, place du Trocadéro, 75116 Paris, France. Emails: pascal.sellier@mnhn.fr, philippe.chambon@mnhn.fr.
3. CNRS, CEPAM (UMR 7264), Université Nice-Sophia Antipolis, 24, avenue des Diables Bleus, 06357 Nice cedex 4, France. Email: olivier.langlois@cepam.cnrs.fr.
4. 'La vie individuelle, quel que soit le type de société, consiste à passer successivement d'un âge à un autre et d'une occupation à une autre. ... C'est le fait même de vivre qui nécessite les passages successifs d'une société spéciale à une autre et d'une situation sociale à une autre: en sorte que la vie individuelle consiste en une succession d'étapes dont les fins et commencements forment des ensembles de même ordre: naissance, puberté sociale, mariage, paternité, progression de classe, spécialisation d'occupation, mort' (van Gennep 1909 [1960], 3–4).
5. 'Pour imaginer la situation du préhistorien devant le fait religieux, il suffit de se représenter un être intelligent débarquant d'un autre système sidéral (ignorant que l'homme est religieux) et mis en présence d'un calice non décoré et d'une coupe à champagne, d'un couteau de boucher et de celui d'un sacrificateur. Quel moyen aurait-il de reconstituer, même vaguement, le sens du sacrifice?' (Leroi-Gourhan 1971, 21; authors' translation provided in the text).
6. Thesis project (E. M., supervisor P. S.), Université Paris 1-Panthéon-Sorbonne, ED112: Diversité Biologique et Archéologie de la Mort: une Approche Populationnelle et Culturelle du Néolithique Soudanais (Haute-Nubie).
7. Kadruka Neolithic project: SFDAS (Section Française de la Direction des Antiquités du Soudan)/NCAM (National Corporation for Antiquities and Museums)/QSAP (Qatar-Sudan Archaeological Project)/CNRS (Centre National de la Recherche Scientifique), director O. L., archaeo-anthropology P. C. and P. S.

References

Arkell, A. J. 1949. *Early Khartoum. An Account of the Excavation of an Early Occupation Site Carried out by the Sudan Government Antiquities Service in 1944-5*. Oxford: Oxford University Press.

Duday, H., Courtaud, P., Crubézy, É., Sellier, P. and Tillier, A.-M. 1990. L'anthropologie 'de terrain': reconnaissance et interprétation des gestes funéraires. *Bulletins et Mémoires de la Société d'Anthropologie de Paris* (n.s.) 2(3–4), 29–49.

Honegger, M. 2001. Fouilles préhistoriques et prospection dans la région de Kerma. Les fouilles archéologiques de Kerma (Soudan). *Genava* 49, 221–228.

Honegger, M. 2004. Settlement and cemeteries of the Mesolithic and Early Neolithic at el-Barga (Kerma region). *Sudan and Nubia* 8, 27–32.

Ledermann, S. 1969. *Nouvelles Tables-Types de Mortalité* (INED Travaux et Documents 53). Paris: Institut national d'études démographiques/Presses Universitaires de France.

Leroi-Gourhan, A. 1971. *Les Religions de la Préhistoire. Paléolithique*. Paris: Presses Universitaires de France.

Moorrees C. F. A., Fanning E. A. and Hunt E. E. 1963a. Age variation of formation stages for ten permanent teeth. *Journal of Dental Research* 42, 1490–1502.

Moorrees C. F. A., Fanning E. A. and Hunt E. E. 1963b. Formation and resorption of three deciduous teeth in children. *American Journal of Physical Anthropology* 21, 205–213.

Reinold, J. 1982. La nécropole néolithique d'El Kadada au Soudan central; les inhumations d'enfants en vase, in Vercoutter, J., Thill, F. and Geus, F. 1985. *Mélanges Offerts à J. Vercoutter*, 279–289. Paris: Éditions Recherche sur les Civilisations.

Reinold, J. 1986. La nécropole néolithique d'El Kadada au Soudan central: quelques cas de sacrifices humains, in Krause, M. (ed.), *Nubische Studien*, 159–169. Mainz: Philipp von Zabern.

Reinold, J. 1994. Le cimetière néolithique KDK.1 de Kadruka (Nubie Soudanaise): premières résultats et essai de corrélation avec les sites du Soudan central, in Bonnet, C. (ed.), *Études Nubiennes* (vol. 2), 93–100. Geneva: Satigny.

Reinold, J. 2000. *Archéologie au Soudan. Les Civilisations de Nubie*. Paris: Editions Errance.

Reinold, J. 2001. Kadruka and the Neolithic in the Northern Dongola Reach. *Sudan and Nubia* 5, 2–10.

Reinold, J. 2007. *La Nécropole Néolithique d'El Kadada au Soudan Central. Volume 1: Les Cimetières A et B (NE-36-O/3-V-2 et NE-36-O/3-V-3) du Kôm Principal* (Fouilles de la Section Française de la Direction des Antiquités du Soudan 1). Paris: Éditions Recherche sur les Civilisations.

Sadig, A. M. 2013. Reconsidering the 'Mesolithic' and 'Neolithic' in Sudan, in Shirai, N. (ed.), *Neolithisation of Northeastern Africa* (Studies in Early Near Eastern Production, Subsistence, and Environment 16), 23–42. Berlin: Ex Oriente.

Salvatori, S. and Usai, D. 2001. First season of excavation at Site R12, a Late Neolithic cemetery in the Northern Dongola Reach. *Sudan and Nubia* 5, 11–20.

Salvatori, S. and Usai, D. 2002. The second excavation season at R12, a Late Neolithic cemetery in the Northern Dongola Reach. *Sudan and Nubia* 6, 2–7.

Salvatori, S. and Usai, D. 2007. The Sudanese Neolithic revisited. *Les Cahiers de Recherches de l'Institut de Papyrologie et d'Égyptologie de Lille (CRIPEL)* 26, 323–333.

Salvatori, S. and Usai, D. (eds.) 2008. *A Neolithic Cemetery in the Northern Dongola Reach: Excavation at Site R12* (Sudan Archaeological Research Society Publication 16; BAR International Series 1814). Oxford: Archaeopress.

Scheuer, L. and Black, S. 2004. *The Juvenile Skeleton*. London: Elsevier and Academic Press.

Sellier, P. 1996. La mise en évidence d'anomalies démographiques et leur interprétation: population, recrutement et pratiques funéraires du tumulus de Courtesoult, in Piningre, J.-F. (ed.), *Nécropoles et Société au Premier Âge du Fer: le Tumulus de Courtesoult (Haute-Saône)* (D.A.F. 54), 188–202. Paris: Éditions de la Maison des Sciences de l'Homme.

Sellier, P. 2016. Différents types de sépulture ou différentes étapes d'une même séquence funéraire? Un exemple démonstratif de chaîne opératoire mortuaire chez les anciens Marquisiens. *Bulletins et Mémoires de la Société d'Anthropologie de Paris* 28, 45–52.

Simon, C. 1997. Premiers résultats anthropologiques de la nécropole de Kadrouka, KDK 1 en Nubie Soudanaise. *Cahier des Recherches de l'Institut de Papyrologie et d'Égyptologie de Lille* 17, 37–53.

Valentin, F., Rivoal, I., Thevenet, C. and Sellier, P. (eds.) 2014. *La Chaîne Opératoire Funéraire: Ethnologie et Archéologie de la Mort* (Travaux de la Maison René Ginouvès, 18). Paris: Éditions de Boccard.

van Gennep, A. 1909 [1960]. *The Rites of Passage* (transl. Vizedom, M. B. and Caffee, G. L.). Chicago, IL: Chicago University Press.

Wengrow, D., Dee, M., Foster, S., Stevenson, A. and Bronk-Ramsey, C. 2014. Cultural convergence in the Neolithic of the Nile Valley: a prehistoric perspective on Egypt's place in Africa. *Antiquity* 88, 95–111.

Chapter 5

Children's Burials in the Eneolithic Cemetery of Sultana-Malu Roşu, Romania

Catalin Lazar,[1] Ionela Craciunescu,[1] Gabriel Vasile[2] and Mihai Florea[1]

Abstract: This paper focuses on Eneolithic child burials discovered in the Sultana-Malu Roşu cemetery, southeastern Romania (*c.* 5000–4000 cal. BC). The associated burial practices may implicitly reflect, through the inclusion of grave goods or other features such as the treatment of the body and the position of the burial within the funerary area, the potential symbolic significance of children and their connection to the household and social groups. Each of these characteristics is a potential active representation of special treatment that may have been applied to children and could therefore be interpreted as a message not so much about the individual identity of the child, but more probably the collective identity of the family or community. Thus, deceased children display evidence of an 'artificial' identity created by adults. The graves will be discussed from the viewpoint of their symbolic potential and their position within society. In addition, possible reasons for the paucity of grave goods and their potential significance, when they do exist in the burials, will be discussed. Differentiation between children and adults, relating to the impact that their deaths may have had on the community and how this is reflected in the funerary ritual, will also be explored.

Keywords: Eneolithic, Romania, cemetery, burial, children, GIS analysis

Introduction

The study of children's burials represents a challenge for archaeologists, in particular for those dealing with prehistoric periods, due to a lack of complementary sources of information, such as historical or literary texts, inscriptions and oral history accounts. Thus, the data derived from archaeological excavations is the only evidence that can help with the process of identifying funerary traditions for children and the nature of their position within past societies. Although children's burials have been included in several studies (e.g. Scott 1999; Sofaer Derevenski 2000; Bacvarov 2008; Lally and Moore 2011), the topic is still open to further debate and new viewpoints.

The fifth millennium BC was the time when the Eneolithic civilisation in southeastern Europe flourished and, as such, it is known as the 'golden fifth millennium' because of the substantial progress made by human communities. Thus, at around 5000 BC, in the territory of northeastern Bulgaria and southeastern Romania, some major changes are visible, including the rise of tell settlements, new building types (symmetrically organised within the settlement area), the development of settlement defence systems (ditches, banks and palisades), the emergence of cemeteries outside settlements, the adoption of new raw materials (e.g. copper, gold and graphite), and changes to ceramic and lithic technologies (Todorova 1986; Bailey 2000; Anthony 2010; Popovici 2010; Lazar 2011). These changes, which comprised the adoption of a new settled way of life, economic developments and new ways of environmental exploitation and control, as well as the development of a complex and stratified society, are very well reflected in funerary practices. This is particularly true for cemeteries where the emergence of elites and socio-economic differences are visible, as demonstrated by the wealthy graves from the Varna I cemetery in Bulgaria and other contemporary funerary areas (Bailey 2000; Anthony 2010; Slavchev 2010; Lazar 2011; Borić 2015).

The aim of the current paper is to explore Eneolithic child burials discovered in the Sultana-Malu Roşu cemetery of Călăraşi County in southeastern Romania (c. 5000–4000 cal. BC). The analysis will focus on the study of children's burials in relation to funerary rituals (e.g. body positions, orientations, grave structures, grave goods and offerings), correlated with palaeodemographic data and the spatial location of the burials in the cemetery (GIS analysis). The aim of this integrated approach is to identify the symbolic potential of children in the community and to gain insights concerning their position within this past society.

Geographical and Chrono-Cultural Settings

The Sultana-Malu Roşu cemetery is located in the northern area of the Balkan region of southeastern Romania (Fig. 5.1) on the right bank of the old Mostiştea River, approximately 7 km from the Danube River, near the border with Bulgaria (Lazar 2014). The burial ground is an extramural cemetery typical for Eneolithic communities in the Balkans, with its location in an uninhabited area, not far from the settlement(s). The people from two settlements belonging to two largely chronologically different Eneolithic communities used the same cemetery (see Fig. 5.1; Lazar 2014; Lazar and Voicu 2015). This situation has also been identified in other Eneolithic burial grounds in the region, including those at Durankulak, Varasti-Gradistea Ulmilor and Cascioarele (Lichter 2001; Dimov 2002; Lazar 2011; Borić 2015). During the first phase of activity

5. Children's Burials in the Eneolithic Cemetery of Sultana-Malu Roşu, Romania

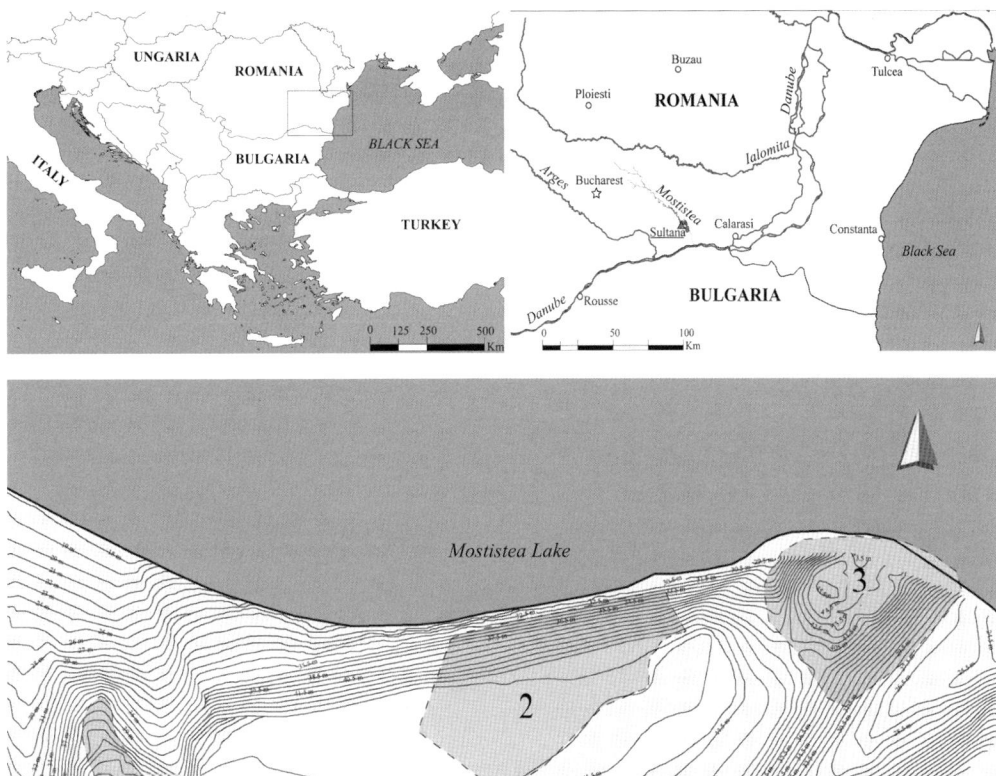

Figure 5.1. The geographical location of the Sultana-Malu Roşu site (top images) and its internal structure of the site (bottom, after Lazar 2014).

(c. 5000–4500 cal. BC), the burial ground was used by the Boian communities who lived in the flat settlement of Sultana-Ghetarie, while in the second phase (c. 4500–4000 cal. BC), it was used by the inhabitants of the Gumelnita tell settlement of Sultana-Malu Roşu (Fig. 5.2).

Generally, following the standard cultural-historical approach of prehistoric studies, archaeologists from Romania and Bulgaria consider both cultural sequences (Boian and Gumelnita) to be components of two distinct cultural complexes (Boian-Maritsa-Karanovo V and Kodjadermen-Gumelnita-Karanovo VI). The conventional view has been that the first cultural entity influenced the origin of the second (Todorova 1986; Petrescu-Dâmboviţa 2001a; 2001b; Popovici 2010). Recent aDNA analysis, however, has demonstrated that both populations, Boian and Gumelnita,

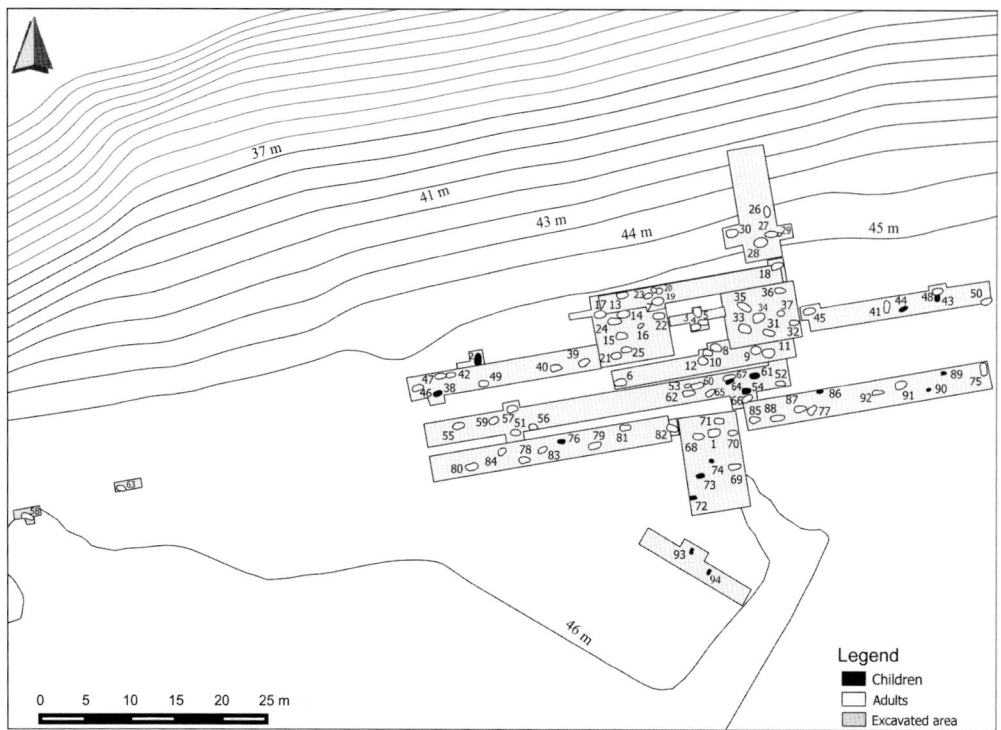

Figure 5.2. Plan of the Sultana-Malu Roşu cemetery with the position of children's graves highlighted.

have similar genetic features that are due to a common origin in southwestern Anatolia from where they arrived in the Balkans after a slow process of migration (Hervella *et al.* 2015).

From a topographical point of view (see Fig. 5.1), the cemetery is located on the highest terrace of Mostiştea Lake, at 150 m (± 1 m) to the west of the Gumelniţa tell settlement (Sultana-Malu Roşu) and 320 m (± 1 m) to the east of the Boian flat settlement (Sultana-Ghețărie). The 94 inhumation graves identified to date were grouped on the terrace edge and along the slopes of Mostiştea Lake. Only 16 graves were those of children (see Fig. 5.2). Most of the graves from the Sultana-Malu Roşu cemetery were single, primary inhumations within simple, ovoid pits, without other traces of funerary constructions. The deceased was laid out in a foetal position on the left side, with an east to west orientation (head to east). Secondary burials and graves with clear evidence for the deliberate removal of certain skeletal elements – especially skulls – have also been documented (Lazar *et al.* 2008; 2009; 2012; Lazar and Voicu 2015). The cemetery's period of use can be estimated to 5071–4450 cal. BC (95.4% probability) based on the AMS radiocarbon data obtained thus far (Lazar and Voicu 2015).

Methods

From a methodological point of view, the archaeological research undertaken on the Sultana-Malu Roşu cemetery involved an interdisciplinary approach. The excavation was undertaken using microstratigraphic methods, coupled with a series of geophysical prospection, a GIS approach for the collection of topographical and archaeological data, aerial research to investigate landscape transformation processes, palaeoecological studies, sampling for various interdisciplinary analysis and also the sieving and flotation of sediment obtained from the investigated features (Lazar et al. 2008; 2009; 2012). The altimetry was measured from a zero point (P0), represented by a terminal located on the terrace at an elevation of 45.1703 metres above sea level, on the basis of the 1975 Black Sea elevation system (Lazar et al. 2009).

Age-at-death of the juveniles was estimated on the basis of dental, cranial and post-cranial osteological features, namely the sequence of formation and eruption of the deciduous teeth (Ubelaker 1989) and the chronology of formation and resorption of the deciduous canines and molar roots (Moorrees et al. 1963). Age-at-death was also estimated on the basis of morphological and metrical characteristics when teeth were not present and as a complementary approach (Fazekas and Kósa 1978; Scheuer and Black 2004). Individuals were assigned to a biological age category as defined by Buikstra and Ubelaker (1994, 9), but modified – perinatal infants and neonatal infant (before or around the birth), infant (0–3 years), child (3–12 years), adolescent (12–20 years), young adult (20–35 years), middle adult (35–50 years) and old adult (50+ years).

The GIS study focused on the analysis of the relationships between the cemetery graves to identify spatial trends based on archaeological and anthropological data. The analytical method seeks to discover clusters of graves and does not focus on individual burials. For this purpose *Multi-Distance Spatial Cluster Analysis: Ripley's K-Function* was used (Deweirdt et al. 2012; Sayer and Wienhold 2013). This type of analysis illustrates the occurrence of statistically significant clustering of the feature centroid when the neighbourhood size changes (Dixon 2002). It has been demonstrated that the size of the study area can influence the results (Conolly and Lake 2006, 164–5) and the ArcMap 10 spatial statistic Ripley's *K*-function was applied to the data. It was performed using the Monte Carlo simulation, and 999 permutations were used to generate 95% confidence intervals. Also, to visualise and explore the clustered areas of data the *Kernel Density Estimation (KDE)* using GRASS GIS (plugin in QGIS 2.8) was applied. This type of analysis enables the comparison of data attributes according to the frequency of distributions of point data changes over the study area (Baxter and Beardah 1997, 349–50). The results of the Ripley's *K*-function analysis were plotted, and the clusters of graves could be visualised on an overlaid cemetery plan. Ripley's *K* analysis was carried out in ArcMap 10 because this software has a greater range of spatial analytic tools than QGIS 2.8. This approach combines the different burial attributes (e.g. age categories, grave goods, distances and clusters) recorded in the archaeological investigation.

Results

The Sultana-Malu Roşu cemetery contained 16 child graves (Table 5.1). To these can be added the burials of two adult females (20–35 years) that were associated with children's bones (see Fig. 5.2) – a fragment of the right side of a mandible derived from a nine- to ten-year-old child was found in the thoracic region of the adult skeleton buried in Grave 21, while Grave 88 was that of a pregnant adult female (Lazar *et al.* 2008; 2009; 2012; Lazar 2014; Lazar and Voicu 2015).

Grave Structure

All of the graves from the Sultana-Malu Roşu cemetery consisted of ordinary pits devoid of plaster lining or any traces of related constructions. As was the case for

Table 5.1. Details of the child burials from the Sultana-Malu Roşu cemetery. *Abbreviations used in relation to body position – C-DU=crouched dorsal upwards; C-LLS=crouched lateral on the left side; PC-LLS=probable crouched lateral on the left side; U=undetermined.

Grave no.	Pit		Body		Grave goods	Animal offerings	Age	
	Shape	Orientation	Position	Orientation			Years	Category
2	oval	N–S	U	N–S		•	1	infant
38	oval	NE–SW	C-LLS	NE–SW			4.5–5.5	child
43	oval	N–S	C-DU	N–S	•		1.5–2.5	infant
44	oval	ENE–WSW	PC-LLS	ENE–WSW			1–1.5	infant
54	oval	E–W	C-LLS	E–W			4.5–5.5	child
61	oval	NE–SW	C-LLS	E–W			10	child
64	oval	NE–SW	PC-LLS	NE–SW	•		2.5–3.5	infant/child
72	oval	E–W	U	E–W	•		0	neonatal infant
73	oval	NE–SW	C-LLS	NE–SW	•		4.5–5.5	child
74	oval	E–W	U	E–W			0.6	infant
76	oval	E–W	PC-LLS	E–W			0.5–1.5	infant
86	oval	E–W	U	E–W	•		0	neonatal infant
89	oval	E–W	U	E–W			0–0.5	perinatal infant
90	oval	NE–SW	PC-LLS	E–W			0	neonatal infant
93	oval	NE–SW	PC-LLS	NE–SW			0	neonatal infant
94	oval	NE–SW	U	NE–SW	•		0.5–1.5	infant

adult graves, the burial pits for children were predominantly ovoid in shape (Fig. 5.3). The dimensions of the child pits differed depending on the size of the body (Lazar *et al.* 2008; 2009; 2012), but in some instances they were larger than the preserved skeleton required (e.g. Graves 2, 72, 86, 89, 90, 93 and 94). In most of the cases, the orientation of the pits was along an east–west axis with few exceptions. The altitudes for the bases of the children's pits ranged between 44.13 and 44.76 metres above sea level, with some having been dug in a layer considered to represent a palaeosoil from the prehistoric period (s.u. T1003) and others in the loess layer (s.u. T1004). The elevations of the tops of the children's grave pits ranged from 44.13 to 44.74 metres above sea level. In cases where it was possible to identify the original depths of the pits (n=5) these were found to range from 0.22 m to 0.32 m beneath the prehistoric layers (s.u. T1003 and T1002). No surface marks have been identified to date but it should not be assumed that they had not originally existed, particularly since there was no evidence for the inter-cutting of graves (Lazar *et al.* 2012).

Bone Preservation

Unfortunately, the skeletons of most of the children from the Sultana-Malu Roşu cemetery were poorly preserved and high levels of fragmentation were evident (see

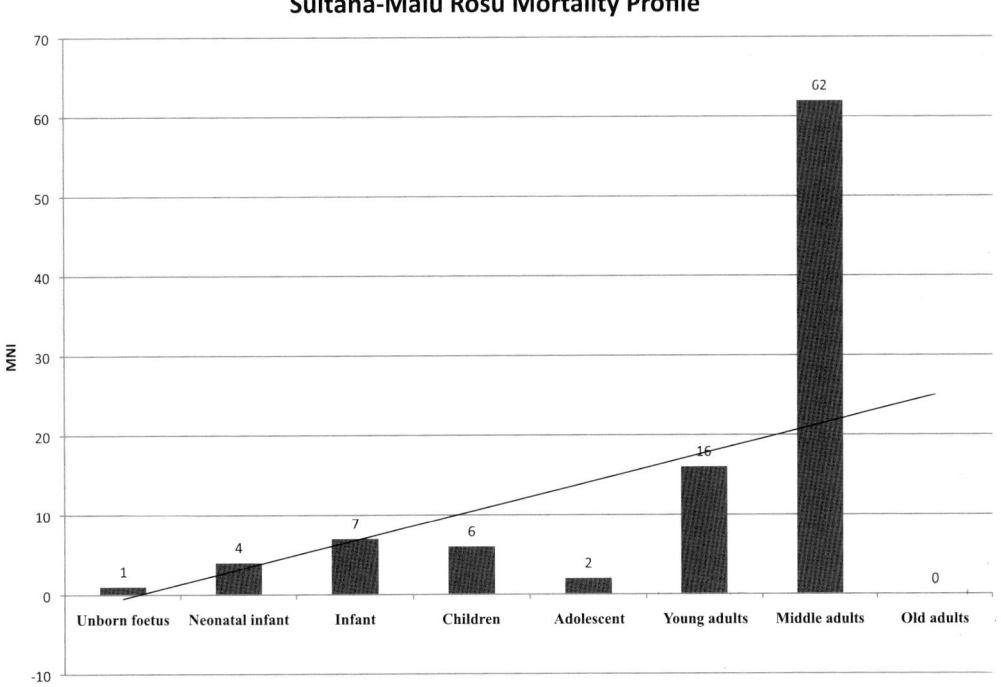

Figure 5.3. Graph showing the mortality profile of the Sultana-Malu Roşu cemetery.

Fig. 5.3). Furthermore, some bones were missing as a result of post-depositional disturbances (e.g. animal burrows) and soil conditions not conducive to their preservation. Added to these extrinsic factors, certain intrinsic properties of immature bone, including a high organic and low inorganic content, lack of fusion, higher porosity, lower bone density, as well as pathological and metabolic conditions that may cause a decrease in bone mineralisation, can contribute to issues of poor preservation (Nielsen-Marsh et al. 2002; Bello and Andrews 2006; Manifold 2012). As such, three of the children's burials contained only skulls or skulls associated with fairly poorly preserved post-cranial remains (n=8). In five instances, the skeletons were relatively well preserved.

The pattern of preservation of the osseous remains from the cemetery are indicative that juvenile bones are less well preserved and less well-represented than adult bones, which were typically represented by 60–90% of the skeleton. The skull and long bones were the best represented anatomical elements, with the remainder of smaller bones having frequently been absent. The poorly preserved and incomplete nature of the juvenile skeletons meant that it was almost impossible to identify any pathological conditions or injuries amongst the children at Sultana-Malu Roșu.

Palaeodemographic Data

The child graves (n=16) identified in the Sultana-Malu Roșu cemetery represent 17% of all individuals recovered from the excavations at the burial ground. The others age groups represented in the cemetery (Fig. 5.3) comprised adolescents (2%), young adults (20%) and middle adults (61%). A minimum numbers of individuals (MNI) of 99 was identified for the cemetery – these individuals were largely recovered from individual graves in addition to three graves that contained the remains of two individuals (Graves 21 and 88 already mentioned above, and Grave 28 that contained the remains of two adult males). The mortality profile is indicative of a high rate of death among young and middle adults, followed by infants, children and neonates.

The age-at-death data for the children from Sultana-Malu Roșu is presented in Table 5.1. Seven individuals of the analysed group were infants, while a further four were neonatal infants (Graves 72, 86, 90, 93), and a perinatal infant (Grave 89) was also represented. A total of four children were also present (Graves 38, 54, 61, 73) and, in another case (Grave 64), it was difficult to determine whether the individual was an infant or a young child due to their poor state of preservation.

Body Treatment

Regardless of age (adults or children), most of the graves in the Sultana-Malu Roșu cemetery followed the same pattern – individuals were laid out in a foetal position (lateral, dorsal or ventral), on the left side (rarely on right side), in a normal anatomical order, with the head positioned in an easterly direction (Lazar et al. 2008; 2009; 2012). Unfortunately, in the case of the individual child burials,

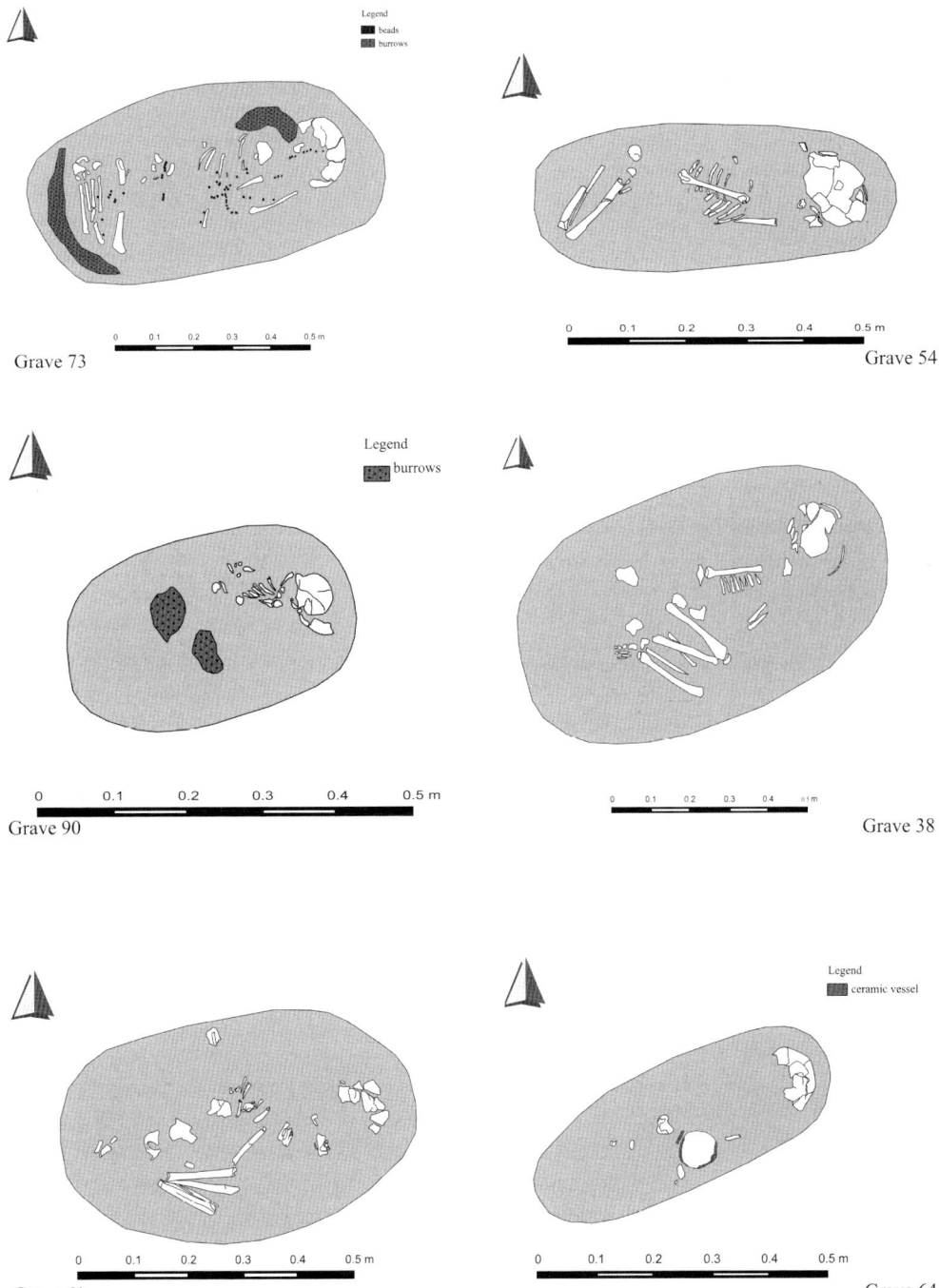

Figure 5.4. Examples of children's burials from the Sultana-Malu Roşu cemetery – Graves 73, 54, 90, 38, 61 and 64.

it was difficult to determine the position of the body since the graves contained such poorly preserved remains (see above), and it was not possible to do this for six individuals. In cases of relatively well-preserved individuals, or those with a medium level of preservation, however, it was possible to determine the position in which the body had originally been laid (see Table 5.1). A total of nine children had been buried lying on their left sides with the legs flexed to the left (Graves 38, 44, 54, 61, 64, 73, 76, 90, and 93), while a single child had been buried lying on the back with the legs flexed upwards (Grave 43; Fig. 5.4). Based on these data it can be concluded that the majority of children received the same funerary treatment as the adults buried in the cemetery.

The orientation of the individuals in the Sultana-Malu Roșu cemetery was determined on the basis of the orientation of the funerary pits (Lazar et al. 2008; 2009; 2012). Most often the bodies of the children were found to have followed the same orientation as that of the funerary pit (see Table 5.1 and Fig. 5.4). The positioning of the skull and/or the location of other fragmented anatomical elements within the funerary pit was used for determining the orientations of poorly preserved skeletons. In a notable majority of cases the children had been buried with the head in an easterly direction (87.5%; 14/16), the only exceptions having been the individuals in Graves 2 and 43 who were orientated along a north–south axis.

Grave Goods

Artefacts were rarely found in children's graves in the Sultana-Malu Roșu cemetery from both a quantitative and qualitative perspective. Only six graves contained objects (see Table 5.1 and Fig. 5.5). In addition, in Grave 44 two *Spondylus* beads were discovered in the fill of the funerary pit after sieving of the sediment. These can only be considered as grave inclusions, however, because they came from the upper area of the burial pit and it seems probable that their original location had been disturbed as a result of various post-depositional occurrences.

The children's graves were not very wealthy in terms of the quantities of objects they contained, with most individuals having been associated with a small number of items (e.g. Grave 43 – two *Dentalium* beads, three marble beads and a stone pendant; Grave 64 – a small ceramic pot; Grave 72 – four bone beads; Grave 86 – a bone awl; Grave 94 – five malachite beads). A notable exception was Grave 73, however, which contained almost 500 beads made from marble (n=1), malachite (n=3), *Spondylus* shells (n=173) and *Lithoglyphus naticoides* snails (n=322). In terms of the quality of the funeral goods, the children's graves contained artefacts made of exotic raw materials, including a stone pendant (Grave 43), marble beads (Graves 43 and 73), malachite (Graves 73 and 94), *Spondylus* shells (Grave 73) and *Dentalium* shells (Grave 43). Objects made from local raw materials, however, were also present and included a bone awl (Grave 86), a ceramic pot (Grave 64), beads made from *Lithoglyphus naticoides* snails (Grave 73) and bone (Grave 72).

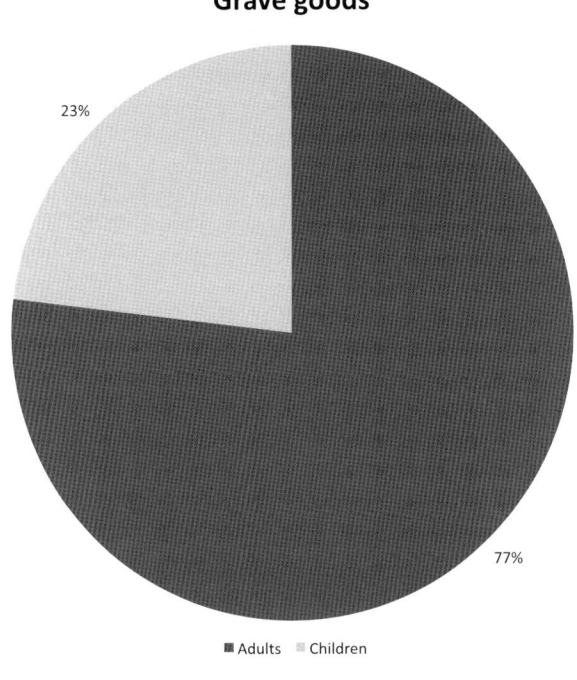

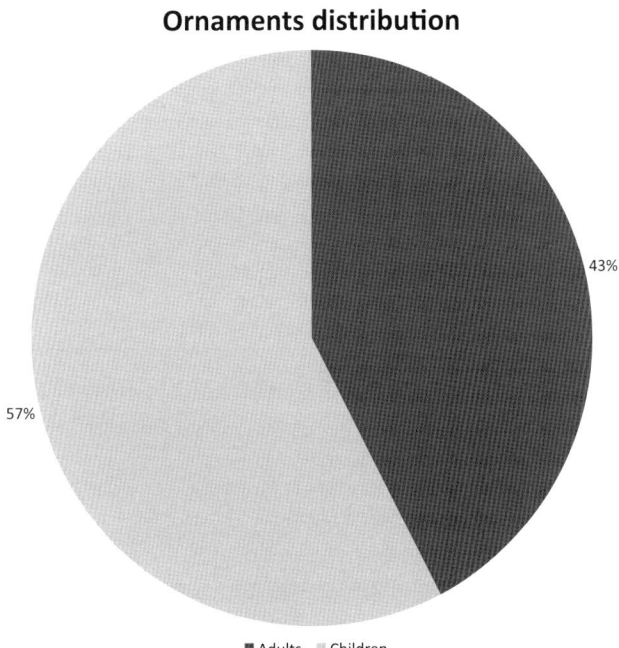

Figure 5.5. Distribution of grave goods (top) and ornaments (bottom) by age at the Sultana-Malu Roşu cemetery.

The position of the grave goods in relation to the children's bodies is indicative that they were deliberately buried in association with the dead. In most cases the pendants and beads were found at the neck and head areas (e.g. Graves 43, 72, 73 and 94), and it is probable that they had been worn as necklaces by the deceased. In Grave 73 several *Spondylus* and *Lithoglyphus naticoides* beads were found at the pelvic area (n=30), and they may have been sewn or attached as decorative items on clothes or formed a belt. The small ceramic pot recovered from Grave 64 was deposited at the pelvic area, and the bone awl retrieved from Grave 86 was found beneath the neonatal infant skull. Therefore, ornaments predominate in the grave goods associated with child burials in the Sultana-Malu Roşu cemetery (67% of the children graves with inventory contains ornaments), a situation that was very similar to that identified for adult graves, but children graves contain more pieces in quantitative terms – see Fig. 5.5 (Lazar et al. 2012).

Animal Offerings

Grave 2 was the only burial to contain animal remains and these comprised a left sheep metatarsal and right horn (Lazar et al. 2008; 2012). The child skeleton was in a very poor state of preservation and it was not possible to make any observations concerning the positioning of the animal bones in relation to the body.

GIS Analysis

The first step in the GIS analysis of the child burials in the cemetery was to create a plan of the burial ground in AutoCAD in which each grave was individually layered on the basis of its unique identification number. The cemetery plan was imported into QGIS 2.8 and associated with a detailed database that contained information on the two-dimensional position of every grave and burial attributes, such as the age groups established and the presence of the furnished graves (Wheatley and Gillings 2002, 19–21). A pair of x, y coordinates of the grave's central point was sufficient for recording the position of each burial assemblage (Šmejda 2004, 57).

Ripley's K-function analysis indicated significant clustering for the burials after 2 m (Fig. 5.6). The burials that contained grave goods were more dispersed, and did not show evidence for significant clustering. The results were plotted with the help of the Kernel density function (Fig. 5.7) and the density of burial events appears to be heterogeneous in different zones of the funerary space. The central area includes a clustered area on the right side (A), and another on the left side (B); they are separated by a clear spatial gap, and provide the strongest evidence for deliberate organisation at the burial ground. Most of the children's graves from the Sultana-Malu Roşu cemetery (Graves 54, 61, 64, 72, 73, 74, 93, 94) are found on the right side (A), while Grave 21, which comprised the burial of an adult female associated with a fragment of child mandible, is located on the left side (B). The western part (C) has

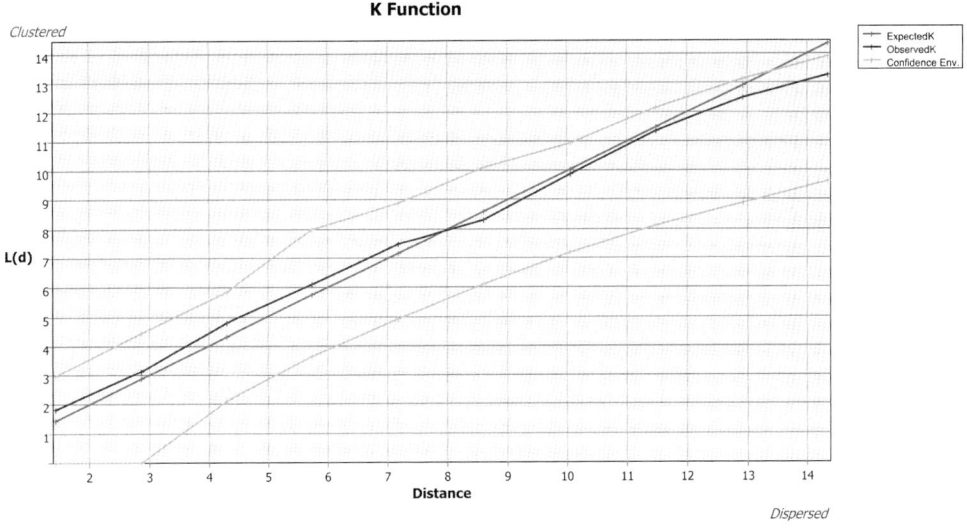

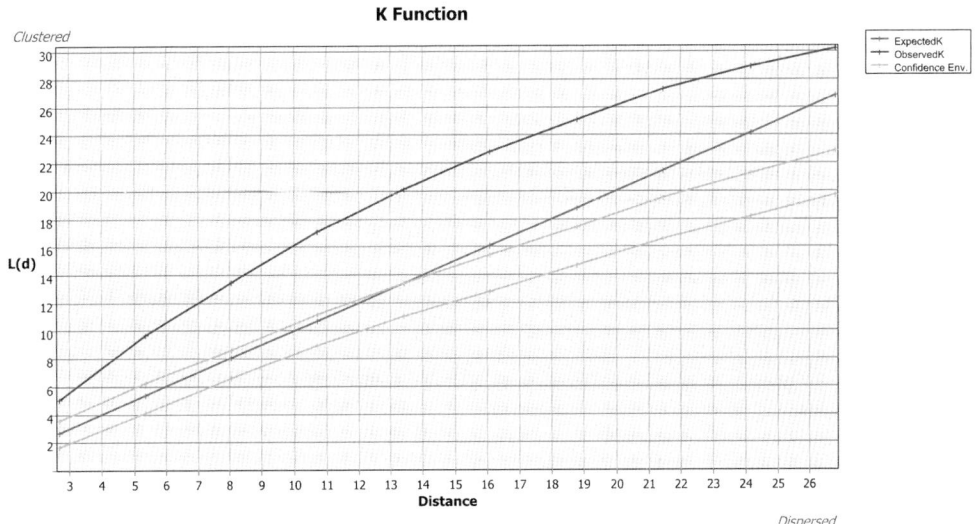

Figure 5.6. Ripley's K-function analysis graphs that indicate significant clustering after 2 m (top) and significantly dispersed graves containing grave goods (bottom).

a lower density of burials, and contains only three children's graves (Graves 2, 38, and 76). The eastern zone of the cemetery contain only a few dispersed graves (D), but includes an area that contains foetal burials (Graves 86, 89, and 90). In addition, Grave 88 that contained an adult female with an unborn foetus in her abdomen was buried in this area (see Fig. 5.2).

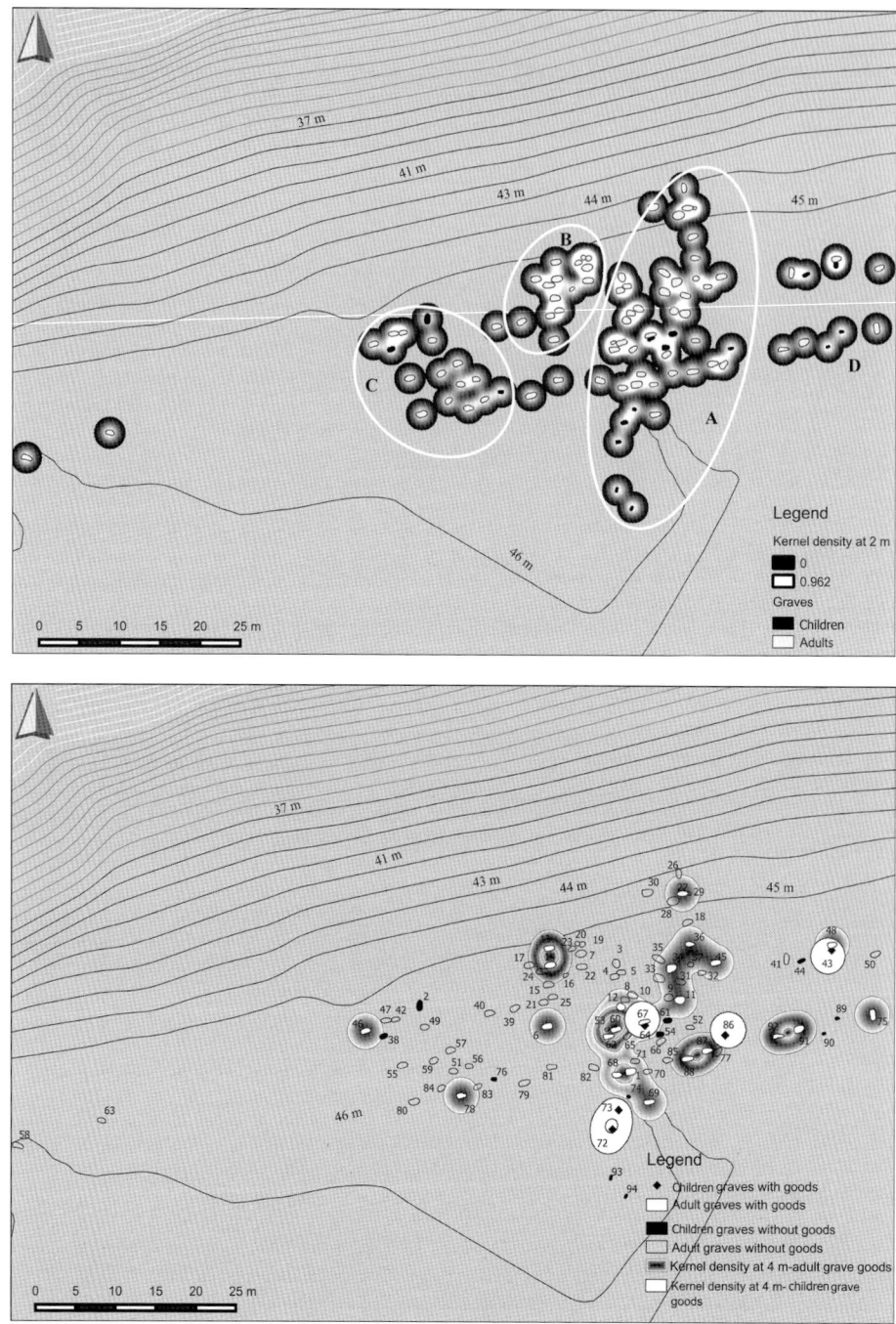

Figure 5.7. Kernel density maps (derived from the Ripley's K-function analysis graphs) that were indicative of burial clustering after 2 m (top) and dispersed graves that contained grave goods after 4 m (bottom).

Discussion

The Sultana-Malu Roşu burial ground is a typical Eneolithic extramural cemetery similar to other examples from the Balkans in relation to its location to nearby settlement(s), the organisation and structure of the burials and the mortuary rituals applied to the deceased (Lichter 2001; Lazar 2011; 2014; Borić 2015). GIS analysis of the location of the child burials has provided some interesting data. The areas of higher density of graves in the middle of the cemetery (A and B) may be understood as a *function of time* (Šmejda 2004), during which a particular part of the cemetery was used for the disposal of the deceased (see Fig. 5.7). These areas contain the earliest burials and probably formed the original core of the cemetery. As time passed these initial graves become surrounded by other burials as the cemetery developed. On the other hand, mapping of the available radiocarbon data for the cemetery (n=16), indicates that the cemetery developed from east to west, which is not surprising since the community that founded the burial ground had a settlement at the eastern limit of the Mostiştea Lake terrace. Also, the presence of graves that belonged to both the Boian and Gumelnita settlements/communities in the central area of the burial ground (A and B) supports this interpretation. It is possible that the spatial development of the cemetery is related to the distribution of child burials and that it can provide insights concerning their social importance within the communities that used the funerary space. Thus, the presence of the richest child burial (Grave 73) in Area A, where the richest adult graves (e.g. Grave 1) are also situated, may be of significance. In addition, the largest cluster of children's graves (n=8) in the cemetery is also found in Area A (see Fig. 5.7). Two small areas of significant aggregation after 4 m in relation to the distribution of grave goods were identified in the central area of the cemetery and overlap with the area of the highest concentration of burials. These included four child burials that were associated with artefacts. Another interesting observation is the distribution of child graves with smaller pits in a distinctive manner and with a sizeable area around them (e.g. Graves 74, 89, and 90) in Areas A and D of the cemetery, which apparently may reflect a special 'isolated' location although further research is required to test this theory.

Integration of juvenile age categories in the spatial analysis resulted in the observation of a number of patterns. It is interesting that in the area of the cemetery in almost all cases the immature individuals are always buried in the proximity of one or more adult graves, both male and female (see Fig. 5.2). It is only in the case of Graves 72, 73, 93, and 94 that the children's graves are apparently isolated, but this situation is due to the stage of excavation of the southern perimeter of the cemetery. Area D of the cemetery includes only the remains of neonatal and infant individuals. Interestingly, the burial of a pregnant adult female (Grave 88) was also discovered in this area. The remains of infants (0–3 years) were found in Areas A, C, and D, but occurred with highest frequency in Area A (n=4), while children (3–12 years) were only buried in Areas A and C, with a higher frequency in Area A (n=3).

It is very odd that no child burials were discovered in Area B, with the exception of Grave 21, which was that of an adult female who had a fragment of the right side of a mandible of a nine- to ten-year-old child associated with her thoracic region (Lazar et al. 2009; 2012).

Overall, no major differences regarding grave structure or funeral treatment (body position and orientation) were apparent between adults and children in the Sultana-Malu Roşu burial ground. The few cases of atypical orientation (e.g. N–S), that do not seem to fit with the overall pattern for the cemetery (Lazar et al. 2012), are also encountered among small numbers of adult burials. The body position (body deposited lying on the back in a crouched position with the legs flexed upwards) of the child from Grave 43, however, appears to represent a unique situation not identified to date among adult burials. Concerning the funerary inventory, children were buried with fewer grave goods than adults (see Fig. 5.5), but this situation reflects only that less children's graves have been recovered than those of adults. In terms of the quantity and quality of grave goods included in the burials, however, some of the children's burials (e.g. Grave 73) are in no way inferior to richer adult graves (Lazar and Voicu 2015). The distribution of artefacts in the children's graves from the Sultana-Malu Roşu cemetery is unequal and may be related to family and kinship groups, gender, expressions of wealth or social status. Alternatively, it may have been due to the symbolic significance of immature individuals in the community, and their connection to particular households or social groups. The determination of associations between children and gender on the basis of grave goods was almost impossible because most of the burials contained ornaments (beads, pendants and necklaces), which were characteristic of both men and women. Only a ceramic pot recovered from Grave 64 could be considered to have been associated with a female gender identity because, until now, ceramic vessels have only been discovered in female graves (e.g. Grave 6) (Lazar and Voicu 2015).

The proportion of children's burials identified in the Sultana-Malu Roşu cemetery is similar to the levels that have been discovered in contemporary cemeteries in Romania and Bulgaria (Lichter 2001; Yordanov and Dimitrova 2002; Kogălniceanu 2008; Lazar 2012; Borić 2015), and generally fits within a normal profile of infant mortality of 15–30% (Scott 1999, 90). It is possible to explain the relatively small proportion of children compared to adults in a number of ways. The burial ground at Sultana-Malu Roşu has not been fully excavated, as is the case for all Balkan Eneolithic cemeteries studied to date (Lazar and Voicu 2015), and it is possible that further child burials are present within the unexcavated areas of the site. The data obtained from other burial grounds of the fifth millennium BC, however, is more suggestive that a process of selective burial was in operation (Borić 2015). Evidence derived from the broader archaeological context of Eneolithic communities in the Balkans seems to support this assertion and child burials have been recovered within both flat and tell settlements in Romania and Bulgaria. Such burials are generally associated with the domestic space, and usually occur beneath the floors of buildings or between houses (Lichter

2001; Bacvarov 2003; Kogălniceanu 2008; Popovici 2010; Lazar 2012). This form of intramural burial occurred throughout the entire Neolithic in southeastern Europe and beyond (Borić 2015). This particular practice is another dimension of the funerary rituals used for the children of these Eneolithic communities and it may be viewed as a continuation of a tradition of considerable longevity.

Conclusions

Archaeological excavations in the Sultana-Malu Roşu cemetery are an ongoing project and the data present here is based on research that was undertaken between 2002 and 2015. The investigation included an area of 848.21 sq. m of the terrace surface that had been used as a burial ground by both the Boian and Gumelnita communities. Undoubtedly, future research will result in the discovery of new graves, including those of children that will enrich the interpretations present in this paper.

The available data is suggestive that the Sultana-Malu Roşu burial ground, along with other Eneolithic cemeteries from the Balkans, demonstrates a standardisation of mortuary practices in the fifth millennium BC. These reveal a complex and structured society with strong funerary traditions and specific burial customs, which are probably a reflection of the eschatological concepts of these communities, accumulated across many generations over a period of more than 1000 years. Moreover, funerary behaviours represent an extended form of identity, both personal and collective, and are a *metaphorical tool* that allows the living to express particular or group biographies through the dead. The children in the Sultana-Malu Roşu cemetery were apparently governed by the same funerary rules as the adults, but with some particular features, especially concerning the location of the graves within the burial ground. Each of these characteristics is a possible active representation of special treatment that was applied to children and can be carefully understood as a message regarding not so much the individual identity of the child, but more likely the collective identity of the family, group or community. Thus, the deceased children display an 'artificial' identity, created by the adults that survived them. By applying the same funerary rules, the adult members exploit the symbolic potential of a tragic event (i.e. the death of a child) and ensure a perpetuation of their conservative traditions. As is the case for the present day, it may have been the case that cemeteries in the Eneolithic were public spaces which made them ideal places for the building and rebuilding of traditions by the living through the dead. This metaphorical reconstruction enabled the adults to express a different kind of identity through their dead children.

Acknowledgements

This work was undertaken through the Partnerships in Priority Areas Program – PN II, developed with the support of MEN – UEFISCDI, project no. PN-II-PT-PCCA-2013-4-2302.

Notes

1. National History Museum of Romania, Calea Victoriei, No. 12, Sector 3, 030026, Bucharest, Romania. Emails: lazarc@arheologie.ro, mihaimfs@yahoo.com, ionela.craciunescu@gmail.com.
2. 'Vasile Pârvan' Institute of Archaeology, Henri Coandă Street, No. 11, Sector 1, 010667, Bucharest, Romania. Email: gsvasile@yahoo.com.

References

Anthony, D. W. 2010. The rise and fall of old Europe, in Anthony, D. W. and Chi, J. Y. (eds.), *The Lost World of Old Europe. The Danube Valley, 5000-3500 BC*, 29–57. New York: Princeton University Press.

Bacvarov, K. 2003. *Neolitni Pogrebalni Obredi. Intramuralni Grobovi ot Blgarskite Zemi v Konteksta na Yoguoiztočna Evropa i Anatoliya*. Sofia: Bard.

Bacvarov, K. (ed.) 2008. *Babies Reborn: Infant/Child Burials in Pre- and Proto-history* (BAR International Series 1832). Oxford: Archaeopress.

Bailey, D. W. 2000. *Balkan Prehistory. Exclusion, Incorporation and Identity*. London: Routledge.

Baxter, M. J. and Beardah, C. C. 1997. Some archaeological applications of Kernel Density Estimates. *Journal of Archaeological Science* 24, 347–354.

Bello, S. and Andrews, P. 2006. The intrinsic pattern of preservation of human skeletons and its influence on the interpretation of funerary behaviours, in Gowland, R. and Knüsel, C. (eds.), *Social Archaeology of Funerary Remains*, 1–13. Oxford: Oxbow Books.

Borić, D. 2015. Mortuary practices, bodies, and persons in the Neolithic and Early–Middle Copper Age of south-east Europe, in Fowler, C., Harding, J. and Hofmann, D. (eds.), *The Oxford Handbook of Neolithic Europe*, 927–958. Oxford: Oxford University Press.

Buikstra, J. E. and Ubelaker, D. H. (eds.) 1994. *Standards for Data Collection from Human Skeletal Remains* (Arkansas Archaeological Survey Research, Series No. 44). Fayetteville, AR: Arkansas Archeological Survey.

Conolly, J. and Lake, M. W. 2006. *Geographic Information Systems in Archaeology*. Cambridge: Cambridge University Press.

Dimov, T. 2002. Entdeckung und erforschung der prähistorischen gräberfelder von Durankulak, in Todorova, H. (ed.), *Durankulak, Band II. Die Prähistorischen Gräberfelder, Teil 1*, 24–34. Sofia: Publishing House Anubis Ltd.

Dixon, F. M. 2002. Ripley's K-function, in El-Shaarawi A. H. and Piegorsch W. W. (eds.), *Encyclopedia of Environmetrics, Volume 3*, 1796–1803. Chichester: Wiley.

Deweirdt, E., Mayer, P., Méniel, P., Metzler, J., Petit, C. and Bourgeois, J. 2012. L'analyse spatiale des nécropoles révistitée. *Archäologisches Korrespondenzblatt* 42, 185–204.

Fazekas, I. G. and Kósa, F. 1978. *Forensic Fetal Osteology*. Budapest: Akadémiai Kiadó.

Hervella M., Rotea M., Izagirre N., Constantinescu M., Alonso S., Ioana M., Lazar C., Ridichie F., Soficaru A. D., Netea M. G. and de-la-Rua, C. 2015. Ancient DNA from south-east Europe reveals different events during Early and Middle Neolithic influencing the European genetic heritage. *PLoS ONE* 10, e0128810. doi:10.1371/journal.pone.0128810.

Kogălniceanu, R. 2008. Child burials in intramural and extramural contexts from the Neolithic and Chalcolithic of Romania: The problem of 'inside' and 'outside', in Bacvarov, K. (ed.), *Babies Reborn: Infant/Child Burials in Pre- and Proto-history* (BAR International Series 1832), 101–111. Oxford: Archaeopress.

Lally, M. and Moore, A. (eds.). 2011. *(Re)Thinking the Little Ancestor: New Perspectives on the Archaeology of Infancy and Childhood* (BAR International Series 2271). Oxford: Archaeopress.

Lazar, C. 2011. Some observations about spatial relation and location of the Kodjadermen-Gumelnița-Karanovo VI extra muros necropolis, in Mills, S. and Mirea, P. (eds.), *The Lower Danube in Prehistory:*

Landscape Changes and Human Environment Interactions – Proceedings of the International Conference, Alexandria 3-5 November 2010, 95–115. Bucharest: Renaissance.

Lazar, C. (ed.) 2012. *The Catalogue of the Neolithic and Eneolithic Funerary Findings from Romania*. Bucharest: Cetatea de Scaun.

Lazar, C. 2014. The Eneolithic necropolis from Sultana-Malu Rosu (Romania) – a case study, in Oosterbeek, L. and Fidalgo, C. (eds.), *Mobility and Transitions in the Holocene* (BAR International Series 2658), 67–74. Oxford: Archaeopress.

Lazar, C. and Voicu, M. 2015. The distortion of archaeological realities through objects: A case study, in Kogălniceanu, R., Gligor, M., Curcă, R. and Stratton, S. (eds.), *Homines, Funera, Astra 2. Life beyond Death in Ancient Times (Romanian Case Studies). Proceedings of the International Symposium on Funerary Anthropology, 23-6 September 2012, '1 fDecembrie 1918' University (Alba Iulia, Romania)*, 67–77. Oxford: Archaeopress.

Lazar, C., Andreescu, R., Ignat, T., Florea, M. and Astaloș, C. 2008. The Eneolithic cemetery from Sultana-Malu Roşu (Călăraşi County, Romania). *Studii de Preistorie* 5, 131–152.

Lazar, C., Andreescu, R., Ignat, T., Mărgărit, M., Florea, M. and Bălăşescu, A. 2009. New data about the Eneolithic cemetery from Sultana-Malu Roşu (Călăraşi County, Romania). *Studii de Preistorie* 6, 165–199.

Lazar, C., Voicu, M. and Vasile, G. 2012. Traditions, rules and exceptions in the Eneolithic cemetery from Sultana-Malu Roşu (Southeast Romania), in Kogălniceanu, R., Curcă, R., Gligor M. and S. Stratton (eds.), *Homines, Funera, Astra. Proceedings of the International Symposium on Funerary Anthropology, 5-8 June 2011, '1 Decembrie 1918' University (Alba Iulia, Romania)* (BAR International Series 2410), 107–118. Oxford: Archaeopress.

Lichter, C. 2001. *Untersuchungen zu den Bestattungssitten des Südosteuropaischen Neolithikums und Chalkolithikums*. Mainz am Rhein: Philipp von Zabern.

Manifold, B. M. 2012. Intrinsic and extrinsic factors involved in the preservation of non-adult skeletal remains in archaeology and forensic science. *Bulletin of the International Association for Paleodontology* 6, 51–69.

Moorrees, C. F. A., Fanning, E. A., and Hunt Jr., E. E. 1963. Formation and resorption of three deciduous teeth in children. *American Journal of Physical Anthropology* 21, 205–213.

Nielsen-Marsh, C. M., Gernaey, A., Turner-Walker, A., Hedges, R., Pike, A. and Collins, M. 2002. The chemical degradation of bone, in Cox, M. and Mays, S. (eds.), *Human Osteology in Archaeology and Forensic Science*, 439–454. Cambridge: Cambridge University Press.

Petrescu-Dâmbovița, M. 2001a. Eneoliticul timpuriu, in Petrescu-Dîmbovița, M. and Vulpe, A. (eds.), *Istoria Românilor, vol. I. Moștenirea Timpurilor Îndepărtate*, 148–154. Bucharest: Editura Academiei Române.

Petrescu-Dâmbovița, M. 2001b. Eneoliticul dezvoltat, in Petrescu-Dîmbovița, M. and Vulpe, A. (eds.), *Istoria Românilor, vol. I. Moștenirea Timpurilor Îndepărtate*, 154–168. Bucharest: Editura Academiei Române.

Popovici, D. N. 2010. Copper Age traditions north of the Danube River, in Anthony, D. W. and Chi, J. Y. (eds.), *The Lost World of Old Europe. The Danube Valley, 5000-3500 BC*, 91–111. New York: Princeton University Press.

Sayer, D. and Wienhold, M. 2013. A GIS-investigation of four early Anglo-Saxon cemeteries: Ripley's K-function analysis of spatial grouping amongst graves. *Social Science Computer Review* 31, 71–89.

Scheuer, L., and Black, S. 2004. *The Juvenile Skeleton*. London: Academic Press.

Scott, E. 1999. *The Archaeology of Infancy and Infant Death* (BAR International Series 819). Oxford: Archaeopress.

Slavchev, V. 2010. The Varna Eneolithic cemetery in the context of the Late Copper Age in the east Balkans, in Anthony, D. W. and Chi, J. Y. (eds.), *The Lost World of Old Europe. The Danube Valley, 5000-3500 BC*, 192–210. New York: Princeton University Press.

Šmejda, L. 2004. Potential of GIS for analysis of funerary areas: Prehistoric cemetery at Holešov, distr. Kroměříž, Czech Republic, in Šmejda, L. and Turek, J. (eds.), *Spatial Analysis of Funerary Areas*, 57–68. Plzeň: Department of Archaeology, University of West Bohemia.

Sofaer Derevenski, J. (ed.). 2000. *Children and Material Culture*. London: Routledge.

Todorova, H. 1986. *Kamenno-mednata Epokha v Bulgariya. Peto Khilyadoletie Predi NovataEra*. Sofia: Izdatepstvo Nauka i Izkustvo.

Ubelaker, D. H. 1989. *Human Skeletal Remains - Excavation, Analysis, Interpretation*. Washington DC: Taraxacum.

Wheatley, D. and Gillings, M. 2002. *Spatial Technology and Archaeology: The Archaeological Applications of GIS*. London: Taylor and Francis.

Yordanov and Dimitrova, B. 2002. Results of an anthropological study of human skeletal remains of the prehistoric necropolis in the vicinity of the village of Durankulak, in Todorova, H. (ed.), *Durankulak, Band II. Die Prähistorischen Gräberfelder, Teil 1*, 326–347. Sofia: Publishing House Anubis Ltd.

Chapter 6

Late Chalcolithic Skeletal Remains and Associated Mortuary Practices from Çamlıbel Tarlası in Central Anatolia

Jayne-Leigh Thomas[1]

Abstract: Recent excavations at the Late Chalcolithic site of Çamlıbel Tarlası in Central Anatolia have uncovered numerous burials and human bones from secondary contexts within the settlement. The skeletal remains reveal that the majority of individuals were young children and infants; only four adult skeletons were found. Numerous pathological and dental conditions were discovered and work is underway to fully explore the range of pathologies exhibited. In correlation with other contemporary sites in the region, it would appear that the mortuary practice found at Çamlıbel Tarlası reflects the tradition of burying juvenile and infant remains in pots or under the floors of the settlement, with the majority of adult individuals buried extramurally.

Keywords: Anatolia, juvenile remains, Late Chalcolithic, mortuary studies

Introduction

During the 2007–9 field seasons, the Chalcolithic site of Çamlıbel Tarlası was excavated by archaeologists from the University of Edinburgh as a smaller project of the Boğazköy Expedition run by the Deutsches Archäologisches Institut. This project was undertaken in an attempt to explore the prehistoric periods in the region prior to the development and rise of the Hittite kingdom (Schoop 2011, 54). The exploration of this area began with excavations at Alişar Höyük in the 1930s and continued at Chalcolithic settlements such as Arslantepe and Çadir Höyük (Steadman *et al.* 2007, 386; Steadman *et al.* 2008, 47; Frangipane 2011, 968). The discovery and subsequent excavations of Çamlıbel Tarlası has provided new insight into the prehistoric settlement structures and cultural stability in this region during the Chalcolithic period.

Little osteological research has been done on human remains from the region; however, preliminary analyses of the remains have revealed significant information regarding population demographics and mortuary practices. This article discusses the osteological analysis of the individuals from Çamlıbel Tarlası and sheds new light on the mortuary practice and burial customs of prehistoric people in Central Anatolia during the fourth millennium BC.

Study Area

Located 2.5 km from the Hittite capital of Hattuşa, Çamlıbel Tarlası is situated in the narrow side valley of the Karakeçili Deresi, on a low plateau between two ridges of basaltic bedrock (Fig. 6.1; Schoop 2008, 150; 2011, 54; 2015, 47). Geomorphological studies show that the current surroundings of Çamlıbel Tarlası are badly damaged from erosion; however, the site would have been no more than 50 m × 50 m in size and would have been utilised for small-scale agriculture (Marsh 2010, 135; Schoop 2015, 47). This rural settlement is represented by four phases of activity, spanning a time period of 120 years during the fourth millennium BC, with phases of ephemeral use occurring between the second, third, and fourth construction phases (Table 6.1; Schoop 2009, 67; 2015, 50).

Original evidence of seasonal human activity at Çamlıbel Tarlası was revealed by the discovery of groups of circular pits consisting of insulation layers comprised of potsherds, stones and clay (Schoop 2015, 50). Surrounded by layers of ash and fragments of iron ore, it has been suggested that these pits may have been utilised as bowl furnaces for metal production (Schoop 2015, 53, 63). The discovery of an ore deposit some two kilometres upstream from Çamlıbel Tarlası further supports

Table 6.1. Phasing at Çamlıbel Tarlası (after Schoop 2011, 55; 2015, 49).

Phase	Characteristic features
Third Phase of Ephemeral Use (TPEU): in plough zone, adult burials	
ÇBT IV	Boundary wall, Flagstone House, courtyard with copper slag, crucibles, incision-decorated pottery
Second Phase of Ephemeral Use (SPEU): non-residential use, bowl furnaces, water flows through site	
ÇBT III	Longhouses, 'Burnt House', incision-decorated pottery, crucibles, copper slag
First Phase of Ephemeral Use (FPEU): non-residential use, bowl furnaces, water flows through site	
ÇBT II	Permanent architecture, many subadult burials
ÇBT I	Non-residential use, possibly seasonal, water course, bowl furnaces, copper ore, adult burial
Virgin soil / bedrock	

6. Late Chalcolithic Skeletal Remains and Associated Mortuary Practices

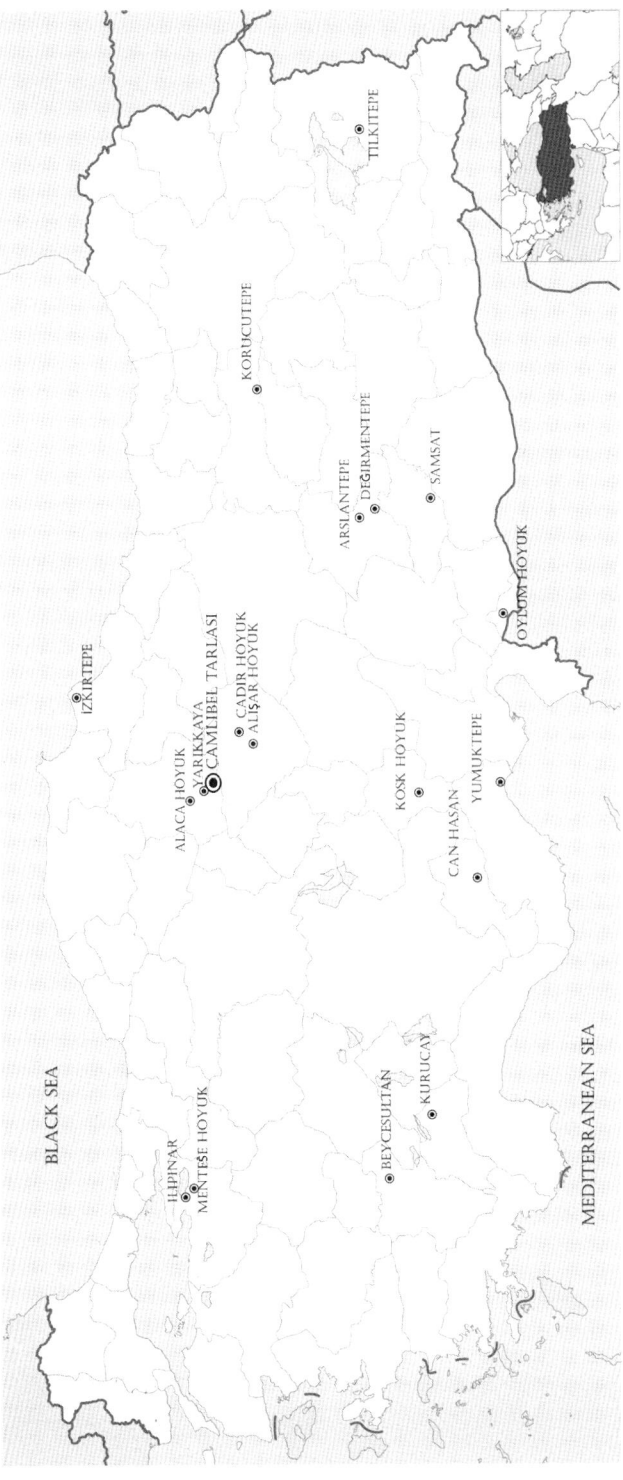

Figure 6.1. Location of Çamlıbel Tarlası and other Chalcolithic sites in Central Anatolia (Turkey Map by NeoRetro, used under CC BY-SA 3.0/ Modified from the original by J.-L. Thomas).

this theory (Schoop 2015, 48). Located close to these pits, large amounts of black ash containing seeds, cereals and flint debitage were found; it is thought that this material is related to agricultural activity taking place on the plateaus surrounding the site (Schoop 2015, 51). After this phase, small structures were built in clusters with large clay ovens and many of the juvenile remains were discovered in juxtaposition to the walls of individual buildings (Schoop 2015, 53). Additional bowl furnaces and fragments of ore were found, indicating a continuation of metallurgical activities at the site (Schoop 2015, 53). During the third phase of habitation at Çamlıbel Tarlası, new long-house buildings and terrace walls were constructed (Schoop 2015, 55). The final phase at Çamlıbel Tarlası is characterised by another period of construction on the site, with smaller rectangular structures and spacious courtyards (Schoop 2015, 56).

Materials and Methods

The human remains were discovered in various conditions. The majority of the infant skeletons were in poor condition, while those of the older children and adults were in a moderate state of preservation. In total, the remains of 19 individuals were recovered from both pot burials and contracted burials. Of the 17 burial contexts, two graves contained the remains of two individuals of varying ages. A total of 68 isolated finds, represented by individual bones or small bone assemblages, were discovered from secondary contexts (Pickard *et al.* 2016, 297). Within the isolated finds the remains of a further eight additional individuals were represented. Methods of recording and assessing age, sex, and pathology of an individual are based on the osteological standards proposed by van Beek (1983), Buikstra and Ubelaker (1994), Bass (1995), Byers (2002) and Scheuer and Black (2004). The following age categories were applied – neonate (prior to birth), perinate (at the time of birth); infant (0–3 years); child (3–12 years); adolescent (13–20 years) and adult (21+ years).

Results and Discussion

The following section details the information revealed from the osteological analysis of the skeletal remains from both burial and secondary contexts (Table 6.2). The majority of individuals from Çamlıbel Tarlası were buried within their own grave or pot, with only two of the graves holding the remains of more than one individual. This is similar with findings from the Chalcolithic sites of Ilıpınar and Menteşe, where all remains were buried separately, except for one grave at Ilıpınar containing the bodies of a young adult female and a perinate (Roodenberg and Alpaslan-Roodenberg 1999, 5). At the site of Korucutepe, Late Chalcolithic graves were discovered in tombs; one of the tombs contained a double burial, the remains being from an adult male and an adult female (van Loon 1978, 10–11; Welton 2010, 121). At the Late Chalcolithic cemetery of Tilkitepe, three sets of double graves were found with the remains of two individuals, in each case an adult and a child (Özgüç 1948, 11; Welton 2010, 122).

6. Late Chalcolithic Skeletal Remains and Associated Mortuary Practices

Table 6.2. Individuals from the grave and secondary contexts (Thomas 2011, 74).

Grave Number	Age at Death	Type of Burial	Pathology
12	30–40 years	Contracted inhumation	None
17	12–15 months	Contracted inhumation	None
8	20–30 years	Contracted inhumation	Moderate dental attrition
4	8–10 years	Contracted inhumation	Severe dental attrition
11	7–9 years	Contracted inhumation	Occipital deformation, Wormian bones
13	6–8 years	Contracted inhumation	Occipital deformation, Wormian bones; moderate attrition
5	6–8 years	Contracted inhumation	Moderate-severe attrition
14	4–5 years	Contracted inhumation	Occipital deformation, Wormian bones; slight attrition
3	2–4 years	Contracted inhumation	Occipital deformation
9	18–24 months	Contracted inhumation? (Disturbed)	None
7	3–5 years	Jar burial	Enamel hypoplasia; slight attrition
16	15–18 months; neonate	Jar burial	Occipital deformation, inflammatory lesions
2	0–3 months; 16–24 gestational weeks	Jar burial	Slight attrition
15	28 gestational weeks–3 months	Jar burial	None
1	9–15 months	Jar burial	None
6	18–24 months	Jar burial	None
15	28 gestational weeks–3 months	Jar burial	None
10	8–12 gestational weeks	Jar burial	None
Isolate	6 ± 2 months	Secondary contexts	None
Isolate	Perinate	Secondary contexts	None
Isolate	Perinate; perinate	Secondary contexts	None
Isolate	25–35 years	Secondary contexts	None
Isolate	Adult (21+ years)	Secondary contexts	None
Isolate	~5 years	Secondary contexts	None
Isolate	~1.5 years	Secondary contexts	None

The difference with the double burials at Çamlıbel Tarlası is that both contained the remains of very young children and both were in vessels, as opposed to flexed burials, where the arms and legs were bent. At the site of İzkirtepe where nearly 700 burials were discovered, 52 graves were discovered to contain the remains of more than one individual and several of these graves contained more than two individuals (Welton 2010, 114). However, the double burials consisted of adult individuals buried with children and, in most cases, the remains of the second individual comprised fragmentary bones that were mixed in with the primary internments, rather than a complete burial (Welton 2010, 114). It may be that the individuals included in the Çamlıbel Tarlası double burials were related or that, having passed away at the same time, they were buried together for symbolic purposes.

Age-at-Death

Age-at-death was established for all 28 individuals analysed, ranging from neonate to approximately 40 years of age. As shown in Table 6.2, the remains of only four adults were present, with the other skeletons deriving from individuals aged less than 10 years. The discovery of primarily juvenile remains from this site compares directly with other Chalcolithic sites in the region as adult individuals tend to be rare. At the site of Alişar Höyük (Table 6.3), for example, the majority of the remains were from juveniles or infants, with only a few adults having been recovered (von der Osten 1937, 32; Welton 2010, 117). Some 31 juveniles were retrieved from the site of Değirmentepe near the Euphrates River during salvage excavations. No adults were found at this site and it was presumed they were buried in a separate location (Özbek 2001, 240). The Chalcolithic sites of Can Hasan, Kösk Höyük and Çadir Höyük also contained large numbers of juvenile, while adult burials were extremely rare at the site of Beycesultan, and it was hypothesised that they were buried extramurally (Lloyd and Mellaart 1962, 23–6; French 1968, 45; Gorny *et al.* 1999, 152–3; Özkan *et al.* 2004, 195–6; Welton 2010, 117–8). The idea that adults in the region were buried extramurally is reflected in the mortuary practice exhibited at Çamlıbel Tarlası, where the majority of remains uncovered were those of juveniles and infants.

Pathologies

Table 6.2 indicates the individuals with assigned pathologies. Eight individuals exhibit varying degrees of dental attrition on a range of teeth, including heavy attrition on deciduous teeth, indicating wear at young ages (Irvine *et al.* 2014, 6). Previous studies from Anatolia and the Levant focusing on dental attrition have concluded that substantial dental wear on affected teeth was attributed to non-masticatory processes or abrasive diets (Alpaslan-Roodenberg 2011, 32; Irvine *et al.* 2014, 7). Dental wear found on the Çamlıbel Tarlası children has been interpreted as a result of a diet high in carbohydrates; this interpretation is substantiated by stable isotope analyses performed on the Çamlıbel Tarlası population which have shown a reliance on C3 plants such as cereals and lentils (Pickard *et al.* 2016, 303). With high carbohydrate diets the most common pathology is dental caries, however, of which there is an unusually low prevalence within the Çamlıbel Tarlası juvenile population (Irvine *et al.* 2014, 7). This has been interpreted as a result of the additional consumption of varying percentages of protein, and quantities of milk and dairy products which tend to offer protection against caries (Hillson 2008, 313; Ullinger 2010, 29; Irvine *et al.* 2014, 8). This theory is supported by the discovery of churns at Çamlıbel Tarlası and the nearby Chalcolithic site of Yarikkaya where pottery sherds contained fatty acid residues, which were interpreted as having derived from milk or dairy products (Schoop 1998, 28–9; Sauter *et al.* 2003, 20; Bartosiewicz and Gillis 2011, 78; Pickard *et al.* 2016, 303).

The presence of heavy dental attrition in addition to other dental lesions indicates the use of teeth in non-masticatory processes such as the habitual holding of objects between the teeth or the biting down hard on objects during production activities (Irvine *et al.* 2014. 8–10). Similar dental conditions have been found on skeletal remains from the Chalcolithic sites of Ilıpınar and Menteşe Höyük. Roodenberg and Alpaslan-Roodenberg (1999, 6) discuss how dental lesions were discovered in very young individuals and suggest that food preparation was probably the cause of dental lesions and varying stages of dental attrition. Individuals from both of these sites also displayed grooves on the front teeth and it has been suggested that these indentations are likely to have arisen as a result of non-masticatory processes (Roodenberg and Alpaslan-Roodenberg 1999, 6; Alpaslan-Roodenberg 2001, 6).

Five individuals show signs of artificial cranial deformation, with a form of circular banding having been used to produce the distortion. Similar results have been found on the juvenile remains from the Chalcolithic site of Değirmentepe near the Euphrates River in Anatolia. Artificial cranial deformation was discovered in 13 individuals, ranging from one month to 13–14 years of age (Özbek 2001, 241). The type of banding found was determined to be circular with either one or two bandages having been applied to create the deformation and Özbek (2001, 238) discusses how this cultural practice may be indicative of long distance trade within the region and the need for defined ethnic markers. At this time, additional research is underway to assess the full extent of cranial deformation within the Çamlıbel Tarlası sample and further information may shed light on the utilisation of this cultural practice within Chalcolithic Anatolia.

Mortuary Practice

All of the graves found from the architectural layers of Çamlıbel Tarlası were those of young children; these remains were found in two mortuary groups, with babies and young infants buried in large pottery vessels in narrow pits (Fig. 6.2) and older children in flexed positions, with the change in burial practice occurring at 4–5 years of age (Schoop 2015, 58–9). The pottery vessels used for the burial were closed with

Figure 6.2. Juvenile burial, aged 4–5 years, in a contracted position (Grave 14) (Thomas 2011, 74).

Figure 6.3. Infant burial, aged 15–18 months, in a pottery vessel. The grave also included the remains of a neonate (Thomas 2011, 74).

Table 6.3. List of Anatolian Chalcolithic sites summarising burial discoveries.

Site	Burial Type	Age of individuals	Period
Çamlıbel Tarlası	Inhumations in ephemeral & settlement layers; jar burials under house floors	Adults, children; children, infants	Late Chalcolithic
Alişar Höyük	Jar burials, cist graves, inhumations under house floors	Primarily children, infants; few adults	Late Chalcolithic
Ilıpınar	Flexed inhumations	Adults	Late Chalcolithic
Kuruçay	Jar burials., inhumations under house & street levels	Children, infants; adults	Late Chalcolithic
Oylum Höyük	Inhumations, jar burials	Adults, infants	Late Chalcolithic
Tilkitepe	Flexed inhumations; jar burials	Adults; infants	Late Chalcolithic
Alaca Höyük	Inhumations, cist graves	Adults, children	Late Chalcolithic
Can Hasan	Jar burials, inhumations under house floors	Children	Middle Chalcolithic
Çadir Höyük	Jar burials under house floors	Children	Middle Chalcolithic
Köşk Höyük	Inhumations beneath house floors	Infants	Early Chalcolithic
Menteşe Höyük	Flexed inhumations	Adults	Early Chalcolithic
Yumuktepe	Inhumations under house floors; inhumations within settlement layers	Children; adults	Chalcolithic
Samsat	Jar burials	Infants	Chalcolithic
Arslantepe	Flexed inhumations under house floors	Adults	Chalcolithic
Değirmentepe	Jar burials	Children, infants	Chalcolithic
Beycesultan	Earthen graves, jar burials	Infants	Chalcolithic

fragments of another ceramic vessel, usually a bowl, and wedged into the side of the grave (Schoop 2015, 58). Another hole was then created by smashing the base of the burial jar once placed in its final location (Schoop 2015, 58). Within the jars, the skull of the individual was often found to be at the bottom of the vessel; however, the postcranial material did not remain in its primary position, so further speculation on the body's position within the vessel is not possible (Schoop 2015, 58). Older children were placed in a contracted position with their heads in a southerly direction and the faces turned towards the east (Fig. 6.3; Schoop 2015, 59). The four adult graves were recovered from layers of ephemeral activity, positioned in a contracted position and facing the east (Schoop 2015, 59).

Similar burial practices were evident at Alişar Höyük (Table 6.3), with individuals being found as inhumations or interred in pots or small cists (von der Osten 1937, 43–4, 51; Özgüç 1948, 23; Welton 2010, 117). The majority of the burials contained the remains of children and infants buried under the floors of houses with bodies, often in a flexed position on their right side (von der Osten 1937, 32; Özgüç 1948, 23; Welton 2010, 117). At the site of Alaca Höyük, Late Chalcolithic cist graves and inhumations were discovered, with the bodies being placed in a flexed position (Özgüç 1948, 11, 42; Welton 2010, 117).

Welton (2010, 117) reports that the burial of juvenile individuals within settlements became common during the Chalcolithic period in Central Anatolia. Within the early levels of Can Hasan, the remains of 12 juveniles were found with whole vessels and later levels produced the remains of juveniles buried beneath the floors of the houses (French 1968, 50; Welton 2010, 117). Jar burials containing the remains of children were discovered from under the floors of Late Chalcolithic houses at Çadir Höyük and over 30 graves containing infant remains were recovered from the floors of buildings at Köşk Höyük (Gorny et al. 1999, 152–3; Özkan et al. 2004, 195–6; Welton 2010, 118).

Adults from other Chalcolithic sites in the region are generally found in simple earth burials, placed in a flexed position (Welton 2010, 118). At the site of Ilıpınar, over 40 graves were recovered in a flexed position within earthen graves (Roodenberg 2001, 351; Welton 2010, 118). This is similar to the situation at Çamlıbel Tarlası where the two adult individuals recovered had been buried in simple inhumation graves and were found in a flexed position.

From the site of Kuruçay dating to the Late Chalcolithic, 50 jar burials and five inhumation burials were discovered beneath the floors of the courtyard and street areas of the settlement. The jar burials contained the remains of infants and children and were closed with stones (Duru 1996, 120; Welton 2010, 119). The inhumations contained the remains of adult individuals and Duru (1996, 121) suggests that the majority of adults were probably buried extramurally (Welton 2010, 120). The burial practice found at Kuruçay is nearly identical to that found at Çamlıbel Tarlası, with numerous children found in pots beneath the settlement and only a few adult remains found as simple inhumations. The difference at Çamlıbel Tarlası is that numerous children were found in a flexed position in burials, rather than being found exclusively in vessels.

At the site of Menteşe Höyük near the modern town of Yenişehir, skeletal material was recovered from simple inhumations in a flexed position on the right side (Alpaslan-Roodenberg and Maat 1999, 41–2; Roodenberg and Alpaslan-Roodenberg 2008, 13; Welton 2010, 120). Yumuktepe produced several intramural burials found beneath the floors of houses within the settlement, consisting of primarily juvenile remains; however adult remains were also found in a flexed position within the settlement (Garstang 1953, 110–1; Welton 2010, 120). At the site of Samsat, infant remains were found within jar and pit burials and the site of Arslantepe contained the remains of individuals placed in flexed positions beneath the floors of houses (Özgüç 1988, 294; Frangipane 2001, 972; Welton 2010, 120).

At the Late Chalcolithic cemetery of Oylum Höyük, several individuals were found in a crouched position and other adults and children were discovered within vessels or simple pits (Özgen and Helwing 2003, 64). Numerous intramural graves were recovered at Tilkitepe with individuals placed in a flexed position and several infants found within vessels (Özgüç 1948, 11; Korfmann 1982, 35; Welton 2010, 121). At Chalcolithic Değirmentepe, infants and small children were found in pots and vessels (Özbek 2001, 240). No adult remains were recovered from the site and it was hypothesised that the adults were either buried extramurally or in a portion of the site which had been obliterated due to flooding (Özbek 2001, 240). The idea of adult extramural interment during the Chalcolithic has also been further explored at Beycesultan where infant burials were found in earthen graves and jars and adult remains are rare (Lloyd and Mellart 1962, 23–6; Welton 2010, 117).

Conclusion

In total, the remains of 24 juveniles and four adult individuals were discovered at Çamlıbel Tarlası. Age-differential burial practices were performed, with infants and young children buried in pots and older children found in simple pit burials. Few adult burials were found at this location and it is generally recognised that adults were buried extramurally or in unexcavated areas of the region. The burial techniques utilised at Çamlıbel Tarlası compare directly with other contemporary sites in the region, where young children and infants are found interred in pots and beneath the floors of houses and adult burials are rare.

The resulting data from the osteological analysis of the remains from Çamlıbel Tarlası has provided important information regarding demographics and burial rites for Central Anatolia during the Late Chalcolithic period. Despite the small sample size new information has been revealed which, with future research, will shed new light on cultural and social aspects of the Çamlıbel Tarlası population.

Acknowledgements

The author would like to thank Dr Ulf-Dietrich Schoop, the Director of the Çamlıbel Tarlası excavations, for access to the skeletal material and photographs. His advice and support is greatly appreciated.

Note

1. Office of the Native American Graves Protection and Repatriation Act, Indiana University, Student Building 318, 701 E. Kirkwood Ave., Bloomington, Indiana, USA 47405. Email: thomajay@indiana.edu.

References

Alpaslan-Roodenberg, S. 2001. Newly found human remains from Menteşe in the Yenişehir Plain: The season of 2000. *Anatolica* 27, 1–14.

Alpaslan-Roodenberg, S. 2011. A preliminary study of the burials from Late Neolithic – Early Chalcolithic Aktopraklik. *Anatolica* 37, 17–43.

Alpaslan-Roodenberg, S. and Maat, G. 1999. Human skeletons from Menteşe Höyük near Yenişehir. *Anatolica* 25, 37–52.

Bartosiewicz, L. and Gillis, R. 2011. Preliminary report on the animal remains from Çamlıbel Tarlası, Central Anatolia. *Archäologischer Anzeiger* 2011, 76–79.

Bass, W. 1995. *Human Osteology: A Laboratory and Field Manual.* Columbia, MO: Missouri Archaeological Society.

Buikstra, J. and Ubelaker, D. 1994. *Standards for Data Collection from Human Skeletal Remains* (Arkansas Archeological Survey Research Series No. 44). Arkansas: Arkansas Archeological Survey.

Byers, S. 2002. *Introduction to Forensic Anthropology: A Textbook.* Boston, MA: Allyn and Bacon.

Duru, R. 1996. *Kuruçay Höyük II: Results of Excavations 1978-88, The Late Chalcolithic and Early Bronze Age Settlements.* Ankara: Türk Tarih Kurumu.

Frangipane, M. 2011. Aslantepe-Matalya: A prehistoric and early historic center in Eastern Anatolia, in McMahon, G. and Steadman, S. (eds.), *The Oxford Handbook of Ancient Anatolia*, 968–992. Oxford: Oxford University Press.

French, D. 1968. Excavations at Can Hasan, 1967: Seventh preliminary report. *Anatolian Studies* 18, 45–53.

Garstang, J. 1953. *Prehistoric Mersin: Yümük Tepe in southern Turkey.* Oxford: Clarendon Press.

Gorny, R., McMahon, G., Paley, S., Steadman, S. and Verhaaren, B. 1999. The 1998 Alişar regional project season. *Anatolica* 25, 149–184.

Hillson, S. 2008. Dental pathology, in Katzenberg, M. and Saunders, S. (eds.), *Biological Anthropology of the Human Skeleton*, 301–340. Hoboken, NJ: John Wiley & Sons.

Irvine, B., Thomas, J.-L. and Schoop, U.-D. 2014. A macroscopic analysis of human dentition at Late Chalcolithic Çamlıbel Tarlası, north-central Anatolia, with special reference to dietary and non-masticatory habits. *Interdisciplinaria Archaeologica* 5, 1–12.

Korfmann, M. 1982. *Tilkitepe: Die ersten Ansätze prähistorischer Forschung in der Östlichen Türkei.* Tübingen: Verlag Ernst Wasmuth.

Lloyd, S. and Mellaart, J. 1962. *Beycesultan, Vol. I: The Chalcolithic and Early Bronze Age Levels.* London: British Institute of Archaeology at Ankara.

Marsh, B. 2010. Geoarchaeology of the human landscape at Boğazköy-Hattuša. *Archäologischer Anzeiger* 2010, 201–207.

Özbek, M. 2001. Cranial deformation in a subadult sample from Değirmentepe (Chalcolithic Turkey). *American Journal of Physical Anthropology* 115, 238–244.

Özgen, E. and Helwing, B. 2003. On the shifting border between Mesopotamia and the west: Seven seasons of joint Turkish-German excavations at Oylum Höyük. *Anatolica* 29, 61–85.

Özgüç T. 1948. *Die Bestattungsbräuche im Vorgeschichtlichen Anatolien.* Ankara: Veröffentlichungen der Universität von Ankara.

Özgüç, T. 1988. Haberler-Kazıları 1987. *Belleten* 52, 291–294.

Özkan, S., Faydalı, E., Öztan, A. and Erek, M. 2004. 2002 Yılı Kösk Höyük Kazıları. *Kazı Sonuçları Toplantısı* 25, 195–204.

Pickard, C., Schoop, U.-D., Dalton, A., Sayle, K., Channel, I., Calvey, K., Thomas, J.-L., Bartosiewicz, L. and Bonsall, C. 2016. Diet at Late Chalcolithic Çamlıbel Tarlası, north central Anatolia: An isotopic perspective. *Journal of Archaeological Science* 5, 296–306.

Roodenberg, J. 2001. A Chalcolithic cemetery at Ilıpınar in Northwestern Anatolia, in Boehmer, R. M. and Maran, J. (eds.), *Archaologie Zwischen Asien und Europa*, 351–355. Rahden: Verlag Marie Leidorf.

Roodenberg, J. and Alpaslan-Roodenberg, S. 1999. *The Neolithic in Eastern Marmara: Two Examples of Settlement*. Leiden: The Netherlands Institute for the Near East.
Roodenberg, J. and Alpaslan-Roodenberg, S. 2008. Ilıpınar and Menteşe: Early settlement in the eastern Marmara region, in Bailey, D., Whittle, A. and Hofmann, D. (eds.), *Living Well Together?: Settlement and Materiality in the Neolithic of South-east and Central Europe*, 8–16. Oxford: Oxbow Books.
Sauter, F., Puchinger, L. and Schoop, U.-D. 2003. Studies in organic archaeometry VI: Fat analysis sheds light on everyday life in prehistoric Anatolia: Traces of lipids identified in Chalcolithic potsherds excavated near Boğazkale, Central Turkey. *ARKIVOC* 3, 15–21.
Schoop, U.-D. 1998. Anadolu'da Kalkolitik Çağda Süt Ürünleri Üretimi. Bir Deneme. *Arkeoloji ve Sanat* 87, 26–32.
Schoop, U.-D. 2008. Ausgrabungen in Çamlıbel Tarlası 2007. *Archäologischer Anzeiger* 2008, 148–157.
Schoop, U.-D. 2009. Ausgrabungen in Çamlıbel Tarlası 2008. *Archäologischer Anzeiger* 2009, 56–66.
Schoop, U.-D. 2011. Çamlıbel Tarlası, ein Metallverarbeitender Fundplatz des Vierten Jahrtausends v. Chr. im Nördlichen Zentralanatolien, in Yalçın, Ü. (ed.), *Anatolian Metal 5*, 53–68. Bochum: Deutsches Bergbau-Museum.
Schoop, U.-D. 2015. Çamlıbel Tarlası: Late Chalcolithic settlement and economy in the Budaközü valley (north-central Anatolia), in Steadman, S. and McMahon, G. (eds.), *The Archaeology of Anatolia I. Recent Discoveries (2011-14)*, 46–68. Newcastle-upon-Tyne: Cambridge Scholars Publishing.
Scheuer, L. and Black, S. 2004. *The Juvenile Skeleton*. London: Elsevier Ltd.
Steadman, S., McMahon, G. and Ross, J. 2007. The Late Chalcolithic at Çadır Höyük in central Anatolia. *Journal of Field Archaeology* 32, 385–406.
Steadman, S., Ross, J., McMahon, G. and Gorny, R. 2008. Excavations on the north-central Plateau: The Chalcolithic and Early Bronze Age occupation at Çadır Höyük. *Anatolian Studies* 58, 47–86.
Thomas, J.-L. 2011. Preliminary observations on the human skeletal remains from Çamlıbel Tarlası. *Archäologischer Anzeiger* 2001, 73–76.
Ullinger, J. 2010. Skeletal Health Changes and Increasing Sedentism at Early Bronze Age Bab edh-Dra, Jordan. Unpublished Ph.D. thesis, Ohio State University.
van Beek, G. 1983. *Dental Morphology: An Illustrated Guide*. Edinburgh: Wright.
van Loon, M. 1978. *Korucutepe: Final report on the excavations of the Universities of Chicago, California (Los Angeles) and Amsterdam in the Keban Reservoir, Eastern Anatolia 1968-70, Vol. 2*. Amsterdam: North-Holland Publishing Company.
von der Osten, H. 1937. *The Alişar Höyük, Vol. III*. Chicago, IL: Oriental Institute Publications.
Welton, M. 2010. Mobility and Social Organisation on the Ancient Anatolian Black Sea Coast: An Archaeological, Spatial, and Isotopic Investigation of the Cemetery at İzkirtepe, Turkey. Unpublished Ph.D. thesis, University of Toronto.

Chapter 7

Processed Babies: Early Bronze Age Infant Burials from Bulgarian Thrace

Kathleen McSweeney[1] *and Krum Bacvarov*[2]

Abstract: The recent analysis of over 50 infants from Early Bronze Age (EBA; third millennium BC) jar and pit burials from various sites in Bulgarian Thrace has revealed funerary practices that were previously unidentified. Varying degrees of skeletal articulation indicate that the bodies of most of the babies were in a state of partial decomposition before being placed in the containers. In addition, there was ample evidence of dismemberment, inexplicably missing bones, as well as additional isolated body parts from other individuals. These factors are suggestive of a mortuary practice in which the dead were 'processed' prior to burial. This paper will explore the evidence and the possible reasons for these unusual mortuary practices.

Keywords: Infant jar burials, burial practices, dismemberment, EBA Bulgarian Thrace

Background to the Research

The custom of infant jar burial in prehistoric Southeast Europe and beyond is well documented from as early as the Neolithic. Burial in jars has been recognised from late seventh-millennium Mesopotamia and the early sixth millennium in Anatolia, the Balkans and the Southern Levant.

The area of the Struma and Vardar river valleys, the west Rhodope and Central Macedonia in the early sixth millennium BC – or the Early Neolithic of the southeast Balkan chronology – was the only place in Europe where jar burial has been practised (Bacvarov 2008). One of the most authoritative neolithisation models considers this very territory as the point of first Neolithic penetration as well as a contact zone between the early settlers and their new neighbours in the second phase of the local Early Neolithic (Nikolov 2007). It is thus possible to relate the earliest jar burials in Southeast Europe to these mutual exchange processes and to trace them back to their hypothetic point of origin. In Western Anatolia, however, which is considered the home of Early Neolithic painted pottery cultures in the central Balkans, no jar burials have been found, the closest parallels being the central Anatolian tell sites of Kösk Höyük, Pinarbaşi-Bor and Tepecik-Çiftlik, referred to the Late Neolithic and Early Chalcolithic of the Anatolian chronology.

Jar burial appears to have originated in the northern Levant, sometime in the pre-Hassuna time, and for a relatively short time influenced culture developments as far as the central Balkans; the appearance of this mortuary practice in the Southern Levant followed soon after. The absence of contemporaneous remains in western Anatolia – bridged by the three central Anatolian sites of Kösk Höyük, Pinarbaşi-Bor and Tepecik-Çiftlik – could also hint at Southeast European autonomy; however, this can hardly be substantiated since the five burials in the Struma and Vardar river valleys, the west Rhodope and Central Macedonia share common diagnostics with their Anatolian and Levantine parallels. What is more plausible is that the idea of burial of foetus/infant/child in a ceramic pot, as an element of the social reproduction and cohesion networks, was transferred along the neolithization routes and its expressions were triggered by certain stimuli, most probably natural events, as is demonstrated by the burials' contemporaneity as well as the clustering of sites both in Southeast Europe and Central Anatolia.

In the Later Neolithic and Chalcolithic, jar burial was further developed, reoccurring at various settlement sites as well as cemeteries in Southeast Europe, sending distinct echoes as far as southern Transdanubia. Burial in a ceramic container was to gradually become the dominating burial practice in Anatolia, the Aegean, and the Levant, elaborated in such forms as the pithos burial of adults.

The early development of burial in a ceramic vessel climaxed in the Early Bronze Age, almost completely covering Anatolia as well as the Aegean and the Levant, in both its forms, pithoi- and jar burials. In Southeast Europe, however, the jar burial area drastically shrank down to a small region in Upper Thrace, although the number of graves at the various sites much exceeded the earlier cases, demonstrating once more close relations to Anatolia and the Levant. The Southeast European burials, however, strictly stuck to the original idea of settlement inhumation of foetuses/babies only, and never adopted later elaborations as the Anatolian pithoi burials of adults or the Palestinian ceramic ossuaries; this fact seems to support the theory of the eastern origins of this burial practice together with some more details such as the occasional flint artefacts, found as grave goods – perhaps related to the ritual of cutting the baby's umbilical cord – or the intentional piercing of the burial vessel's bottom or damaging of its mouth rim, both occurring since the first appearance of jar burial in Southeast Europe as well as in the northern Levant.

By the Early Bronze Age (EBA) the practice had spread throughout Mesopotamia, Anatolia and the Aegean and in a small area of Bulgarian Thrace in Southeast Europe, with only one example found in Thessaly (see, for example, Bacvarov 2008; Mishina 2008, and references therein). This paper focuses on a number of infant burials from EBA Thrace.

7. Processed Babies: Early Bronze Age Infant Burials from Bulgarian Thrace

Material Examined

A total of 42 jar burials from five sites, Tell Yunatsite (Nos. 1–4, 8–14, 16–18, 21, 22, 27, 41, 45 and 45c), Tell Ezero (Nos. 1, 2, 4 and 5), Nova Zagora (No. 1), Tell Kran (Nos. 1, 2 and 4–9) and Tell Karanovo (Nos. 1–9) (Fig. 7.1; for background information on the sites see, Georgiev *et al.* 1979; Kancheva-Ruseva 2000; Hiller and Nikolov 2002; Hiller *et al.* 2005, Andreeva 2007; Mishina and Balabina 2007; Andreeva 2011; McSweeney *et al.* 2016) have been considered in this analysis (Table 7.1). The term 'jar burial' is defined here as burial of subadults in a ceramic vessel that can take various forms but is different to cremation or secondary burial in containers. In addition to the jar burials, individuals from four pit burials from Tell Yunatsite were also examined (Nos. 5, 15, 19 and 25).

With the exception of two jar burials from Tell Kran (4 and 9), the skeletal remains had all been removed from their jars at some stage prior to the current skeletal analysis and placed in bags. In one instance (Tell Kran 8) the bones were in bags, labelled

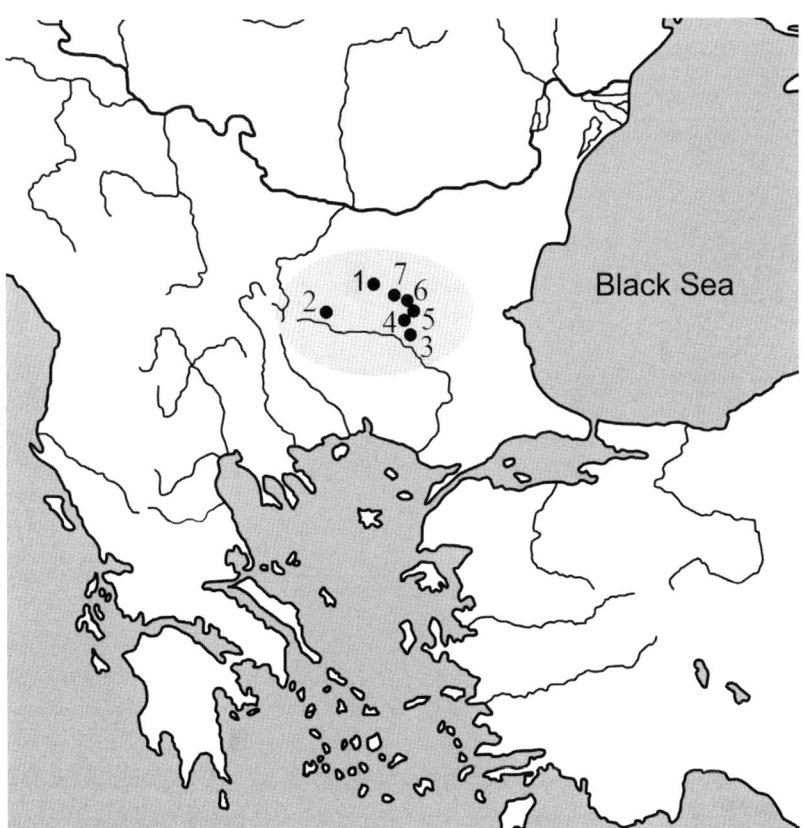

Figure 7.1. Map showing the jar burial distribution area in Early Bronze Age (EBA) Thrace: Kran (no. 1); Yunatsite (no. 2); Galabovo (no. 3); Ezero (no. 4); Dyadovo (no. 5); Nova Zagora (no. 6); Karanovo (no. 7) (prepared by Krum Bacvarov).

Table 7.1. Burials and the number of individuals from each site.

Site	Jar burials	Double burials	Pit burials	No. Individuals
Tell Yunatsite	20	4	4	28
Tell Ezero	4	1		5
Nova Zagora	1			1
Tell Kran	8			8
Tell Karanovo	9			9
Total	42	5	4	51

Figure 7.2. Tell Kran 9, before (left) and after (right) removal of part of the jar. Note the undisturbed contents (photographs by Kathleen McSweeney and Krum Bacvarov).

from spits one to five, suggesting that the contents of the jar had been systematically removed. In the case of Tell Kran 4 the first author excavated some remains that were still in partial articulation in the base of the jar. A bag of disarticulated remains was also present from this context. Although most of the post-cranial bones were accounted for in Tell Kran 4, only the base of the cranium was identified and it is possible that the remainder of the skull had originally been in the jar but had become lost along with the rest of the jar since deposition. In only one instance (Tell Kran 9) did the authors have the opportunity to examine undisturbed remains in an intact jar (Fig. 7.2). This opportunity arose after the analysis of the other burials. The contents of Tell Kran 9 were micro-excavated. Once a layer of bones was uncovered these were removed and then the next layer of bones was exposed, and so on, until the pot was empty of all contents. In all, 13 layers were removed and careful note was taken of bones in articulation. Once all of the bones were removed, the soil matrix was dry and wet-sieved. Despite very careful excavation, several small bones and unerupted teeth were retrieved during sieving, which suggests that some small bones from jars previously excavated could have been missed. The first author then osteologically examined the bones.

The Sites

The locations of the sites from which the baby burials originated are shown in Figure 7.1. Most of the five Thracian sites considered in this paper are multi-layered tell sites, with the exception of Nova Zagora, which is a flat site. Tell Yunatsite is a large tell, with 9 m of cultural deposits, and was first excavated as early as 1939. It yielded horizons throughout prehistoric and historical periods from the Early Chalcolithic through to the Middle Ages. A total of 28 infant burials dating to the EBA were found on this site. The remains examined as part of this study included all of the jar burials and four pit burials. The EBA horizons at Tell Ezero, a 10 m high tell, which also had Neolithic and Chalcolithic levels, produced four infant burials, at least two of them jar burials, from excavations in the 1950s, and a further 10 burials, three of which were in jars, from excavations between 1961 and 1971. Only four of the jar burials from this site formed part of the current project. Kran is a small tell, only 5 m high, in Upper Thrace, of Late Neolithic and EBA date. Eight of the jar burials included in the study were from this site. A further nine jar burials in the sample came from Tell Karanovo, a large tell, some 12.4 m high, with cultural levels from the Neolithic, Chalcolithic and EBA. All of the jar burials were found under the floors of houses. The site of Nova Zagora, the only flat site to date to produce jar burials, yielded six jar burials in association with houses, although five of these were poorly preserved (Bacvarov 2008). Only one jar burial from this site is included in the study.

The Jars

The jars used for burial were of various sizes and shapes. Those used at Tell Yunatsite are described as 'jugs, bowls, pots, with or without lugs, or even bottom parts of broken vessels' (Bacvarov 2008, 64). At Tell Ezero infants were buried in large pots, *c.* 37 cm in height, while at Nova Zagora the fragmented jars were estimated to be about 30 cm high. In many cases lids on the jars were reported; for example, at Tell Karanovo, one of the jars was sealed with a conical bowl (Bacvarov 2008, 65), while at Tell Yunatsite it was said that the 'vessels were sealed with lids' (Mishina 2008, 145) and a clay plate was used at Tell Ezero (Bacvarov 2008, 65). In almost all cases the burials were found either under the floors of houses or in close association with houses.

Condition of the Remains

While a few of the skeletons were almost complete (for example, Tell Yunatsite 45c), the majority had some bones missing. A few consisted of only a small number of bones. In about 50% of the burials the hand and foot bones were missing (see below). Those bones that were present were generally in an excellent state of preservation with many complete bones.

Age-at-Death

The assessment of age-at-death among the baby burials was based on bone dimensions (Fazekas and Kosa 1978), dental development (van Beek 1983) and skeletal development (Scheuer and Black 2004). Details of age-at-death are shown in Figure 7.3. The majority (69%; 35/51) of the babies were aged between 38 and 40 foetal weeks, i.e. full term foetuses. Unfortunately, it is not possible to establish from the skeletal remains whether these had been still or live births. Twelve of the babies had not reached full term and were perhaps more likely to have been still born. Most of these pre-term babies were aged from 32 to 37 intrauterine weeks. The youngest was 22 prenatal weeks. It is almost certain that this latter individual had been stillborn, or died very soon after birth, as foetuses of less than 28 weeks in the absence of the intervention of modern medicine are not viable due to the underdevelopment of the internal organs (Lewis 2007, 94). There were a few burials of older children. One of the babies had probably lived for a few weeks, a further two were between three and six months of age and one was an older child of about 18 months. This older child, two of the full-term foetuses and a 28-week foetus were from the pit burials.

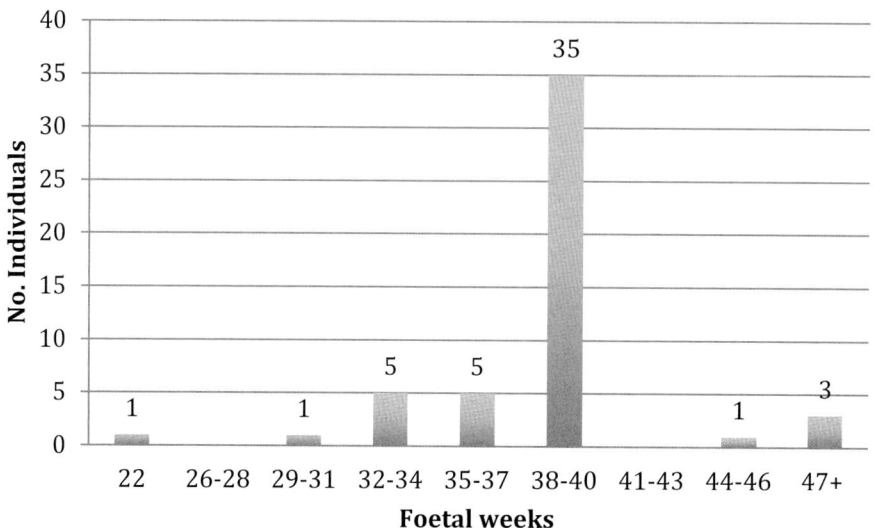

Figure 7.3. Age at death in weeks. The age ranges used are adapted from Mays (2010) (prepared by Kathleen McSweeney).

Cause of Death

There was very little evidence of pathology associated with the babies but this is not in itself significant. Most disease processes initially affect the soft tissues and take time to progress to the skeleton and it is unlikely that the majority of the babies, if born live, lived for more than a few days. Most neonates who do not survive the birth process are affected by 'congenital anomalies, prematurity, low birth weight [and/or] birth trauma' (Lewis 2007, 84). One perinate (Tell Yunatsite 3) had abnormal reactive bony growth on the internal surfaces of its frontal and parietal bones. This may be related to a condition that developed during the intra-uterine period and is very likely to be related to the reason why this child did not survive. Such lesions may be connected to meningeal infections, nutritional deficiencies or trauma among children (Lewis 2004). In the case of Tell Yunatsite 3, a 38- to 40-week-old foetus, peri-mortem fractures of the frontal bone were apparent. Whether these were related to trauma prior to death or had occurred post-mortem is not possible to say with any degree of certainty.

Double Burials

There were five double burials: four from Tell Yunatsite (jar burials 2, 4, 8 and 14) and one from Tell Ezero (burial 4). It is not clear whether the individuals had been deposited in the jars together, or at different times. In two cases, the babies were of different ages and the bones of the individuals were found bagged separately. The separation of the remains may have occurred during excavation if there was a clear stratigraphic difference between the two sets of remains, or at the time of a previous anthropological examination. In a further two cases (Tell Yunatsite 4 and 14), the bones of the two individuals of similar ages were completely mixed, double burial being identified by various duplicated bones. Interestingly, both of the individuals in Tell Yunatsite 14 were aged 32 intrauterine weeks, suggesting that these may possibly have been twins. In one double burial of individuals of different ages (approximately one and three postnatal months, respectively), Tell Ezero 4, clear differences in colour and texture indicated that different taphonomic processes had affected the bones, suggesting either that deposition had occurred at different times, or that the corpses had been previously deposited elsewhere and eventually redeposited in the jar. Stratigraphic information, i.e. excavating the jars in spits, could have helped clarify these questions, but this appears not to have been carried out.

Completeness of the Remains

Nine individuals (17.6%) (from Tell Yunatsite, Tell Ezero, Tell Kran and Tell Karanovo) were largely complete, while another eight (15.7%) were less than 50% complete. Among the other 34 individuals there was ample evidence of inexplicably missing bones. Twenty-three babies that were otherwise in good condition were missing one or both scapulae or clavicles. Tell Kran 9, for example, which was largely complete and in good condition, had the right scapula but not the left and the left clavicle

Table 7.2. Individuals with missing skeletal elements (excludes the eight individuals that were less than 50% complete).

Missing scapulae or clavicles	23
Missing ribs/Spine	4
Missing one or more pelvic bones	20
Missing hands/feet	24

but no right. In Tell Ezero 4, a double burial with bones in good condition, both babies were missing the right scapula and one had no clavicles. Excluding the skeletons that were less than 50% complete, four babies had no, or only a few, ribs and/or vertebrae. For example, Tell Yunatsite 13, a fairly complete skeleton, had most of the spinal elements present but no ribs (and no hand or foot bones) and Tell Yunatsite 19, a partial skeleton that consisted largely of the upper body had most of the rib cage intact but no spine. Tell Ezero 1 was a full skeleton apart from the pelvis. Altogether 20 individuals (excluding the eight partial skeletons) had incomplete pelvic girdles. Hand and foot bones were markedly under-represented. In only three cases were most of the hand and foot bones present (Tell Yunatsite 45c, Tell Kran 9 and Tell Karanovo 7), while in 24 cases (again excluding the eight partial skeletons) no, or only one or two, hand or foot bones were observed (Table 7.2).

Additional Bones

Eight jar burials contained extra human bones. For example, Tell Yunatsite 3 had a foetal humerus with the bones of an 18-month-old baby. Tell Yunatsite 5 had an extra cervical vertebra. Tell Yunatsite 27 had an adult hyoid with a 38-week-old foetus, Tell Yunatsite 45c had a hand bone of an older child with a full-term foetus, Tell Ezero 4 had two adult rib fragments with a 38-week-old foetus, and there were two cases of extra zygomatic (cheek) bones (Tell Yunatsite 13 and Tell Karanovo 2). Tell Yunatsite 13 also had an additional right humerus.

Articulation versus Disarticulation

Several skeletons showed indications of strict, or semi-strict articulation, along with some disarticulated bones (Tell Yunatsite 16 and 45c; Tell Kran 4, Tell Kran 8 and 9 and Tell Karanovo 1). In Tell Yunatsite Burial 45c it could be established from photographs taken at various stages in the excavation of the contents of the jar, which was conducted on site, that most of the bones were in general articular order. In many cases the only articulated parts were the ribs and spine. The micro-excavation of Tell Kran 9 indicated evidence of both articulation and disarticulation at all excavation levels. For example, while the right radius and ulna (lower arm bones) were in strict articulation and the left radius and left humerus were in anatomical position, the left ulna was found some distance away from the rest of the remains and not associated with any other bones (Fig. 7.4). However, the possibility that this had moved during the process of decomposition after deposition in the jar must be borne in mind (Duday 2009). Other skeletal areas found in articulation in Tell Kran 9 were the lower part of the skull and upper spine, several parts of the spinal column, including the

7. Processed Babies: Early Bronze Age Infant Burials from Bulgarian Thrace

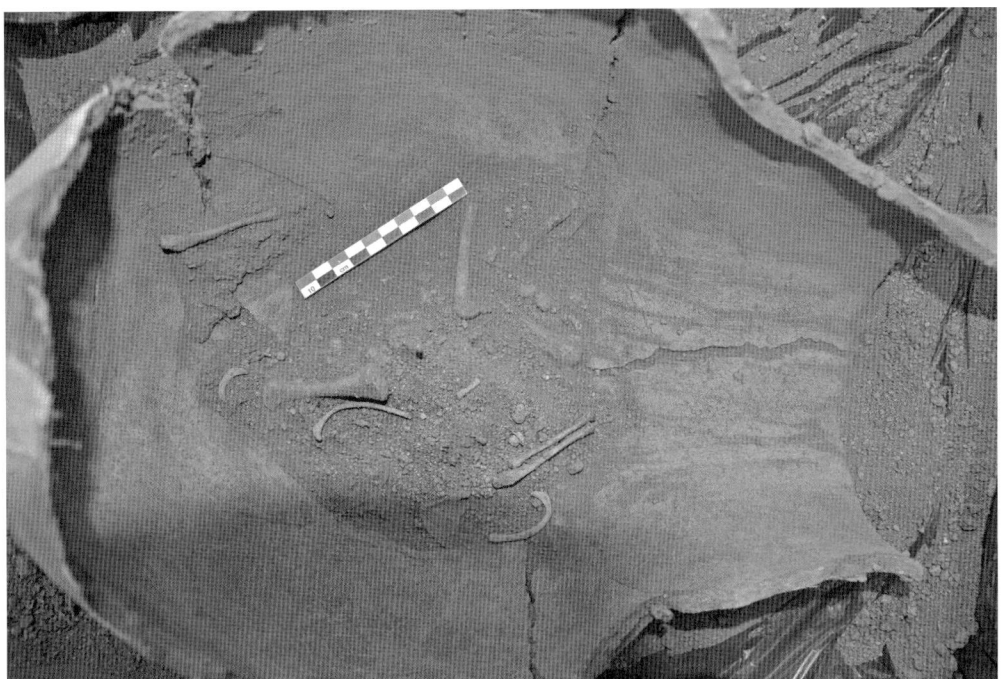

Figure 7.4. Articulation versus disarticulation (Tell Kran 9). Note the isolated and disarticulated left ulna at the top of the picture and the articulated right ulna and radius at the bottom of the picture (photograph by Kathleen McSweeney and Krum Bacvarov).

thoracic, lumbar and sacral areas. In addition, an articulated pelvis was found in close association with the lower spine and most of the ribs were in the correct anatomical order, as was the right leg. In other burials evidence of articulation came from groups of bones held together by the soil matrix.

State of Decomposition

In order to interpret the significance of the presence of both articulated and disarticulated remains within a single burial context it is important to understand the process of decomposition and how this affects the state of articulation (Duday 2009). Soon after death the soft tissues, including the skin and organs, are the first parts of the body to decompose. Decay progresses until, finally, only the skeleton, which consists of both bones and cartilage, is left; the cartilage then gradually decomposes until only the bones remain. Subadults have relatively more cartilage than adults, although most of the cartilage gradually starts to ossify with age. In adulthood, some cartilage remains around the joints, the rib cage and the spinal column. In addition to environmental factors, the rate of skeletonisation varies depending on the amount of cartilage present in each anatomical area. That around the labile joints (e.g. hands and feet) decays first and then, progressively, the remaining cartilage decomposes. The last area of cartilage to decompose is normally that surrounding the ribs and spine, because there is relatively more of it in these areas, precisely the anatomical areas most commonly found in articulation within the jars. There is some evidence to support the perseverance of the thorax in the decomposition of the body. Duday (2009, 10–11) observes a state of almost complete skeletonisation, 'with a small element of the thorax still retaining its anatomical connection and some strips of tissue' when describing a series of Japanese vignettes depicting decomposition of the body and, according to Rodriguez and Bass (1985, 851), a field study collecting data on decomposition rates of human cadavers showed that, 'the sternal area (mesosternum and costal cartilages) of the cadavers in our study withstood major decomposition'. The combination of both articulated and disarticulated bones within the same individual therefore suggests that the bodies were already partly decomposed prior to placement in the jars.

Evidence of Dismemberment

There was considerable evidence for dismemberment in the form of gouges, slices and peeling of bone at the joints (Table 7.3). The bones from 18 jar burials from all five sites and two of the pit burials from Tell Yunatsite showed some evidence of this, mostly at the major joints, i.e. the shoulders, elbows, hips and knees, but also at the smaller joints at the wrists and ankles. A few cuts on limb bone shafts and ribs were also noted. The following examples highlight the extent and nature of the dismemberment.

There was evidence of dismemberment in the form of gouges at the knee ends of both femurs in Tell Karanovo 7, with corresponding damage at the proximal (knee) ends of the tibiae. This individual also showed indications of disjointing at the elbows and wrists and a diagonal slice at the right humerus, probably caused by detachment at the right shoulder. In Tell Ezero 1 dismemberment at the knee end of the left femur was suggested by peeling visible on the bone surface, probably caused by pulling apart at the joint. This is a process that, according to White *et al.* (2012, 62), is particularly seen on immature bones (note they were describing such bony changes in relation to cannibalism, although this practice is not suggested here). A cut mark was also visible at the shoulder joint of the right scapula. In Tell Kran 9 there was a gouge at the lateral end of the left clavicle, associated with a missing left scapula. The presence of a number of other lesions indicates dismemberment in the pelvic area/abdomen and thorax. These included small cut marks on the anterior surface of two lumbar vertebrae and perimortem damage, possibly deliberate, at the left pelvic bone, the left femoral head and four left ribs.

Table 7.3. Number of cases where there was clear evidence of dismemberment. (This excludes the many cases where there was perimortem damage to the articular ends of bones but no clear evidence that this was deliberate).

Area of dismemberment	No. cases
Shoulders	6
Elbows	3
Knees	5
Pelvis	1
Thorax	1
Wrists	1
Ankles	1

The single individual from Nova Zagora had been disjointed at the shoulders; both scapulae had marked damage at the articulation with the humerus, corresponding damage was evident at the right clavicle (the left clavicle was missing) and the head of the left humerus appears to have been 'pulled off'. The individual from Tell Kran 5, a full-term foetus, had also been disjointed at the shoulder and at the elbows. Only nine of the 24 ribs were represented and some of these had evidence of peeling at the vertebral ends. Only a few bones were present of this individual. Tell Yunatsite jar burial 8 contained the very full remains of two individuals, aged 38 foetal weeks and 40 foetal weeks, respectively, based on long bone dimensions. The larger individual had a deep 'V'-shaped cut at the right acetabulum of the pelvis (Fig. 7.5). The colour of the broken surface clearly indicates that this damage had occurred in antiquity. No other corresponding damage was noted on the adjoining head of the femur. The only way that such damage could have occurred in the past without damaging the femoral head would have been if the right hip joint had been in the process of decomposition. Although a few bones from each individual had been broken in antiquity, there was no clear evidence of dismemberment, such as cut marks, on these.

Plant Remains

The mouth of the Tell Yunatsite 45c jar was sprinkled with einkorn (*Triticum monococcum* L.). Plant remains were also found in the soil matrix of Tell Kran 9 – einkorn (*Triticum monococcum* L.; 1 grain); emmer (*Triticum dicoccum* Shrank; 2 grains); hulless barley (*Hordeum vulgare* var. *nudum*; 1 grain); as well as many small unidentifiable fragments of wheat. It is not certain how the grains arrived in the jars but a 'deliberate scenario' raises some interesting questions about burial practices. It is not clear whether

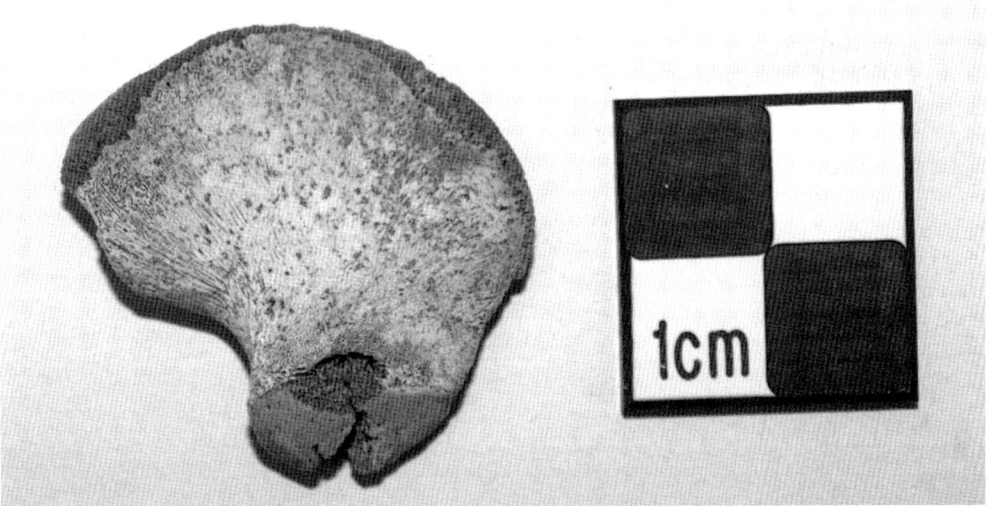

Figure 7.5. Tell Yunatsite 8. Cut in right ilium. Age 38–40 foetal weeks (photograph by Kathleen McSweeney and Krum Bacvarov).

plant remains were present in any of the other jar burials, but their inclusion in the two burials mentioned is suggestive they may have had a role in jar burial practices.

Animal Bone Inclusions

A few burials contained animal bones – seven from Tell Yunatsite, the single burial from Nova Zagora, Tell Kran 6 and 9, and Tell Karanovo 6. These varied in size, number and species and may well have been accidental inclusions. As this cannot be stated for certain, however, they are mentioned here.

Summary of Burial Practices

The osteological examination of the remains suggests the following burial practices. Burial in jars was mainly for foetuses (whether viable or otherwise), although there were also a few perinates and very young infants. Some jars contained double burials, although it is not clear whether these babies had been buried simultaneously, or at different times. Very few of the individuals were complete. Many deposits contained additional isolated bones from different individuals, in some cases children of the same age, in others older children, or adult remains. Some burials contained animal bones, although these varied in species and anatomical area, and the lack of any pattern suggested that these might have been accidental inclusions. The possible deliberate inclusion of grains in two of the jars suggests an interesting further dimension to burial rituals.

From the excavation of one incomplete jar (Tell Kran 5) and the intact jar from Tell Kran 9, where the deposits were undisturbed, and from some excavated bones in which anatomical order was retained because of post-mortem concretion, it was clear that at least some of the remains had still been in articulation at the time of deposition, while others were present out of articulation. This suggests that at least some of the babies were placed in the jars while in a state of partial decomposition. This fact, together with the good condition of many of the remains, indicates that, while a period had elapsed between death and deposition in the jar, this may have been fairly short, perhaps only days or weeks, depending on the season. There are many variables involved in skeletonisation of the body, including temperature, humidity, precipitation, insect activity, body size, and trauma (Mann et al. 1990). Mann and colleagues (1990, 107), for example, reported the almost full skeletonisation of an adult female placed on the ground in hot weather, after only two weeks. We can never be certain of the time lapse involved, however, especially since conditions will have varied depending on the season during which the babies died. The ample evidence for dismemberment suggested that this was an important part of mortuary practice.

Interpretation of the Burial Practices

The combination of bones in good condition along with missing elements indicates that the babies were initially laid out, or perhaps buried, elsewhere. The inclusion of the occasional additional isolated bones may be suggestive of accidental incorporation, perhaps from specific locations used for this purpose. An alternative interpretation that these were deliberate inclusions is a possibility that has been suggested elsewhere, however, where it has been suggested that the baby's body was somehow 'incomplete' and additional bones were required to 'complete' it (Bacvarov and McSweeney 2011). As indicated above, however, there appears to be no pattern to the deposited extraneous bones and perhaps this interpretation is implausible.

It seems quite clear that it was not important for complete skeletons to be placed in the jars. While in most cases the pots contained fairly complete skeletons there were intriguingly missing bones, whose loss could not be explained by poor preservation. The combination of articulated and disarticulated body parts suggests that the babies were partially decomposed before they were placed in the jars, although the evidence for dismemberment before deposition in jars indicates that some bodies parts had not completely decomposed. One consideration is that dismemberment took place to enable the bodies to fit in the jar. However, the fact that two of the four babies from pit burials had also been dismembered, as well as the use of relatively large jars, with mouth diameters large enough for a complete body to be inserted, suggests that this practice was carried out for a different purpose.

Prior to the excavation of Tell Kran 9, the absence of an osteoarchaeologist at the excavation of most of the burials posed several problems in the interpretation of burial practices. While preservation was generally very good, there was a number of missing skeletal elements. It was not clear if these had been originally deposited in the jars and had been lost during excavation. This was a possibility because not all of the pots were intact. Some of the burials had also been the subject of previous anthropological analysis (Buzhilova 2007) and so it is also conceivable that some bone samples had been removed for analysis. Therefore, it could not be stated with any degree of certainty that in all cases the missing bones had never been deposited in the jars. In the case of additional bones, these could have been original inclusions, or the result of post-excavation mixing. Some limited evidence for articulation had survived, but in the majority of cases no such evidence remained. In the case of double burials, the presence or lack of any clear stratification in the pots could have clarified whether these individuals had been buried together, or separately. However, the micro-excavation of Tell Kran 9 conducted by the authors supports the interpretation of partial articulation and hence partial decomposition, as well as the inextricable absence of missing elements. The absence of the left scapula in Tell Kran 9, when all of the other bones were present, mostly intact and in good condition, cannot be explained as having been caused by poor burial conditions, or by post-excavation mixing. It is clear that this bone had never been placed in the jar with the other remains. In keeping with other evidence for widespread dismemberment among the EBA baby burials, the Tell Kran 9 infant had also been disjointed. As Tell Kran 9 was virtually complete, however, what is still not clear is why there are so many partial skeletons in most of the jars. Nor, as there were no additional bones present in this burial, does it explain why some burials contained additional elements

A number of questions arise relating to the aim of these burial practices, including infanticide. There have been claims that infant jar burials are related to sacrificial death (see for example, Orrelle 2008). It could be argued that in EBA Bulgarian Thrace the placing of infant burials under the floors of, or in close association with, houses (Mishina 2008; Orrelle 2008) suggests a ritual connection, and therefore that deliberate killing is a possibility. The vast majority of infants in this study were full-term foetuses. Mays (2010, 82) states that 'perinatal burials may be the result of stillbirths, natural deaths in the immediate postnatal period, or victims of infanticide'. In a comparison of Romano-British with medieval and modern perinatal deaths, Mays concluded that a peak of 38 to 40 gestational weeks in the Romano-British data compared with more even spreads of 20-plus to 47-plus gestational weeks in the medieval and modern periods, suggests that infanticide had been practised in the Romano-British period (Mays 1993; 2010). The mortality profile among the infants in this study (Fig. 7.3) is very similar to that of the Romano-British produced by Mays. However, Halcrow and colleagues (2008) suggest that high levels of mortality in full-term Late Prehistoric Thailand foetuses may simply be due to natural causes. Whether the infants found in the EBA jars were deliberately killed is an argument that we leave open to question at this stage. The sample size is small and we may simply be seeing the deliberate selection of certain individuals who suffered natural deaths. There have been suggestions that infants buried in jars tended to be male (Faerman and Smith 2008) and it is our intention to undertake DNA analysis of the Bulgarian EBA infants in the future.

The study of infant jar burials in Bulgarian Thrace has revealed some fascinating burial practices, suggesting that these babies were subjected to funerary processes never previously identified. The aim of these processes remains unclear, as well as whether the babies suffered natural or un-natural deaths. It is hoped that broader, more detailed research into jar burials, such as DNA analysis to identify the sex of the babies and thus possible selection, and the analysis of more examples will provide greater insight into the processing of babies in Bronze Age Thrace.

Acknowledgements

We would like to thank Dr Tsvetana Popova for the information about plant remains in Tell Kran 9 and Tell Yunatsite 45c.

Notes

1. School of History, Classics and Archaeology, University of Edinburgh, William Robertson Wing, Teviot Place, Edinburgh, EH8 9AG, UK. Email: Kath.McSweeney@ed.ac.uk.
2. National Institute of Archaeology and Museum, Bulgarian Academy of Sciences, 2 Saborna Str., BG-1000 Sofia, Bulgaria. Email: krum.bacvarov@googlemail.com.

References

Andreeva, D. 2007. Prouchvane na plasta ot rannata bronzova epoha v tel Kran, Kazanlashko (predvaritelno saobshtenie). *Problemi i izsledvaniya na trakiyskata kultura* 2, 3–16.

Andreeva, D. 2011. Arheologicheski prouchvaniya na tel Kran, Kazanlashko prez 2009 g. *Problemi i izsledvaniya na trakiyskata kultura* 5, 3–17.

Bacvarov, K. 2008. A long way to the West: Earliest jar burials in southeast Europe and the Near East, in Bacvarov, K. (ed.), *Babies Reborn: Infant/child burials in Pre- and Protohistory* (BAR International Series 1832), 61–70. Oxford: Archaeopress.

Bacvarov, K. and McSweeney, K. 2011. 'Potted babies': A bioarchaeological approach to the study of jar burial symbolism, in Nikolov, V., Bacvarov, K. and Gurova, M. (eds.), *Festschrift for Marion Lichardus-Itten* (Studia Praehistorica 14), 399–408. Sofia: National Institute of Archaeology and Museum.

Buzhilova, A. P. 2007. Antropologiya rannego bronzovogo veka tellya Yunatsite, in *Tell Yunatsite: Epoha bronzy, vol. II/1*, 207–216. Moscow: Vostochnaya literatura.

Duday, H. 2009. *The Archaeology of the Dead: Lectures in Archaeothanatology*. Oxford: Oxbow Books.

Faerman, M. and Smith, P. 2008. Has society changed its attitude to infants and children? Evidence from archaeological sites in the Southern Levant, in Gusi, F., Muriel, S. and Olària, C. (eds.), *Nasciturus: Infans, Puerulus. Vobis Mater Terra. La Muerte en la Infancia*, 211–229. La Roja: Diputació de Castelló.

Fazekas, I. G. and Kosa, F. 1978. *Forensic Foetal Osteology*. Budapest: Akademiai Kiado.

Georgiev G. I., Merpert, N. Y., Katincharov, R. and Dimitrov, D. 1979. *Ezero: Rannobronzovoto selishte*. Sofia: BAN.

Halcrow, S. E., Tayles, N. and Livingstone, V. 2008. Infant death in Late Prehistoric Southeast Asia. *Asian Perspectives* 47, 371–404.

Hiller, S. and Nikolov, V. 2002. *Tell Karanovo 2000-1. Vorläufiger Bericht.* (Schriftenreihe des Instituts für Klassische Archäologie der Universität Salzburg, Reihe I, Heft 17) Salzburg: Paris Lodron Universität Salzburg.

Hiller, S., Nikolov, V. and Lang, F. 2005. *Tell Karanovo 2002-4. Vorläufiger Bericht.* (Schriftenreihe des Instituts für Klassische Archäologie der Universität Salzburg, Reihe I, Heft 18) Salzburg: Paris Lodron Universität Salzburg.

Kancheva-Ruseva, T. 2000. Grobove ot bronzovata epoha v praistoricheskoto selishte v Nova Zagora. *Arheologia* 3–4, 31–34.

Lewis, M. E. 2004. Endocranial lesions in non-adult skeletons: Understanding their aetiology. *International Journal of Osteoarchaeology* 14, 82–97.

Lewis, M. E. 2007. *The Bioarchaeology of Children: Perspectives from Biological and Forensic Anthropology*. Cambridge: Cambridge University Press.

McSweeney, K., Bacvarov, K., Nikolov, V., Andreeva, D. and Bonsall, C. 2016. Infant burials in Early Bronze Age Bulgaria: A bioarchaeological appraisal of funerary behaviour, in Nikolov, V. und Schier, W. (Hrsg.), *Der Schwarzmeerraum vom Neolithikum bis in die Früheisenzeit (6000-600 v. Chr.): Kulturelle Interferenzenin der zirkumpontischen Zone und Kontakte mit ihren Nachbargebieten.* (Prähistorische_Archäologie_in Südosteuropa 30), 383–394. Rahden/Westf.: Marie Leidorf.

Mann, R. W., Bass, W. M. and Meadows, L. 1990. Time since death and decomposition of the human body: Variables and observations in case and experimental studies. *Journal of Forensic Sciences* 35, 103–111.

Mays, S. 1993. Infanticide in Roman Britain. *Antiquity* 67, 883–888.

Mays, S. 2010. *The Archaeology of Human Bones*. London: Routledge.

Mishina, T. 2008. A social aspect of intramural infant burials analysis: The case of EBA Tell Yunatsite, Bulgaria, in Bacvarov, K. (ed.), *Babies Reborn: Infant/child burials in Pre- and Protohistory* (BAR International Series 1832), 137–146. Oxford: Archaeopress.

Museibli, N. 2008. Zahoronenie mladentsev v glinyanyh sosudah u plemen Leilatepinskoy kultury, in *Otrazhenie tsivilizatsionnyh protsesov v arheologicheskih kulturah Severnogo Kavkaza i sopridelnyh teritoriy*. Severo-osetinskiy institute gumanitarnyh I sotsialnyh issledovaniy im, 267–273. V.I. Abaeva Vladikavkazskogo nauchnogo tsentra RAN I pravitelstva Respubliki Severnaya Osetiya-Alania: Vladikavkaz.

Nikolov, V. 2007. Problems of the early stages of neolithization in the southeast Balkans, in Biagi, P. and Spataro, M. (eds.), *A Short walk through the Balkans: The first farmers of the Carpathian Basin and adjacent regions. Proceedings of the Conference held at the Institute of Archaeology UCL on June 20-2, 2005*, 183–188. Societa per la Preistoria e Protostoria della Regione Friuli-Venezia Giulia: Trieste.

Orrelle, E. 2008. Infant jar burials – a ritual associated with early agriculture? in Bacvarov, K. (ed.), *Babies Reborn: Infant/child burials in Pre- and Protohistory* (BAR International Series 1832) 71–78. Oxford: Archaeopress.

Rodriguez, W. C. and Bass, W. M. 1985. Decomposition of buried bodies and methods that may aid in their location. *Journal of Forensic Sciences* 30, 836–852.

Scheuer, L. and Black, S. 2004. *The Juvenile Skeleton*. New York: Elsevier/Academic Press.

Van Beek, G. 1983. *Dental Morphology: An Illustrated Guide*. Bristol: PSG Wright.

White, T. D., Black, M. T. and Folkens, P. A. 2012. *Human Osteology*. Amsterdam: Elsevier/Academic Press.

Chapter 8

'Missing infants': Giving Life to Aspects of Childhood in Mycenaean Greece via Intramural Burials

Katerina Kostanti[1]

Abstract: This paper aims to examine the absence of infants aged under 24 months in Late Helladic communal burial grounds, and provide evidence to demonstrate that they are treated in a different way in comparison with older children and adults. An insight into the age data of the deceased reveals that missing infants are present in domestic contexts, although not in adequate numbers. The latter practice seems to reflect a conscious mixing of the residential and funerary space juxtaposing or, moreover, conciliating life and death. All of the above enable the investigation of various aspects of the social landscape, such as the formation of life stages and different identities, mechanisms of memory and the expression of status and power.

Keywords: intramural burial, childhood, infant, Mycenaean

Introduction

Burial practices have always been considered crucial to the understanding of the structure of the past especially for prehistoric societies in which the symbolic and social worlds of the living play an important, yet elusive, role in the selection of a burial's placement. Since age is among the major factors that structure human societies, the study of burial practices focusing on the age of the deceased opens a new perspective and may bring to light important information about their social identities and relevant social entity. As Triantaphyllou (2016) has noted, at specific sites, age appears to be a fundamental criterion in the selection of an area for the disposal of the deceased. In addition, the fluidity and changeable nature of the boundaries between the social aspects of age, gender, ethnicity and status during life, make it important to consider all of these factors in order to gain clear and solid results.

Age categories are neither clearly defined nor unanimously accepted as such. Childhood comprises several stages of growth including neonates, infants and toddlers, as well as younger and older children (Scott 1999, 4; Triantaphyllou 2001, 36), and we must bear in mind that the *osteological child* and *cultural child* do not always coincide (Ingvarsson-Sundström 2008, 19–23). This is also the case for chronological, biological and social age (Halcrow and Tayles 2008, 192), three factors which define age categories, symbolically expressed through material culture (Triantaphyllou 2016). For our convenience, we will use simplistically the terms *infant* for individuals younger than 24 months and *child* for those aged between two and twelve years of age (Hallager and McGeorge 1992, 39; Schepartz et al. 2009, 166, tab. 10.3; for different opinions see Polychronakou-Sgouritsa 1987, 9; Ingvarsson-Sundström 2008, 25, tab. 2; for an overview of the problem of subadult age categories see Halcrow and Tayles 2008).

The term *intramural*, as a norm, is applied in cases of deliberate co-occurrence of residential architecture and human burials in relation to both space and time, but its content is not always defined with accuracy. In addition, the poor examination of stratigraphic evidence and excavation data may lead to the erroneous attribution of the term to a burial that was incidentally connected to architectural remains. Therefore, specific relevant terminology needs to be developed, which considers aspects, such as the precise relationship of the grave to the architecture, its low or high visibility within the house and its exact function in the process of commemoration (Laneri 2011, 44). An analysis of the corpus of excavation data has identified a range of specific patterns in relation to the spatial arrangement of burials for Mycenaean Greece. These include burials under floors or doors, in corners or beneath hearths, as well as interments in between houses, within ruined domestic units and in deserted settlement sectors. It should be noted, however, that the deposition of the deceased in deserted settlements may confuse the proper use of the term *intramural burial* (Maran 1995, 71). The latter practice requires further study, however, which would enable it to be better understood and properly placed either within or outside the spectrum of intramural burials.

The under-representation of infants and children in the mortuary archaeological record has been repeatedly pointed out (e.g. Cavanagh and Mee 1998, 111; McHugh 1999, 19; Parker Pearson 1999, 103; Scott 1999, 4, 90, 125–6; Triantaphyllou 2001, 37) and explained as a result of age biases in relation to burial customs which necessitate a differential treatment of deceased subadults. In the present paper, I will explore this phenomenon particularly in relation to infants so that this age category can be embedded within the broader Mycenaean perception of childhood. The nature of intramural and extramural burials will be compared in conjunction with the associated skeletal evidence and an attempt will be made to identify aspects of the social personae acquired by infants in Mycenaean society. Furthermore, I will advance some propositions to explain why they were not buried within extramural contexts, but rather were included within the domestic sphere. The presence of human skeletal remains within residential areas has the potential to provide insights concerning the role of the dead in social and communal life.

Infant Burials in Mycenaean Greece

The currently available data is suggestive that the occurrence of intramural burial practices in the Aegean Early Bronze Age (see Table 8.1 for a summary of Aegean Bronze Age chronology) was of an occasional character, whereas it is marked by widespread distribution in the Middle Bronze Age (although the presence of the practice has been recently challenged by Sarri 2016) and it declined over the course of the Late Bronze Age (Cavanagh and Mee 1998, 129; Lanaras 2003; McGeorge 2012, 296). In the Middle Helladic period, adults and juveniles were both buried intramurally (Cavanagh and Mee 1998, 129), with a preponderance of infant and child burials in actual domestic units (Cavanagh and Mee 1998, 24; Ingvarsson-Sundström 2008, 102–3) without any strong evidence for the differential treatment of infants and older children.

Table 8.1. Aegean Bronze Age Traditional Chronology. Late Helladic subperiods (after Cline 2008, fig. 145).

Relative chronology	Absolute chronology (BC)
Early Bronze Age	3200–2000
Middle Bronze Age	2000–1625
Middle Helladic	2000–1625
Late Bronze Age	1625–1125
Late Helladic I	1625–1525
Late Helladic II	1525–1425
Late Helladic IIIA1	1425–1375
Late Helladic IIIA2	1375–1325
Late Helladic IIIB1	1325–1200
Late Helladic IIIC	1200–1125

In the Late Helladic period, the burials that occurred within settlements were mainly those of neonates and infants, and sometimes older children, and the practice exclusively involved inhumation (Kostanti 2009, 110–2; Pomadère 2013) (see Table 8.2 for a summary of infant and child burials mentioned in the text). A few examples of intramural adult burials are known from Lefkandi in Euboea (Musgrave and Popham 1991), Ayios Stephanos in Laconia (Taylour 1972; Taylour and Janko 2008) and Tiryns in the Argolid (Kilian *et al.* 1981, 173, fig. 25), however, and their frequency of occurrence might increase further since research is still ongoing (Fig. 8.1). A comparative analysis of the studied skeletal material from both intramural burials and burials in cemeteries in Mycenaean Greece reveals that infants under 24 months of age are usually absent from organised cemeteries, yet present in domestic contexts, albeit in small numbers (Kostanti 2009, 118–9). In addition, only a small percentage of subadult skeletal remains – those of individuals less than 18 years of age (i.e. 10.6% at Pylos, Schepartz *et al.* 2011, 365; 23.1% at East Lokris, Iezzi 2009, 178, tab. 11.1) – have been uncovered in Late Bronze Age burials. This is in contrast to the expected high child mortality rate for prehistoric times, which might have risen to as much as 15–30% of the total number of the deceased, in analogy to the norm for pre-industrial societies (Scott 1999, 90).

Intramural Infant Burials

During the transitional transformative period from Middle Helladic to Late Helladic I, that led to the birth of the Mycenaean civilisation, neonates (Pomadère 2010, 537)

Table 8.2. Summary of infant and child burials mentioned in the text.

Intramural/Extramural Burial	Site	Chronology	Individuals and Age
Intramural	Midea	LHI–II	1 child of unspecified age
Intramural	Lerna	LHII	1 infant ≤ 12 months 1 child of unspecified age
Intramural	Lefkandi	LHII	2 foetuses or premature neonates 2 infants of 6 months 1 child of 2–4 years
Intramural	Lefkandi	LHIIIC	1 foetus or neonate 2 children of 2–3 years 1 child ≥ 3 years 1 child of 5 years 1 child of 8–9 years 2 infants or children of unspecified age
Intramural	Ayios Konstantinos	LHIIIA2–IIIB	1 infant 3–6 months 1 infant 6–12 months 1 foetus of 8–8.5 months gestation
Intramural	Modi Troizinias	LHIIIC	1 foetus or neonate
Intramural	Ayios Stephanos	LH period	48 foetuses, neonates, infants and children
Intramural	Asine	LH period	8–9 children of unspecified age
Extramural	Argos	LHI–II	Children ≥ 6 years
Extramural	Argos	LHIIB–IIIA1	1 child of 2.5–3.5 years
Extramural	Ancient Agora of Athens	LHIIB–IIIB/C	20 children ≥ 18 months
Extramural	Clauss	LHIIIC	9 children ≥24 months 1 infant of 12 months
Extramural	Deiras	LHIIIA–B	1 child of 3 years
Extramural	Pylos	LH period	Children ≥4 years
Extramural	Asine	LH period	3 neonates
Extramural	Tragana	LHIIIB–C	neonates
Extramural	Modi Lokridos	LHIIIB–C	neonates
Extramural	Ayia Sotira	LHIIIA–B	neonates
Extramural	Kazarma	LHII	neonates
Extramural	Grave III, Grave Circle A Mycenae	LHI	1 infant ≤3 months
Extramural	Grave Circle A Mycenae	?	1 infant ≤ 2 years
Extramural	Grave Circle B Mycenae	MH–LHI	1 child of 2 years 2 children of 5 years

8. 'Missing infants': Giving Life to Aspects of Childhood in Mycenaean Greece

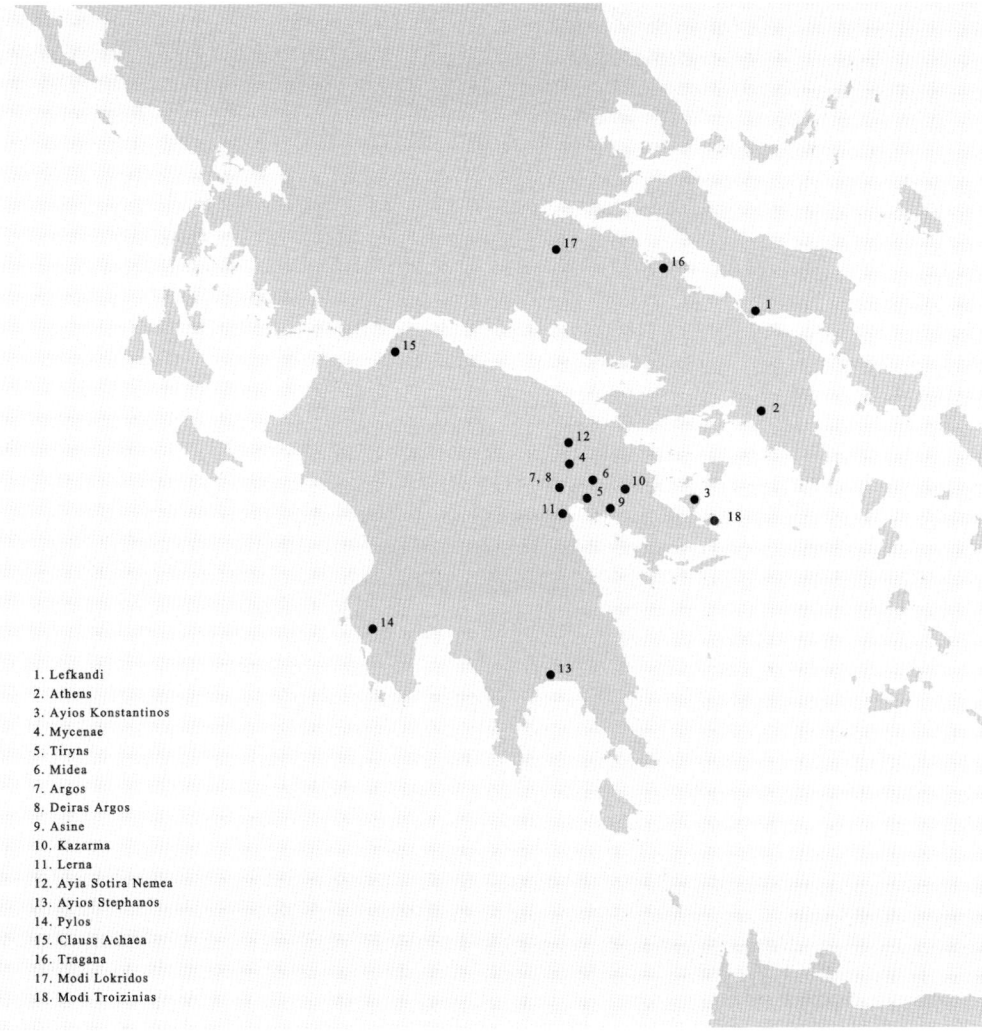

Figure 8.1. Schematic map of mainland Greece showing the sites mentioned in the text (prepared by Katerina Kostanti).

1. Lefkandi
2. Athens
3. Ayios Konstantinos
4. Mycenae
5. Tiryns
6. Midea
7. Argos
8. Deiras Argos
9. Asine
10. Kazarma
11. Lerna
12. Ayia Sotira Nemea
13. Ayios Stephanos
14. Pylos
15. Clauss Achaea
16. Tragana
17. Modi Lokridos
18. Modi Troizinias

and infants of less than one year of age (the upper age limit is not specifiable with certainty) (Lebegyev 2009, 22), tend to be differentiated by their exclusion from organised extramural cemeteries. Instead they were buried in intramural contexts, a practice that continued into the next period for infants under 24 months. During the Late Helladic III (LHIII) period this trend appears to have diminished, however, only to become more common again in the latter stages of the Late Bronze Age when a rise in the number of intramural burials is observed (Kostanti 2009, 118–9). Intramural infant/child burials are more common in the Peloponnese (at least 89 individuals), while in eastern Central Greece and Euboea (at least 25 individuals) their number increases during LHIIIC (Kostanti 2009, 111, tab. 1). In general, intramural burials of infants and children tend to remain unfurnished (53% in the Peloponnese and 75% in eastern central Greece and Euboea) unless furnished with perishable goods (Fig. 8.2) (Kostanti 2009, 112, tab. 2). In rare cases, wealthy child burials have been discovered intra muros as at LHI–II Midea in the Argolid where, it has been suggested, that the infant/child – provided with a vase, a carnelian bead, a container of perishable material with ivory handles and raw material possibly for the production of dye – was the offspring of local elite (Demakopoulou *et al.* 2002, 36–7). The graves were placed under floors, near the hearth, in corners of walls, under staircases, in backyards, and between houses. Sometimes the floor of the grave was paved with pebbles or with a layer of sand. Infants were placed in simple pits, in cist graves, in clay jars or covered with large ceramic sherds, in clay larnakes, or in wooden boxes (Kostanti 2009, 110–2).

At Lerna in the Argolid, two possible, though not convincingly, intramural LHII burials were those of an infant less than one-year-old and a child of unspecified

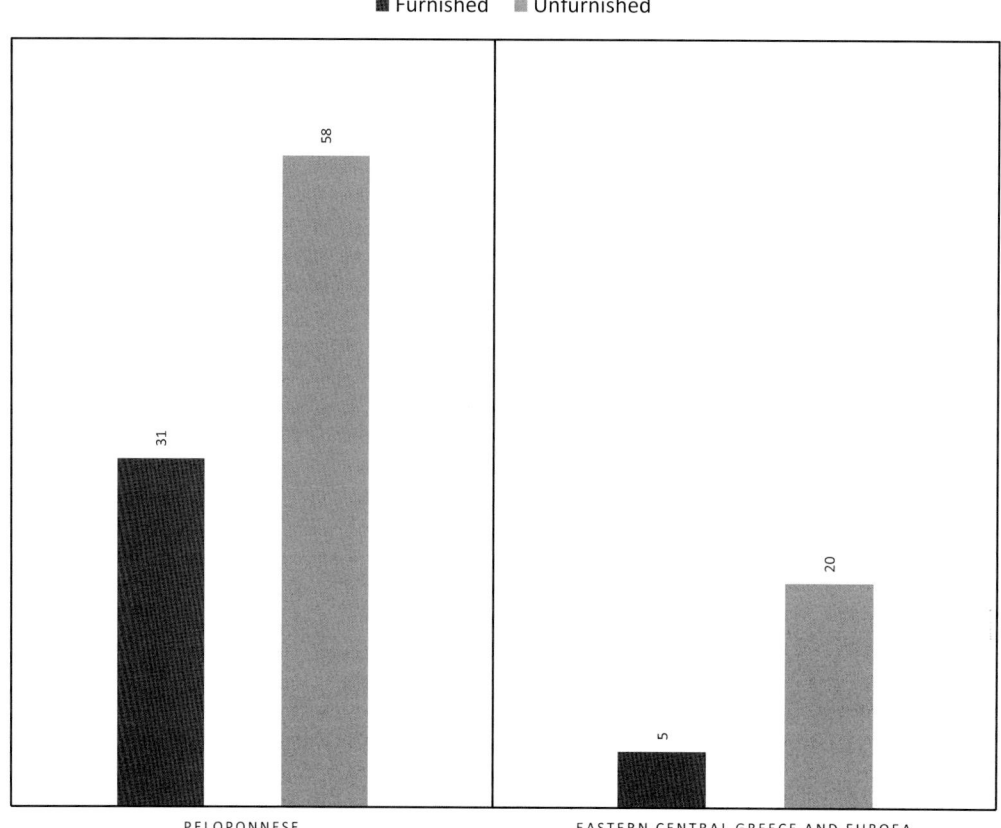

Figure 8.2. Bar chart comparison of furnished and unfurnished intramural child/infant burials in Peloponnese and eastern Central Greece and Euboea.

age (Lebegyev 2009, 20). Infants under 24 months are present in the intramural burials at Ayios Stephanos in Laconia and at Lefkandi in Euboea, although in these two cases adults were also buried intra muros. At Ayios Stephanos the number of intramural burials of children and adults decreases after the LHI period, from 47% (74 burials) of the total 157 burials in the site in the previous Middle Helladic period and the transitional MH/LHI, to 39% (62 burials) during the Late Helladic period; the percentage of neonate and infant burials increases during the LHIIIA1 and decreases once more in later LH phases until the final abandonment of the site in the early stages of LHIIIC (Taylour and Janko 2008, 141–4). At this settlement, the Late Helladic intramural burials of foetuses, neonates, infants and children amount to at least 48 individuals (Kostanti 2009, 39–44; Taylour and Janko 2008, 144). For the LHII period at Lefkandi, the intramural burials of two infants of six months of age, two foetuses or premature neonates, and a child between two and four years were present; for the LHIIIC period of the 15 buried individuals only one was a foetus or neonate, while two were aged between two and three years, one was over three years, one was five years old, one was eight or nine years old and the remains of two infants or children of unknown age were also present (Musgrave and Popham 1991). The LHIIIA2–IIIB burials recovered from the Mycenaean sanctuary at Ayios Konstantinos Methana, dedicated to a deity connected with the watery element, fecundity and the underworld (Konsolaki-Yannopoulou 2002; Tzonou-Herbst 2002, 172), can be considered to be a unique set (Konsolaki-Yannopoulou 2003a, 269–70). The excavation of Room C unearthed a large cist, which contained the remains of two infants aged three to six months and six to twelve months respectively as well as a foetus of eight to eight-and-a-half months' gestation, and was cut into the pavement of a room adjacent to Room A, the main room of the sanctuary (Fig. 8.3). One of the infants was articulated and comprised the last burial to have been made. The burials were richly furnished with vases, figurines, two bronze rings (one of these being too large for an infant, but nevertheless worn symbolically on a finger), necklaces, a seal stone, an amulet and seashells. According to the excavator the grave was used, throughout the active period of the sanctuary, for the burials of foetuses and infants. It was proposed that these individuals may have been from families that belonged to the local elite in an attempt to prevent future infant death and to facilitate the rebirth of the dead infants (Konsolaki-Yannopoulou 2003a, 270–1). On the islet of Modi in Troizinia, under the floor of a LHIIIC storeroom, excavation unearthed four narrow and one T-shaped bone plaques belonging to the casing of a wooden box used as a burial container (Konsolaki-Yannopoulou 2003b, 420–1, fig. 15). Preliminary anthropological study (conducted by Dr Yasar Isçan, Institute of Forensic Sciences, Istanbul University) of the skeletal material, discovered among the remains of the box, indicated that it belonged to a foetus or neonate (Konsolaki-Yannopoulou pers. comm.). Unfortunately, none of the contemporary infant or child skeletal remains derived from extramural burials at the same sites have been subject to anthropological analysis which makes it impossible to conduct a thorough comparative analysis of the data.

Figure 8.3. Plan of the Mycenaean sanctuary at Ayios Konstantinos, Methana. The cist grave in Room C, which contained the remains of two infants aged 3-6 months and 6-12 months respectively as well as a foetus of 8-8.5 months' gestation, is indicated by the arrow (courtesy of Eleni Konsolaki-Yannopoulou).

Extramural Infant Burials

Extramural burials usually do not include the remains of infants under 24 months. At Argos, in the LHI-II period, the anthropologically studied skeletons of extramural burials all derived from individuals older than six years (Lebegyev 2009, 23). The child in one of the richest extramural tombs at LHIIB-IIIA1, Argos, died at the age of two-and-a-half to three-and-a-half years, which might be considered as a liminal age – between infanthood and childhood (Kaza-Papageorgiou 1985, ft. 11). In the LHIIB-LHIIIB/C extramural cemetery of the Ancient Agora at Athens skeletal remains from 20 children, all of whom were older than 18 months, have been unearthed (Immerwahr 1971, 110, 158–246; Kostanti 2009, 82–9). In this assemblage, the anthropologist Lawrence Angel identified the skeleton of a wealthy LHII burial to be that of an 18-month-old infant furnished with exceptionally rich gifts – an ivory comb and hairpin, a glass-paste and amethyst necklace with a gold pendant, a conch shell and 10 vases, one of which was a unique shape which imitated metal prototypes (Immerwahr 1971, 205, ft.1). Among the 10 children buried in the LHIIIC chamber tombs of Clauss in Achaea, the preliminary anthropological study has shown that nine were older than 24 months of age and one was around twelve months of age (Paschalidis 2016). At the cemetery of Deiras the youngest skeleton to be examined was that of a three-year-old child who had been buried in a LHIIIA-B context (Lebegyev 2009, 21). Finally, infants or children under four years of age are absent from the LHI–IIIC Pylian tombs (Schepartz et al. 2011, 365) and it is considered that young children in Pylos were rarely buried in a similar manner to adults (Schepartz et al. 2009, 165–6).

Children, older than two to three years at death, began to be buried with adults in chamber tombs from the LHIIIA onwards (Lebegyev 2009, 24). The recently identified traces of wooden chests associated with some skeletal material inside chamber tombs in the large cemetery of Asine in the Argolid provide rare evidence for the burial of neonates in community cemeteries (Mårtensson 2002; Nordquist and Ingvarsson-Sundström 2005, 157). In the immediate vicinity of buildings of the same settlement, however, the remains of approximately eight or nine children have also been unearthed (Frödin and Persson 1938, 128–9 and 146). Newborn babies were recently recognised in a disarticulated state in LHIIIB–C chamber tombs at Tragana and Modi in eastern Lokris; in the LHIIIA–B cemetery of chamber tombs at Ayia Sotira in Nemea and in the LHII tholos tomb at Kazarma (Triantaphyllou 2016).

Another exception to the rule can be found in the burial of an infant in the extramural Grave III of Grave Circle A at Mycenae whose body appears to have been entirely covered with gold foil (Fig. 8.4) (Karo 1930–3, 62, taf. LIII; Papazoglou-Manioudaki et al. 2010, 161 with different opinions about the number of infants). On the basis of the preserved height of the gold suit, which would have covered a body measuring approximately 50 cm, it has been suggested that the infant was less than three months old (Konstantinidi-Syvridi forthcoming) and may have been holding a gold model of a scale, presumably for symbolic purposes, while a magnificent gold diadem may have adorned its forehead. Further study is needed to help determine why an individual of such a young age had been chosen to be buried in this magnificent grave along with two adult males and a female (Papazoglou et al. 2010, 172–9), the latter of whom has recently been identified as a priestess (Konstantinidi-Syvridi forthcoming). The above question is even more interesting when one considers the rarity of non-adult burials in both Grave Circles (Voutsaki 2012, 178) – in Grave Circle B 21 adults were buried along with three children, one of whom was aged two years while the others were both five years of age (Angel 1973, 379, 383 and 392; Mylonas 1973, 145–6 and 186). In Grave Circle A 15 adults and one subadult were buried; two fragments possibly belonging to the same infant, less than two years of age, have also been identified among the skeletal material, although the possibility that they derive from later burials above Grave Circle A cannot be excluded (Papazoglou-Manioudaki et al. 2010, 213–4, fig. 31). The unique burial of the infant wrapped in gold, when correctly understood, could shed new light on the lives of infants of nobility, with hereditary rights, in this case related to the priestly office (Konstantidi-Syvridi forthcoming). Even in this instance, however, the foil covering the infant is undecorated and the piece covering the face does not bear facial features. This situation differs from the five gold death masks and luxuriously decorated funeral attire discovered in association with adults in Grave Circle A, thereby confirming that age biases applied to burial customs (Voutsaki 2012, 178).

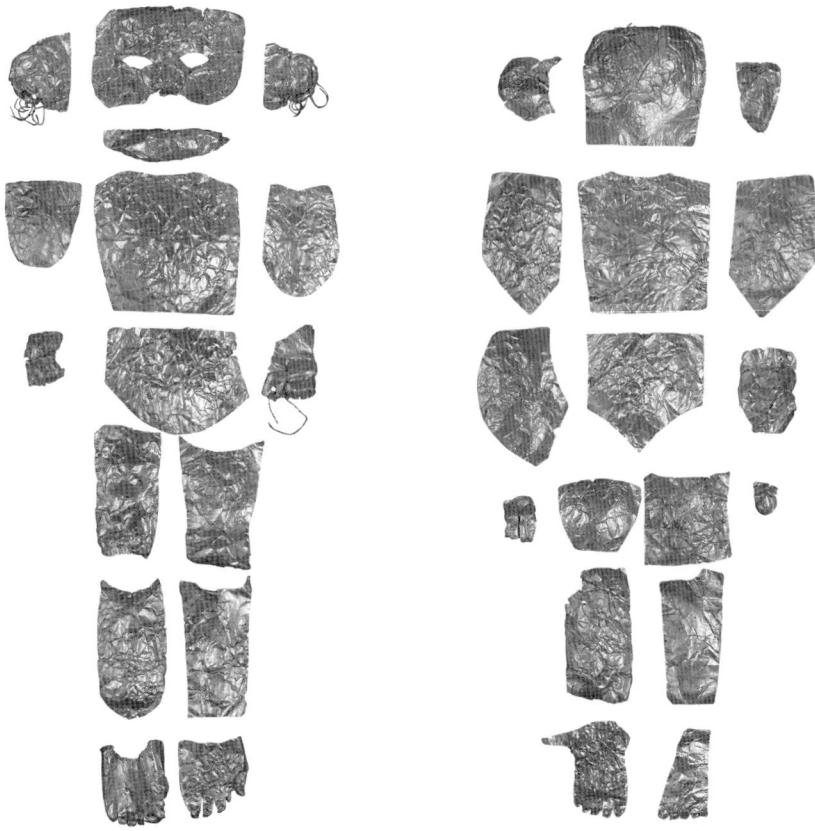

Figure 8.4. The gold suit of the infant in Grave III, Grave Circle A at Mycenae (National Archaeological Museum, Athens © Hellenic Ministry of Culture and Sports/Archaeological Receipts Fund).

Discussion

On the basis of the aforementioned evidence it appears to be the case that infants less than 24 months of age tend to be absent from Late Helladic organised communal cemeteries and – at least some of them – were buried instead in a domestic context (unfortunately precise figures are currently not available). This age mark is not random – the end of the second year of an infant's life is characterised by an increasing tendency towards independence. The infant/child is capable of walking alone and talking, combining words to form simple sentences (Bloom 1993, 4, 149, 264, pl. 7.7); it is not a coincidence that the Latin term *infantia* means inability to speak (Scott 1999, 1). The set of deciduous teeth is almost complete, with the first deciduous molars emerging at 18 months (± six months) and the second at 24 months (± eight months) (Ubelaker 1989, 51, fig. 24). The age at which deciduous teeth have emerged seems to coincide with the age mark between infants and children proposed in the present paper, that is the age of 24 months. As such, the development of teeth must be considered as a milestone in infant growth and age categorisation (for ethnographic examples see Ucko 1969, 270–1), the latter operating as one of the many factors affecting burial customs. It is very interesting that the limit of two years of age maintains its importance in Greece during the following centuries. On inscribed ancient Greek funerary stelae, the adjective *ahoros* (premature) is never, or almost never, attributed to children under the age of two years, as if the death of those individuals was not seen as unnatural and premature (Golden 1990, 83). The importance of this age limit is also discernible in a huge cemetery of jar burials exclusively of infants aged less than 24 months on the island of Astypalaea, dating from the Late Geometric until the Roman period (Hillson 2009, 139).

Children older than 24 months were probably formally entering the Mycenaean community through rites of integration and passage, and they were therefore permitted burial within communal cemeteries. In some settlements, however, a limited number of children older than 24 months continued to be buried intramurally as at Ayios Stephanos and Lefkandi. Since the practice of intramural burials was rather common in the Middle Helladic, a period with simpler social structure than in Mycenaean times (Voutsaki 2012, 164–6), the continuation of the custom for these older children could be explained as an archaic mortuary ritual practised by traditional and conservative groups within the population or at peripheral sites or by less stratified societies. Based on ethnography, it can also be suggested that other factors related to the individuals – such as their having suffered from particular diseases, their physical appearance or a specific cause of death, such as smallpox, lightning or drowning – may have been involved in the continuation of the practice (Ucko 1969, 271; for a short discussion of the possible effects of unusual appearance and pathology see Ingvarsson-Sundström 2008, 105, ft. 457), although there are no such indications in the relevant skeletal material. Similar criteria, which connect some deformities or causes of death with the supernatural realm, may also be applied to the above-mentioned cases of the extramural burial of infants, which was also contrary to common funerary practices.

Taboos and exclusion rules along with rites of separation and reintegration may have existed for stillborns, neonates and infants at different age stages, and possibly for their mothers, at birth and at death, two circumstances often considered to be highly polluting events in preindustrial societies (Van Gennep 1909, 71–92, 218–9; Ingvarsson-Sundström 2008, 106–7, 120–1; McGeorge 2011, 11). The age at which children acquire social status varies among societies and between periods. As other scholars have observed, infants are excluded from community cemeteries throughout prehistory up until modern times (Scott 1999, 126), when infants are still occasionally buried within the house (Ariès 1962, 39).

The change in burial practices at liminal age stages suggests changes in relation to the perceived social status and social persona of the deceased. Through the study of the former it is possible to obtain a more complete image of the multiple social personae of children and, in this case, infants in Mycenaean Greece.

Among infants less than 24 months of age different subcategories emerge which are based on social biases; behaviours which battle between conservatism and innovation, along with religious and cosmological beliefs. Since the biological sex of infants cannot be accurately defined (Ingvarsson-Sundström 2008, 32) this parameter, along with a lack of gender specific burial gifts for children and infants could not be used to further articulate the social identity of infants through burial practices (although *askoi* have been considered indicative of female adult and subadult burials; Mylonas 1973, I, 288).

The striking difference between the generally poorly furnished or completely unfurnished burials and those that were wealthy enables further aspects of childhood to be reconstructed. Rich infant burials may be interpreted through various approaches, usually considered to reflect the prominent social status of their family and the effort to attribute the latter to the deceased infant (Pader 1982, 62; McHugh 1999, 24–5; Cavanagh and Mee 1998, 129, where it is argued that girls receive often rich burial gifts that would have been included in their dowries). The wealthy burials of the two individuals at Athens and Argos probably fall into the above category and the liminal age of the deceased may indicate that they had already entered the next age stage through the appropriate rites of passage and were therefore permitted burial in communal ground. On the other hand, the totally unfurnished burials may not demonstrate just a lower social class, but may additionally stress the marginal spiritual and social state of the infant, which does not need burial gifts for entering the world of beyond, due to its closer connection with the afterlife.

If we accept the noted differences between the part of the population that used simple graves (intramural ones included) and those that used chamber tombs (Lewartowski 1995, 110–1), which has been explained as a reflection of differences in social status and wealth and beliefs about death and the afterlife, the rare presence of infants in chamber tombs may be explained as an attempt of groups of the population to emphasise existing social differences within Mycenaean social structure. The unique burial of infants and a foetus in a Mycenaean sanctuary highlights unexplored relations between juveniles and religious (magical?) rituals and strengthens the idea that the dead belonged within the sphere of the sacred (Cavanagh and Mee 1998, 114).

The occurrence of organised burial grounds versus interments in current or former domestic sectors, and exceptionally in a sanctuary, is an indication that the living did not always seek to separate their world from the realm of beyond. Various interpretative propositions have been put forward concerning the exclusion of infants from communal cemeteries and their inclusion (or a percentage of them) within the domestic sphere. These include the need for the segregation of non-adult persons from the remainder of the dead, as the former did not qualify as active and equal members of society (Wells 1990, 139); making their death an event that impacted upon the household and not the whole community (Lebegyev 2009, 28); or the performance of sympathetic witchcraft in an effort to attract the power of fertility into the house where a child had been buried (Golden 1990, 85), hoping for the rebirth of a healthy offspring in order to preserve the family line (Morris 1987, 63–5; Konsolaki-Yannopoulou 2003a, 268, ft. 89). The latter view is supported in the case of jar burials which, albeit rare, are found only intramurally in Mycenaean Greece. Burials of infants in jars could have nurtured the hope that the dead infant would come back to life through a womb, bearing in mind that the ceramic vessel was perceived as a schematic human/female body (Goodison 1989, 40) and has been associated with beliefs about regeneration (Maniki 2006, 53–4). The vase/womb could provide the infant with the necessary vehicle to enable their rebirth (McGeorge 2011, 12).

The deliberate mixing of the domestic life and the mortuary sphere is associated, among others, with mechanisms for the commemoration of the deceased (Adams and King 2010; for a connection between memory and burials in abandoned and ruined settlement sectors see Sarri 2016), although it is not yet clear if this practice reinforces memory or oblivion. The absence of grave marks may be counterpoised by the everyday presence of the physical remains of the deceased near the household. The performance, or not, of rituals during the interment and the mourning period is a matter to be investigated through careful contextualisation of an intramural burial and its surrounding environment. The coexistence of the living and the dead is open to a twofold explanation – on the one hand, the expression of absolute oblivion, if the grave can no longer be reached as it is hidden and covered up by everyday activities, sounds, odours and worries, and, on the other hand, the attainment of the most profound commemoration by allowing the deceased to continue to 'coexist' spatially and in a sense 'participate' in family and communal life. The exclusion of infants from cemeteries, and the missing percentage of infants in the Mycenaean mortuary archaeological record, should not be explained as an indication that they held lesser social importance during life (Triantaphyllou 2016). As burials are highly symbolic actions, the choice of the location of the grave is not a random act by the living and the individuals buried inside settlements were considered eligible for this particular treatment for specific reasons that may vary in space and time. A new age group emerges from the archaeological data, that of the infants aged less than 24 months of age, which includes several different social identities, related to the status of their families, cosmological beliefs and conservative or innovative behaviours, all of which are important aspects of the ongoing process of gaining an understanding of the social structure of Mycenaean society.

Acknowledgements

I would like to thank E. Murphy and M. Le Roy for inviting me to contribute to this volume; K. Manteli and E. Murphy for improving the English text; G. Fakarou, M. Kontaki, K. Paschalidis, E. Konstantinidi-Syvridi and E. Konsolaki-Yannopoulou for their assistance in many ways; and the anonymous reviewers for constructive observations. This paper forms part of my ongoing doctoral research on the practice of intramural burials in the Aegean Bronze Age which is being undertaken in the Institut für Ur- und Frühgeschichte, University of Heidelberg.

Note

1. Hellenic Ministry of Culture and Sports, Athens National Archaeological Museum. Tositsa 1, 106 82 Athens, Greece. Email: katkostanti@hotmail.com.

References

Adams, R. L. and King, S. M. 2010. Residential burial in global perspective. *Archeological Papers of the American Anthropological Association* 20, 1–16.

Angel, J. L. 1973. Appendix – Human skeletons from Grave Circles at Mycenae, in Mylonas, G. E., *O Tafikos Kyklos B ton Mykinon* (Library of the Archaeological Society at Athens nr. 73), 379–397. Athens: The Archaeological Society at Athens.

Ariès, P. 1962. *Centuries of Childhood. A Social History of Family Life*. New York: Alfred A. Knopf.

Bloom, L. 1993. *The Transition from Infancy to Language. Acquiring the Power of Expression*. Cambridge: Cambridge University Press.

Cavanagh, W. G. and Mee, C. B. 1998. *A Private Place. Death in Prehistoric Greece* (Studies in Mediterranean Archaeology Vol. 125). Göteborg: Paul Åströms Förlag.

Cline, E. H. 2008. Problems of chronology: Egypt and the Aegean, in Aruz, J., Benzel, K. and Evans, J. M., *Beyond Babylon: Art, Trade, and Diplomacy in the Second Millennium B.C.*, 453–454. New York: Metropolitan Museum of Art.

Demakopoulou, K., Divari-Valakou, N. and Schallin, A.-L. 2002. Excavations in Midea 2000 and 2001. *Opuscula Atheniensia* 27, 27–58.

Frödin, O. and Persson, A. W. (eds.) 1938. *Asine. Results of the Swedish Excavations 1922-30*. Stockholm: Generalstabens Litografiska Anstalts Förlag i Distribution.

Golden, M. 1990. *Children and Childhood in Classical Athens*. Baltimore, MD: John Hopkins University Press.

Goodison, L. 1989. *Death, Women and the Sun: Symbolism of Regeneration in Early Aegean Religion* (Bulletin Supplement 53) London: University of London, Institute of Classical Studies.

Halcrow, S. E and Tayles, N. 2008. The bioarchaeological investigation of childhood and social age: problems and prospects. *Journal of Archaeological Method and Theory* 15, 190–215.

Hallager, B. and Mc George, P. J. P. 1992. *Late Minoan III Burials at Khania. The Tombs, Finds and Deceased in Odos Palama* (Studies in Mediterranean Archaeology 93). Göteborg: Paul Åströms Förlag.

Hillson, S. 2009. The world's largest infants' cemetery and its potential for studying growth and development, in Schepartz, L. A., Fox, S. C. and Bourbou C. (eds.), *New Directions in the Skeletal Biology of Greece* (Hesperia Supplement 43), 137–154. Princeton, CA: The American School of Classical Studies at Athens.

Iezzi, C. 2009. Regional differences in the health status of the Mycenaean women of East Lokris, in Schepartz, L. A., Fox, S. C. and Bourbou C. (eds.), *New Directions in the Skeletal Biology of Greece* (Hesperia Supplement 43), 175–192. Princeton, CA: The American School of Classical Studies at Athens.

Immerwahr, S. A. 1971. *The Athenian Agora, XIII: The Neolithic and Bronze Ages*. Princeton, CA: The American School of Classical Studies at Athens.

Karo, G. 1930–3. *Die Schachtgräber von Mykenai*. Munich: Verlag F. Bruckmann A-G.

Kaza-Papageorgiou, D. 1985. An Early Mycenaean cist grave from Argos. *Athenische Mitteilungen* 100, 1–21.

Kilian, K., Podzuweit, C. and Weisshaar, H.-J. 1981. Ausgrabungen in Tiryns 1978, 1979. *Archäologischer Anzeiger*, 149–256.

Konsolaki-Yannopoulou, E. 2002. A Mycenaean sanctuary on Methana, in Hägg, R. (ed.), *Peloponnesian Sanctuaries and Cults. Proceedings of the Ninth International Symposium at the Swedish Institute at Athens, 11-13 June 1994* (Skrifter utgivna av Svenska Institutet i Athen, 4°; 48), 25–36. Stockholm: Svenska Institutet i Athen. Paul Åströms Förlag.

Konsolaki-Yannopoulou, E. 2003a. Taphes nipion sto mykinaiko iero tou Ag. Konstantinou sta Methana, in Konsolaki-Yannopoulou, E. (ed.), *Argosaronikos. Praktika 1ou Diethnous Synedriou Istorias kai Archaiologias tou Argosaronikou, Poros 26-9 Iouniou 1998*, Vol. A, 257–284. Athens: Municipality of Poros.

Konsolaki-Yannopoulou, E. 2003b. I mykinaiki egkatastasi sto nisaki Modi tis Troizinias, in Kyparissi-Apostolika, N. and Papakonstantinou, M. (eds.), *Praktika B' Diethnous Epistimonikou Symposiou 'H Perifereia tou Mykinaikou Kosmou', 26-30 Septemvriou, Lamia 1999*, 417–432. Athens: YPPO-ID' EPKA/Archaeological Receipts Fund.

Konstantinidi-Syvridi, E. forthcoming. O lakkoeidis tafos III tou Kyklou A ton Mykinon: Tafos ton gynaikon i tafos ton iereon? in Kalogerakou, P. and Kountouri, E. (eds.), *Timitikos Tomos gia ton Kathigiti Georgio Styl. Korre*. Athens.

Kostanti, K. 2009. Paidia sto Proistoriko Aigaio. Martyries ton Tafon tis Ysteris Epochis tou Chalkou sto Mykinaiko Elladiko Horo. Unpublished Master's thesis, University of Crete, Rethymno. Available at <http://elocus.lib.uoc.gr/dlib/8/7/9/metadata-dlib-017e10b3255d47787262a07863ed83d9_1265005819.tkl>

Lanaras, B. 2003. Mia intra muros tafi vrefous sto mesokykladiko Akrotiri Thiras. Istoria tou ethimou ton intra muros tafon sto Aigaio. Prospatheia ermineias enos idiaiterou evrimatos, in Vlachopoulos, A. V. and Birtacha, K. B. (eds.), *Argonaftis. Timitikos Tomos gia ton Kathigiti Christo G. Douma apo tous Mathites tou sto Panepistimio Athinon 1980-2000*, 445–460. Athens: I Kathimerini.

Laneri, N. 2011. Defining residential graves. The case of Titris Höyük in southeastern Anatolia during the late IIIrd Millennium BC, in Henry, O. (ed.), *Le Mort dans la Ville. Pratiques, Contextes et Impacts des Inhumations Intra-Muros en Anatolie, du Début de l'Age du Bronze à l'Époque Romaine, Actes des 2e Rencontres d'Archéologie, Istanbul 14-15 Novembre 2011*, 43–52. Istanbul: Institut Français d'Études Anatoliennes Georges Dumézil – CNRS USR 3131.

Lebegyev, J. 2009. Phases of childhood in Early Mycenaean Greece. *Childhood in the Past* 2, 15–32.

Lewartowski, K. 1995. Mycenaean social structure: a view from simple graves, in Laffineur, R. and Niemeier W.-D. (eds.), *Politeia. Society and State in the Aegean Bronze Age. Proceedings of the 5th International Aegean Conference, University of Heidelberg, Archäologisches Institut, 10-13 April 1994*, Vol I. (Aegaeum 12), 103–113. Austin, TX: University of Texas at Austin.

Maniki, A. 2006. I Tafiki Praktiki tou Enchytrismou sto Aigaio tin Proimi Epochi tou Chalkou: To Paradeigma tou Akrotiriou sti Thira. Unpublished Master's Thesis, University of Crete, Rethymno. Available at <http://elocus.lib.uoc.gr/dlib/4/1/3/metadata-dlibf2d2582fde3cd39192a1da99446953b0_1241688173.tkl>

Maran, J. 1995. Structural changes in the pattern of settlement during the Shaft Grave period on the Greek mainland, in Laffineur, R. and Niemeier, W.-D. (eds.), *Politeia: Society and State in the Aegean Bronze Age. Proceedings of the 5th International Aegean Conference, University of Heidelberg, Archäologisches Institut, 10-13 April 1994*, Vol I. (Aegaeum 12), 67–72. Austin, TX: University of Texas at Austin.

Mårtensson, L. 2002. Traces of boxes: linings of wooden boxes in Helladic tombs, in Wells, B. (ed.), *New Research in Old Material from Asine and Berbati in Celebration of the Fiftieth Anniversary of the Swedish Institute at Athens* (Skrifter utgivna av Svenska institutet i Athen 8° 17) 43–48. Stockholm: Paul Åströms Förlag.

Mc George, P. J. P. 2011. Intramural infant burials in the Aegean Bronze Age, in Henry, O. (ed.), *Le Mort dans la Ville. Pratiques, Contextes et Impacts des Inhumations Intra-Muros en Anatolie, du Début de l'Age du Bronze à l'Époque Romaine, Actes des 2e Rencontres d'Archéologie, Istanbul 14-15 Novembre 2011*, 1–19. Istanbul: Institut Français d'Études Anatoliennes Georges Dumézil – CNRS USR 3131.

McGeorge, P. J. P. 2012. The Petras intramural infant jar burial: context, symbolism, eschatology, in Tsipopoulou, M. (ed.), *Petras, Siteia - Twenty-Five Years of Excavations and Studies. Acts of a Two-Day Conference Held at the Danish Institute at Athens, 9-10 October 2010* (Monographs of the Danish Institute at Athens 16), 291–302. Athens: The Danish Institute at Athens.

McHugh, F. 1999. *Theoretical and Quantitative Approaches to the Study of Mortuary Practices* (BAR International Series 785). Oxford: Archaeopress.

Morris, I. 1987. *Burial and Ancient Society. The Rise of the Greek City-State.* Cambridge: Cambridge University Press.

Musgrave, J. H. and Popham, M. 1991. The Late Helladic IIIC intramural burials at Lefkandi Euboea. *Annual of the British School at Athens* 86, 273–296.

Mylonas, G. E. 1973. *O Tafikos Kyklos B ton Mykinon* (Library of the Archaeological Society at Athens 73). Athens: The Archaeological Society at Athens.

Nordquist, G. and Ingvarsson-Sundström, A. 2005. Live hard, die young: mortuary remains of Middle and early Late Helladic children from the Argolid in social context, in Dakouri-Hild, A. and Sherratt, S. (eds.), *Autochthon. Papers Presented to O.T.P.K. Dickinson on the Occasion of his Retirement* (BAR International Series 1432), 156–174. Oxford: Archaeopress.

Pader, E. J. 1982. *Symbolism, Social Relations and the Interpretation of Mortuary Remains* (BAR International Series 130). Oxford: Archaeopress.

Papazoglou-Manioudaki, L., Nafplioti, A., Musgrave, J. H. and Prag, A. J. N. W. 2010. Mycenae revisited Part 3. The human remains from Grave Circle A at Mycenae. Behind the masks: a study of the bones of Shaft Graves I–V. *Annual of the British School at Athens* 105, 157–224.

Parker Pearson, M. 1999. *The Archaeology of Death and Burial.* Stroud: Sutton Publishing.

Paschalidis, K. 2016. Ties of affection. Burials of parents and children in the Mycenaean cemetery of Clauss, near Patras, in Papadopoulou-Chrysikopoulou, E., Chrysikopoulos, V. and Christakopoulou, G. (eds.), *Achaios. Studies Presented to Professor Thanasis I. Papadopoulos*, 207–218. Oxford: Archaeopress Archaeology.

Polychronakou-Sgouritsa, N. 1987. Paidikes Tafes sti mykinaiki Ellada. *Archaiologikon Deltion* 42A, 8–29.

Pomadère, M. 2010. De l'indifferenciation à la discrimination spatiale des sépultures? Variété des comportements à l' égard des enfants morts pendant l' HM-HRI, in Philippa-Touchais, A., Touchais, G., Voutsaki, S. and Wright, J. (eds.), *MESOHELLADICA. La Grèce Continentale au Bronze Moyen. Actes du Colloque International Organisé par l'École Française d'Athènes, en Collaboration avec l'American School of Classical Studies at Athens et le Netherlands Institute in Athens, Athènes, 8-12 Mars 2006* (BCH Supplement 52), 525–539. Athens: École Française d'Athènes, Editions De Boccard.

Pomadère, M. 2013. Entre zône d' habitat et cimetière clos, l'évolution spatiale des aires funéraires du début du Bronze récent au premier Age du Fer, in Darcque, P., Etienne, R. and Guimier-Sorbets, A.-M. (eds.), *PROASTEION, Recherches sur le Périurbain dans le Monde Grec* (Travaux de la Maison de l'Archéologie et de l'Ethnologie, René-Ginouvès 17), 225–243. Paris: Editions de Boccard.

Sarri, K. 2016. Intra, extra, inferus and supra mural burials of the Middle Helladic period: spatial diversity in practice, in Dakouri-Hild, A. and Boyd, M. (eds.), *Staging Death: Funerary Performance, Architecture and Landscape in the Aegean*, 117–138. Berlin: DeGruyter.

Schepartz, L. A., Miller-Antonio, S. and Murphy, J. M. A. 2009. Differential health among the Mycenaeans of Messenia: status, sex and dental health at Pylos, in Schepartz, L. A., Fox, S. C. and Bourbou, C. (eds.), *New Directions in the Skeletal Biology of Greece* (Hesperia Supplement 43), 155–174. Athens: The American School of Classical Studies at Athens.

Shepartz, L. A., Papathanasiou, A., Miller-Antonio, S., Stocker, S. R., Davis, J. L., Murphy, J. M. A., Malapani, E. and Richards, M. 2011. No seat at the table? Mycenaean women's diet and health in Pylos, Greece, in Schepartz, L. A. (ed.), *Anthropology à la Carte: The Evolution and Diversity of Human Diet*, 359–374. San Diego, CA: Cognella Academic Publishing.

Scott, E. 1999. *The Archaeology of Infancy and Infant Death* (BAR International Series 819). Oxford: Archaeopress.

Taylour W. D. 1972. Excavations at Ayios Stephanos. *Annual of the British School at Athens* 67, 205–270.

Taylour W. D. and Janko, R. 2008. *Ayios Stephanos. Excavations at a Bronze Age and Medieval Settlement in Southern Laconia* (Annual of the British School at Athens Supplement 44). London: The British School at Athens.

Triantaphyllou, S. 2001. *A Bioarchaeological Approach to Prehistoric Cemetery Populations from Central and Western Greek Macedonia* (BAR International Series 976). Oxford: Archaeopress.

Triantaphyllou, S. 2016. Constructing identities by ageing the body in the prehistoric Aegean: the view through the human remains, in Mina, M., Triantaphyllou, S. and Papadatos, Y. (eds.), *An Archaeology of Prehistoric Bodies and Embodied Identities in the Eastern Mediterranean. Convergence of Theory and Practice*, 160–168. Oxford: Oxbow Books.

Tzonou-Herbst, I. N. 2002. A Contextual Analysis of Mycenaean Terracotta Figurines. Unpublished Ph.D. thesis, University of Cincinnati.

Ubelaker D. H., 1989. *Human Skeletal Remains: Excavation, Analysis, Interpretation*. Washington: Taraxacum.

Ucko P. 1969. Ethnography and the archaeological interpretation of mortuary remains. *World Archaeology* 1, 262–280.

Van Gennep, A. 1909. *Les Rites de Passage*. Paris: Éditions A. and J. Picard (1981 reprint).

Voutsaki, S. 2012. From value to meaning, from things to persons: the Grave Circles of Mycenae reconsidered, in Urton, G. and Papadopoulos, J. (eds.), *The Construction of Value in the Ancient World* (Monographs of UCLA), 160–185. Los Angeles, CA: Kotsen Institute.

Wells, B. 1990. Death at Dendra. On mortuary practices in a Mycenaean community, in Hägg, R. and Nordquist, G. C. (eds.), *Celebrations of Death and Divinity in the Bronze Age Argolid. Proceedings of the 6th International Symposium at the Swedish Institute at Athens, 11-13 June 1988*, 125–140. Stockholm: Paul Åströms Förlag.

Chapter 9

Bronze Age Child Burials in the Southern Trans-Urals (21st–15th Centuries cal. BC)

Natalia Berseneva[1]

Abstract: This study concerns the Sintashta, Petrovka and Alakul' cultural groups. They are currently dated from the 21st to the 15th centuries cal. BC, and the sites are located in the steppe part of the southern Trans-Urals. One of the most remarkable traits of the burial grounds is that subadults constitute from 50% to 80% of all the burials. From a comparison of the variations apparent in children's burial rite, we can generally conclude that the place of children in the social structure was quite important in these cultures, despite the fact that the spatial organisation of the burials and grave goods varied substantially. We also can suppose that the initial age of labour and gender socialisation in these cultures occurred at approximately the same age (after 3–5 years). The age at which gender-distinctive clothing appears was apparently different, however, and sometimes started as early as one year old in the Alakul' society but later for children in the Sintashta and Petrovka societies.

Keywords: Bronze Age, South Urals, Sintashta, Petrovka and Alakul' cultures, child burials, socialisation

Introduction

The territory under study is the southern part of the Ural Mountains, namely the Trans-Urals (Fig. 9.1). The Ural area can be defined in terms of its geographic location as a natural boundary between Europe and Asia. In terms of administrative divisions, it covers several provinces of the Russian Federation as well as the northwestern part of Kazakhstan. It is characterised by a great expansive landscape and environmental diversity: steppe, forest-steppe, forests and mountains. The beginning of the Bronze Age in the steppe zone was marked by the emergence of bronze metallurgy and food-producing economies based primarily on livestock husbandry.

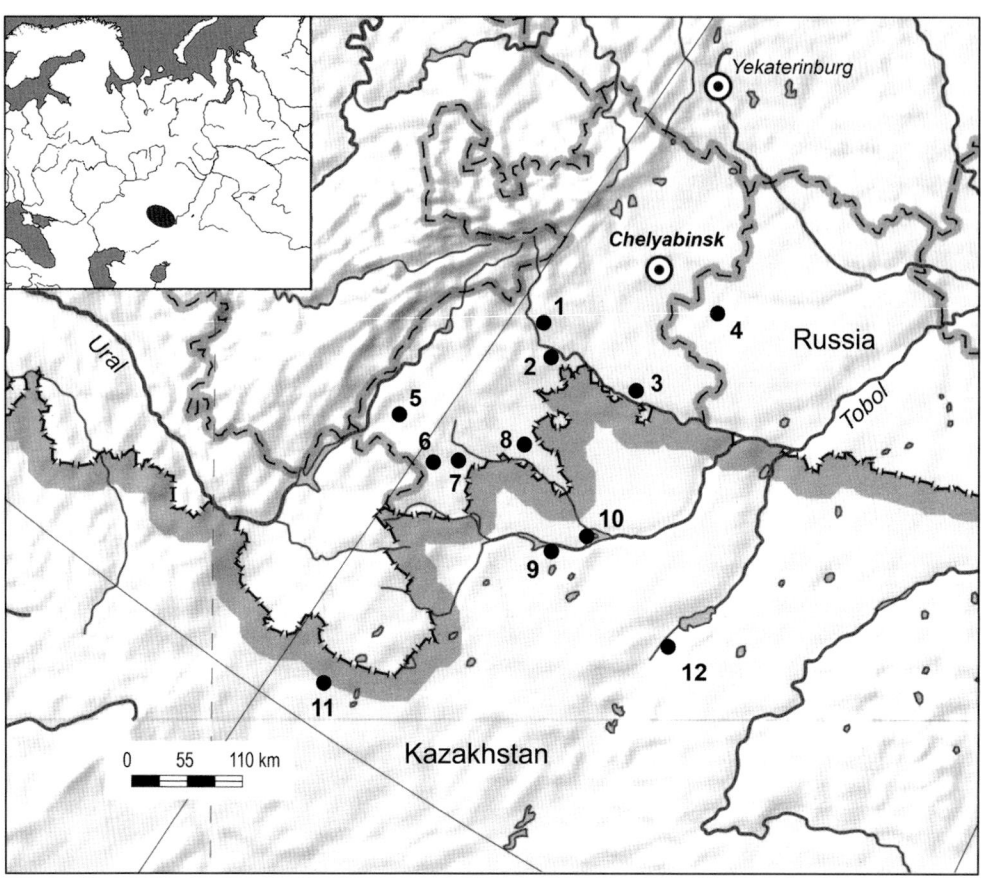

Figure 9.1. Map of key Sintashta, Petrovka and Alakul' burial grounds. 1 – Stepnoye-1, Stepnoye VII; 2 – Krivoe Ozero; 3 – Troizk-7; 4 – Alakul'; 5 – Bolshekaraganskyi; 6 – Sintashta; 7 – Kamennyi Ambr-5; 8 – Kulevchi VI; 9 – Lisakovskyi I; 10 – Halvay III; 11 – Tasty-Bytak I; 12 – Bestamak.

The first period of the Bronze Age (the first half of the third millennium BC) is connected with the Yamnaya (Pit-Grave) 'family' of cultures in the steppe zone of the Urals.[2] The Yamnaya sites are very well represented by barrows (kurgans), but only a few such mounds are known on the eastern slopes of the Ural Mountains. In general, during the third millennium BC the territory appears to have been poorly settled. Beginning about the 21st century cal. BC, the southern Trans-Urals was occupied by the Sintashta (up to 18th century cal. BC), Petrovka (the 19th–18th centuries cal. BC), Alakul' (17th–15th centuries cal. BC) and to a smaller extent Srubnaya (18th–15th centuries cal. BC) populations (Table 9.1).[3] According to the opinion of many archaeologists, the Sintashta–Petrovka–Alakul' continuum constitutes a sequence of kindred cultures (Koryakova and Epimakhov 2007). Their food economies were based on livestock breeding and no firm traces of plant-based agriculture have been found.

9. Bronze Age Child Burials in the Southern Trans-Urals (21st–15th Centuries cal. BC)   127

Table 9.1. Summary of the main characteristics of the different Bronze Age cultures.

	Sintashta period (21–18 cal. BC)	Petrovka period (19–18 cal. BC)	Alakul' period (17–15 cal. BC)
Settlements	Fortified settlements, round or rectangular in form	Fortified settlements	Open settlements
Burial grounds	Barrow groups or flat cemeteries, sporadic burials at settlements	Barrows in multi-phase cemeteries, sporadic burials at settlements	Barrow groups or flat cemeteries, sporadic burials at settlements
Burials	Both multiple and individual inhumation graves for all ages	Single inhumation burials are predominant for children. Multiple tombs – for adults	Single inhumation burials are predominant for children. Multiple tombs – for adults
Grave goods	Sintashta ceramics, tools, ornaments, weaponry, horse harness and traces of chariots	Petrovka ceramics, ornaments, weaponry, tools, horse harness	Alakul' ceramics, ornaments
Animal sacrifices	Abundance of whole sacrificed animals (horses, sheep, calves, dogs) in graves	Sacrificial deposits 'head-extremities', paired sacrifices of whole horse carcasses	Sacrificial sheep deposits 'head- extremities' or separate bones to a small extent
Key sites (burial grounds)	Kamennyi Ambr-5, Bolshekaraganskyi, Krivoe Ozero, Stepnoye-1, Halvay III, Bestamak	Troizk-7, the Petrovka part of the Krivoe Ozero, Kulevchi VI and Stepnoye VII	Stepnoye VII, Kulevchi VI, Alakul', Lisakovskyi I, Tasty-Butak I
Publications	Epimakhov 2005; Zdanovich 2002; Vinogradov 2003; Kupriyanova 2016; Shevnina and Logvin 2015; Shevnina and Voroshilova 2009	Vinogradov 1984, 2003; Kupriyanova and Zdanovich 2015; Kostyukov and Epimakhov 1999	Kupriyanova and Zdanovich 2015; Vinogradov 1984; Sal'nikov 1952; Usmanova 2005; Sorokin 1962

As such, these people can be considered to have been settled pastoralists. The end of the Bronze Age is currently dated to between the 13th and the 10th centuries cal. BC. This period is primarily represented by unfortified settlements and burial sites are extremely rare.

It is important to note that in the South Ural steppes a large number of child burials (>50% of burials) is typical only for the Sintashta period and especially (up to 80%) for the Petrovka – Srubnaya/Alakul' periods (Figs. 9.2 and 9.3). In this study the juveniles are described according to the following age categories: infant – birth to two years; child – 2–15 years; adolescent – 15–17 years; adult – more than 17 years of age (Lewis 2011, tab. 1; Table 9.2). All of the age identifications were conducted

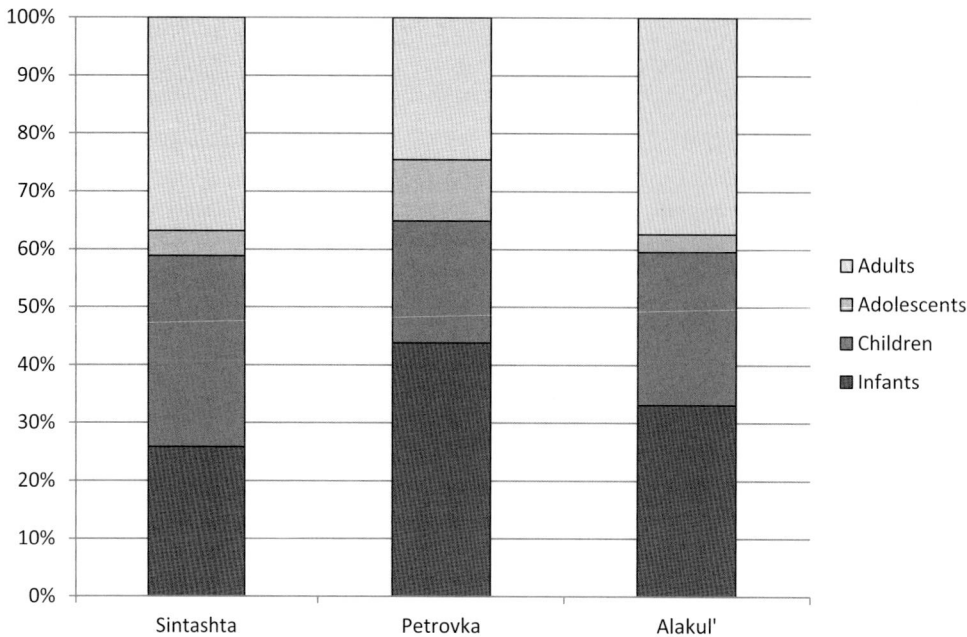

Figure 9.2. The representation of juveniles of different ages in the Sintashta, Petrovka and Alakul' cultures.

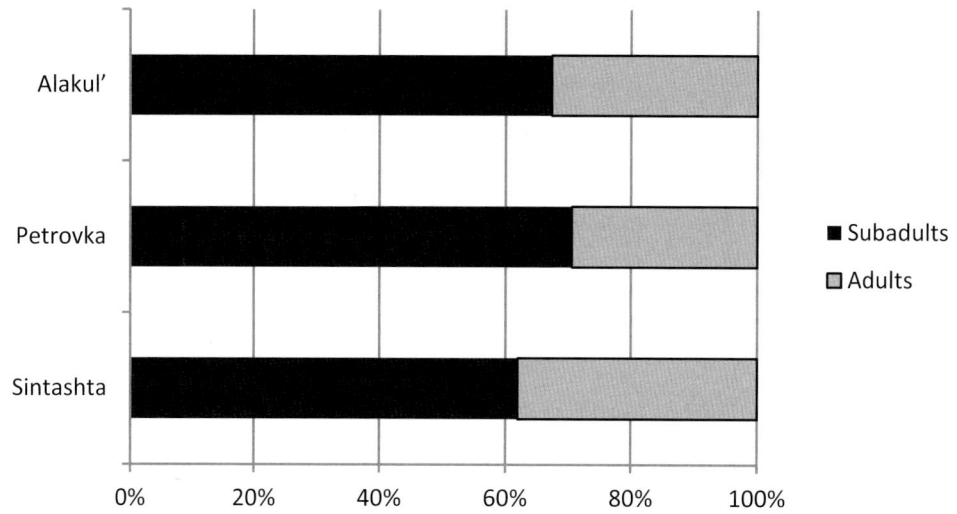

Figure 9.3. Overall proportions of adults and subadults in the Sintashta, Petrovka and Alakul' cultures.

by anthropologists, but it should be noted that it was difficult to assign individuals to these categories for all sites because of the use of different age divisions by the researchers and poor preservation of bone remains. Unfortunately, palaeopathological

Table 9.2. Age-at-death profiles of the Sintashta, Petrovka and Alakul' Cultures.

	Subadults			Adolescents	Adults	Total
	Infants	Children				
	<2 years	2–10 years	10–15 years	15–17 years	>17 years	
Sintashta	54 (25.8%)	56 (26.8%)	13 (6.3%)	9 (4.3%)	77 (36.8%)	209 (100.0%)
Petrovka	25 (43.9%)	8 (14.0%)	4 (7.0%)	6 (10.5%)	14 (24.6%)	57 (100.0%)
Alakul'	76 (33.1%)	52 (22.6%)	9 (4.0%)	7 (3.0%)	86 (37.3%)	230 (100.0%)

investigations have only been conducted for some Sintashta cemeteries. So, on the basis of palaeoanthropological observations of individuals from Barrow 25 in the Bolshekaraganskyi cemetery, the population was relatively healthy without signs of chronic or periodic diseases and no evidence of violent death was apparent (Lindstrom 2002, 163). Individuals buried in the Kamennyi Ambr-5 cemetery also appeared to have been relatively healthy (Epimakhov 2005). The people buried in the Krivoe Ozero cemetery, however, were considered to have been relatively unhealthy with evidence of dental enamel hypoplasia, periodontal disease (adults) and mastoiditis (Rykushina 2003, 359–60). Regrettably, we have no such data for the Petrovka and Alakul' collections, with the exception of the results of an investigation of dental diseases based on the materials of the Bestamak and Lisakovskiy burial grounds (Ventresca Miller et al. 2014).

This study is based on data derived from well-documented and published Bronze Age sites excavated in the southern Trans-Urals (see Table 9.1). The information affords the opportunity to investigate and reconstruct the mortuary rites of children and identify the different stages of their life course based on the materials of these kindred cultural groups who lived in the southern Trans-Ural area at the end of the Middle and into the Late Bronze Age. It is also possible to explore the dynamics of change apparent in the burial rituals of the children in these societies.

Child burials

Sintashta Culture

In the Sintashta burial grounds subadults (individuals up to 15 years old) constitute more than 50% of all the dead (for details see Berseneva 2010) (Fig. 9.4). The burial rites for Sintashta children were the same as for adults, and the internal space of child burials was organised much in the same way (Fig. 9.5). The body position on the left side (rarely on the right) in a contracted position, with the hands near the face, was also the same. Most children were buried in non-individual tombs (62%)

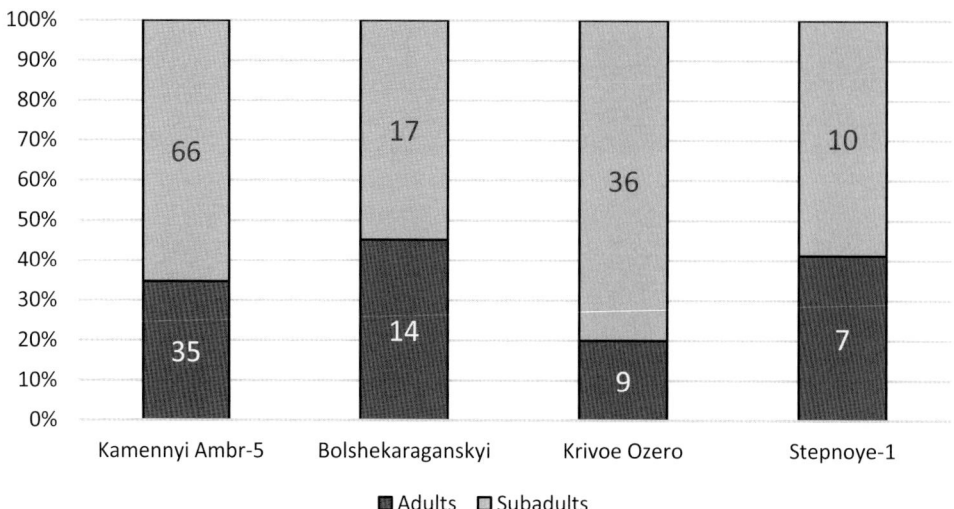

Figure 9.4. The proportions of adults and subadults at key Sintashta sites.

along with other subadults or adults in various compositions (with older adolescents, adult males, adult females or both adult males and females; Fig. 9.6). There are no signs of any obvious patterns. The majority of individual child pits were smaller in size than those of adults, but there were also large tombs with multiple burials of children, some of which contained as many as eight people who had been buried simultaneously (Epimakhov 2005). A small number of Sintashta graves (10 cases, nearly 3.0%), that contained pairs of individuals lying in a position facing each other, have also been uncovered. Usually a child was buried in such a position along with an adult. An interesting example of this type is Grave 8 from Barrow 2 at the Kamennyi Ambr-5 cemetery. Here a male, aged 22–26 years old, 'embraced' an eight-year-old child, while two other subadults (both 5–6 years old) lay near the 'pair' (see Fig. 9.6).

Sintashta grave goods usually included many categories – weaponry, horse trappings, clothing attachments and ornaments, as well as tools and objects linked to metallurgy. Adults were usually accompanied by grave goods but 45.5% of the subadults were provided solely with pottery. Children were usually given sheep and dogs as animal sacrifices. The burials of subadults less than three years old did not contain any tools or bronze items. Their grave goods chiefly constituted ornaments, mainly beads, so-called amulets (drilled fangs of small predators), pottery and also astragali that were often found in large numbers. Some children from 3–14 years of age were buried with weaponry which was quite intriguing. It should be noted that the number of weapons recovered from Sintashta child graves was relatively small, but weapons are totally absent at pre-Sintashta (Yamnaya) and post-Sintashta (Petrovka, Alakul' and Srubnaya) sites and are extremely rare in the steppe cultures of Bronze Age Eurasia as a whole (Jones-Bley 1994). Weapons in child graves were found in five cemeteries – Kamennyi Ambr-5 (arrowheads; Epimakhov 2005); Stepnoye-1

9. Bronze Age Child Burials in the Southern Trans-Urals (21st-15th Centuries cal. BC)

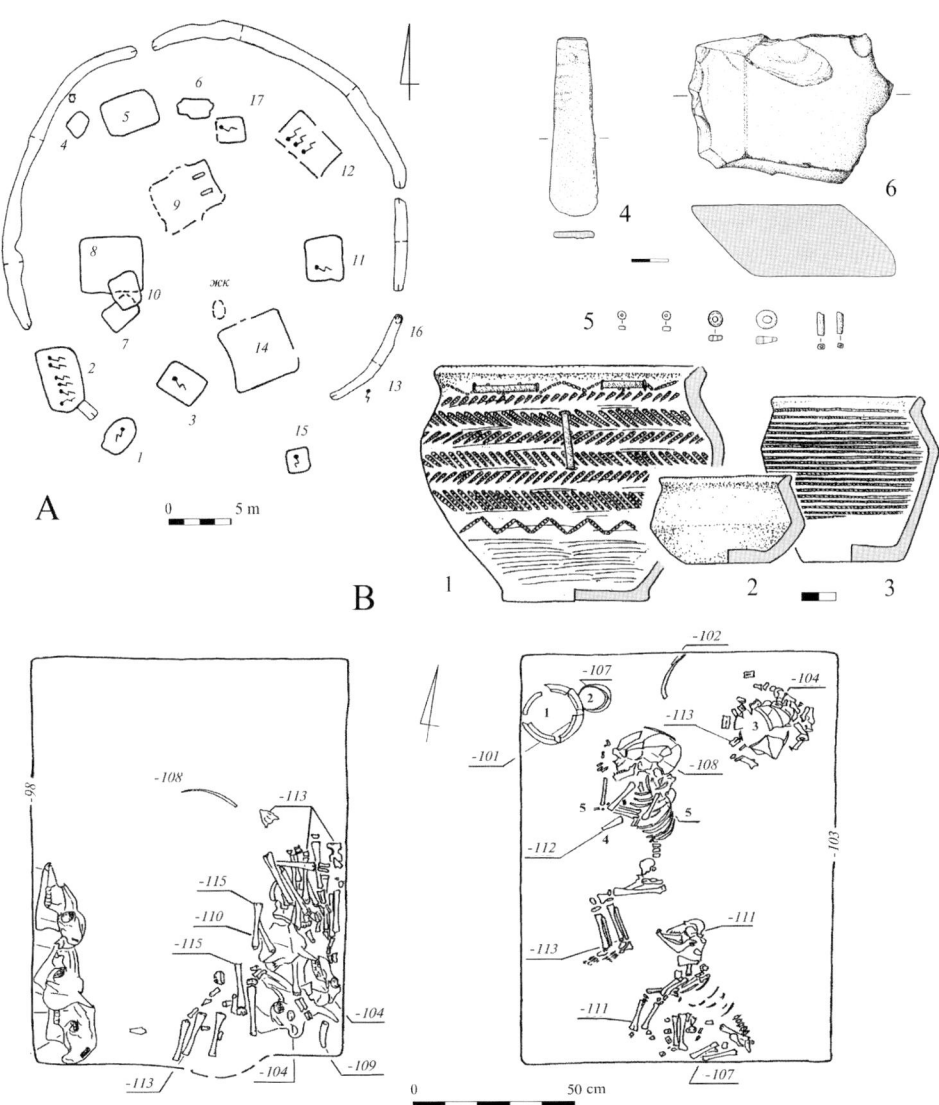

Figure 9.5. Kamennyi Ambr-5 cemetery (Sintashta Culture). A - Barrow 4. Plan. B - Barrow 4. Grave 15. Plan and grave goods: 1-3 - ceramic vessels; 4 - bronze adze; 5 - beads; 6 - stone slab - anvil. The burial was that of a subadult (3-7 years old) (after Epimakhov 2005).

(arrowheads and cheek-pieces; Kupriyanova 2016); Bolshekaraganskyi (arrowheads; Zdanovich 2002); Bestamak (a bronze spearhead; Shevnina and Voroshilova 2009) and Halvay III (bronze arrowheads, two bronze spearheads, a bronze battle-axe, see below; Shevnina and Logvin 2015). In total, 74 bronze, stone and bone arrowheads, three bronze spearheads, a bronze battle-axe, the horn parts of a composite bow and four horn cheek-pieces were discovered in the graves of children (Table 9.3).[4]

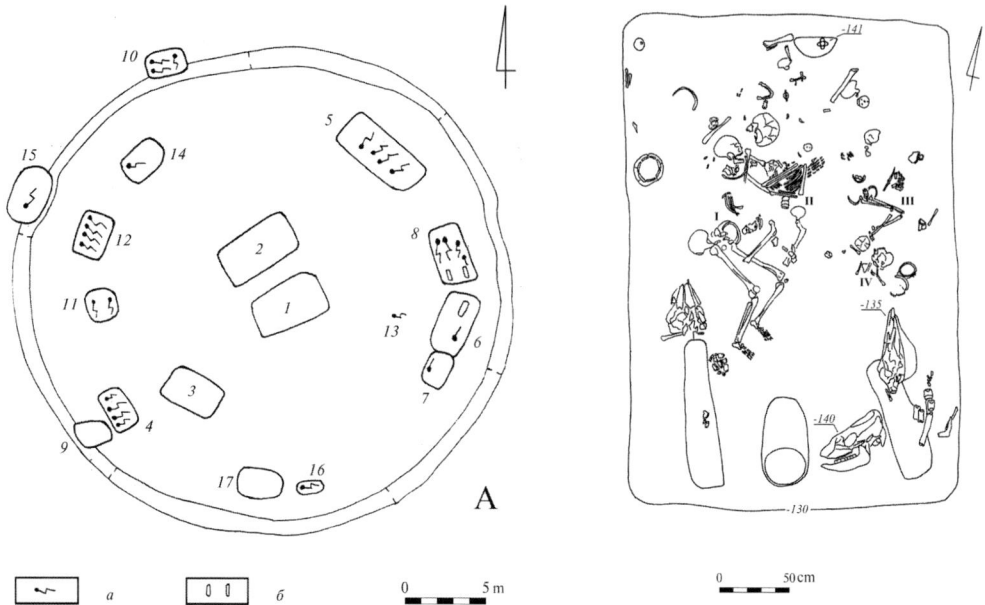

Figure 9.6. Kamennyi Ambr-5 cemetery (Sintashta Culture). A – Barrow 2. Plan. B – Barrow 2. Grave 8. Plan. The associated individuals comprised – I – male (22– 26 years old); II – subadult (8 years old); III – subadult (5 years old); IV – subadult (6 years old) (after Epimakhov 2005, 10, 35-7).

The recently investigated Halvay III burial is especially interesting and it is difficult to believe that the burial was that of a child, but only the teeth of a four- to five-year-old child were discovered in the grave, which appeared to have been undisturbed. The burial pit was very large (3.2 m × 2.6 m, 3 m deep), and it had a massive wooden ceiling supported by posts. The burial produced six ceramic vessels and a rich set of grave goods – two sets of arrowheads, including some made from bronze, two bronze spearheads, a bronze battle-axe, a bronze knife and stone tools (Shevnina and Logvin 2015, 110–1). The authors supposed that the child was buried following 'adult' burial rites (Shevnina and Logvin 2015, 128). Nevertheless, even an adult warrior grave containing evidence of this much wealth is extremely rare. The Halvay III child burial is the only Sintashta grave to contain two bronze spearheads. Only 11 such items have been recovered from the entirety of Sintashta cemeteries excavated to date. Moreover, the presence of two spearheads and a battle-axe together constitutes a unique case. Bronze arrowheads of this very specific form (ornamented with so-called 'herringbone' pattern) are also very rare. Aside from Halvay III, they have only been found in two other cemeteries and it is difficult to interpret such an unusual case. If the huge tomb had been intended for the interment of the child only, then the individual must have been extremely important, perhaps the first-born of a very influential family. It is possible that some of the items were placed in the grave with the expectation that

Table 9.3. Correlation of grave goods and age-at-death of subadults (Sintashta Culture).

Grave Goods	Number of subadults	Age-at-death of subadults (years)
Bronze knife	20	3–15
Bronze awl	8	2–15
Bronze adze	7	3–15
Bronze arrowhead	10	3–15
Beads	22	1–15
Bronze spearhead	2	4–5
Bronze axe	1	4–5
Bone parts of a bow	2	9–10
headdress decoration	3	4/5–14
Stone tools	5	3–15
Cheek-piece	2	1.5 and 12–14
Drilled fang	14	1/2–14
Astragali	22	1/2–11

they would be passed on to someone in the afterlife. Another possibility is that the great pit had been intended for an adult, who had not been buried there for some unknown reason, and not solely for the child.

Petrovka Culture

The Late Bronze Age is represented by the Petrovka and Alakul' Cultures which continue Sintashta cultural traditions. The Petrovka child mortuary ritual is simpler than that of the preceding Sintashta in terms of grave goods, sacrificial deposits of animals and grave pit size, but is very similar in regard to body position and the inclusion of pottery. The main distinction is that Petrovka child burials are predominantly individual (62.9%) and occur in relatively small shallow pits. Multiple child tombs that contain more than three subadults in one grave (as in the Sintashta cemeteries) are not found. Subadults up to 15 years of age constitute at least 70.1% of all those buried (see Table 9.2; Fig. 9.7).

From a demographic point of view, the entire subadult age group appears to be quite unusual. Infants up to two years of age constitute 77.4%, while 52.1% of all juveniles were less than 15 years old, and overwhelmingly the majority of these individuals were neonates or even foetuses. The rest of the burials belong to children aged 8–9 or 10–12 years. Surprisingly, however, the remains of children whose age-at-death was between two and 5–7 years of age were completely absent. Only three individuals aged 12–15 years were represented and all of these subadults had been buried in the primary tombs along with adults. Adults are also represented in extremely low numbers (Stepnoye VII burial ground – 26.5% of all the buried; Krivoe

Figure 9.7. The proportions of adults and subadults at key Petrovka sites.

Table 9.4. Correlation of grave goods and age-at-death of subadults (Petrovka Culture).

Grave Goods	Number of subadults	Age-at-death of subadults (years)
Bronze knife	3	0–10
Astragali	4	0–10
Bronze bracelet	2	8–11
Beads	10	0–11
Shell	1	0–0.5
Hair decorations	2	8–11
Drilled fang	5	1–11

Ozero – 13.6%) (see Fig. 9.7). It is possible that post-mortem selection of the dead took place in Petrovka society or that there was an alternative, archaeologically invisible, way of interment for certain groups of people that is causing distortion to the normal structure of the buried population. Any such 'alternative burials', however, remain to be discovered.

A total of 60.9% of undisturbed (non-robbed) subadult burials contained no grave goods, with the exception of pottery. The burial rite for infants aged up to two years is distinguished by their absolute simplicity. The rectangular pits were small in size and depth (usually 0.5 m × 0.7–1.0 m and 0.30–0.50 m deep) and the remains of the grave ceiling were rarely discovered. The overwhelming majority of the dead were laid on the left side in a flexed position. In terms of grave goods one or, more rarely, two pots were located near the head of the deceased and in some cases (34.8%) small

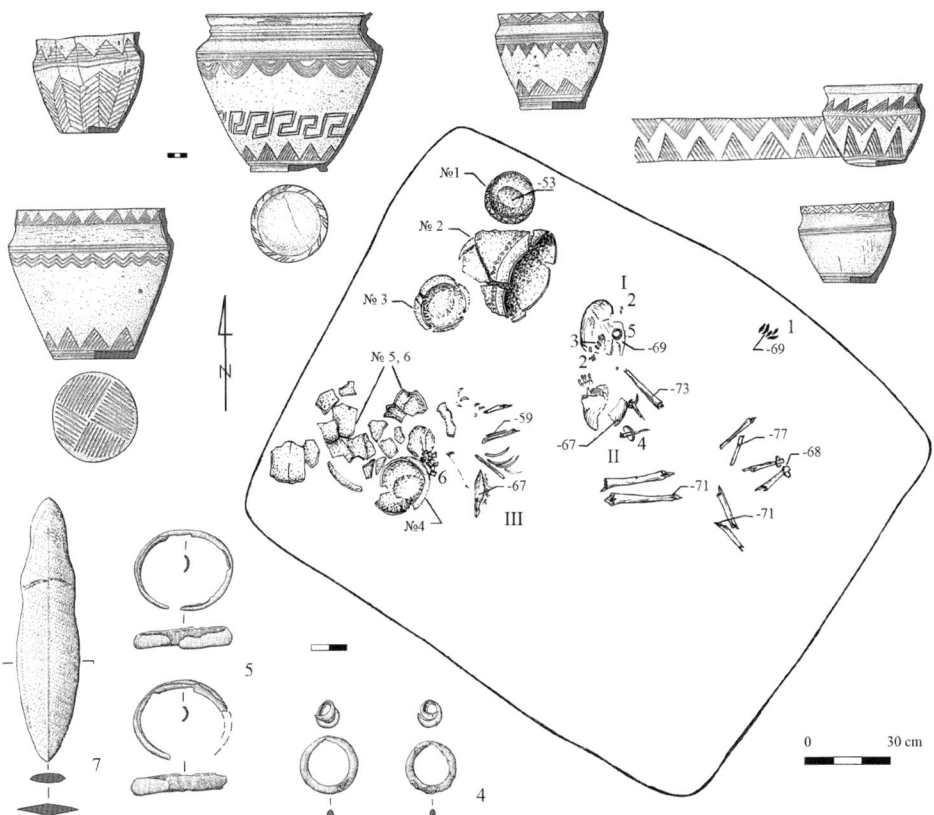

Figure 9.8. Stepnoe VII cemetery (Petrovka Culture). Burial 19. Burial of three children (9–10 years old). 1–6 – ceramic vessels; 1 – pendant from animal fangs; 2 – stone and paste beads; 3 – bronze ornaments; 4 – bronze bracelets; 5 – bronze pendants; 6 – sheep astragali; 7 – bronze knife (after Kupriyanova and Zdanovich 2015).

ornaments (beads, pendants from animal fangs and shells) are present but other grave goods are extremely rare. Animal sacrifices did not occur in the burials of small infants if they were interred in the absence of older children or adults.

The burials of children in the older age group (7–10 years) demonstrate more variety in burial treatment. The size of graves differs in accordance to body proportions and the amount of grave goods increases slightly. These burials produced two bronze knives and, in addition, certain types of 'adult' ornaments, such as bronze bracelets and strings of bronze beads appear on the shoes as decorations or ties. Sets of bone gaming pieces (astragali) are also quite common, but artefacts related to hunting and weaponry are almost absent (Table 9.4). The burials of these older children are sometimes accompanied by sacrificial animal deposits, consisting of the heads and distal parts of sheep.

The older child and adolescent (9 to 17 years) burial rite is almost the same as that of adults in regard to grave goods, as well as the occurrence of clothing attachments and animal sacrifices. Individuals of this age were often interred alongside adults in large collective tombs. In some cases, they were arranged as a pair with an adult in a specific way so that one individual appears to have been 'embracing' the other. This was the case in the Kulevchi VI cemetery (Kurgan 4, Grave 5), where an individual of 9–11 years was 'embraced' by an adult male. There is a strong indication that the subadult may have been female because she was associated with a wealthy set of specific bronze ornaments – a headdress, two bracelets, beads on her ankles and amulets (Vinogradov 1998).

Three children aged 8–10 years old were interred in Grave 19 at the Stepnoye VII burial ground (Kupriyanova and Zdanovich 2015; Fig. 9.8). Two of the children were placed in a 'face to face' position lying on their left and right sides respectively, while the remains of the third individual were discovered some distance away and comprised a small disordered heap. The child, lying on its right side (possibly a girl), was accompanied by a large number of ornaments (bronze bracelets, a complicated headdress, beads on her ankles and strings of fang amulets) similar to the items from Kulevchi VI discussed above. It is interesting that the two other children had no *personal* accessories. Nevertheless, the grave contained a relatively wealthy set of items, including a bronze knife and astragali, as well as six ceramic vessels. Aside from its smaller size, this grave looked very similar to the adult multiple tombs discovered in the burial ground (Kupriyanova and Zdanovich 2015, 50–3).

Alakul' Culture

During the Alakul' period 'infant' cemeteries appeared. These contained up to 40 small shallow graves located at the periphery of barrows where adult individuals were buried. As was the case for the preceding cultures, double and triple graves of small infants (up to two years) are known (see Fig. 9.8). The age structure of the buried population totally differs to that of Sintashta and is more similar to that for the Petrovka Culture (see Table 9.2). Infants up to two years of age constitute around 80% of all those buried in certain cemeteries. Thus, in the Alakul' part of the Stepnoye VII burial ground, the number of such burials reaches 83.6%. At the Kulevchi VI and Alakul' burial grounds a similar situation can be noted (Fig. 9.9). The Alakul' sites located in the northern part of Kazakhstan (Lisakovskyi and Tasty-Butak), however, demonstrate a different age distribution which contains subadults with a broader spread of ages (see Table 9.2).

We can clearly see changes in the nature of burial ritual during the Alakul' period. The number of individual graves sharply increases, in contrast to the number of large multiple tombs which decreases. The treatment of the dead also became different and mortuary ritual is greatly simplified. In both the Sintashta and Petrovka Cultures it is possible a degree of post-mortem selection, which determined who was buried in kurgan cemeteries, was in operation and certain categories of people may have

9. *Bronze Age Child Burials in the Southern Trans-Urals (21st–15th Centuries cal. BC)* 137

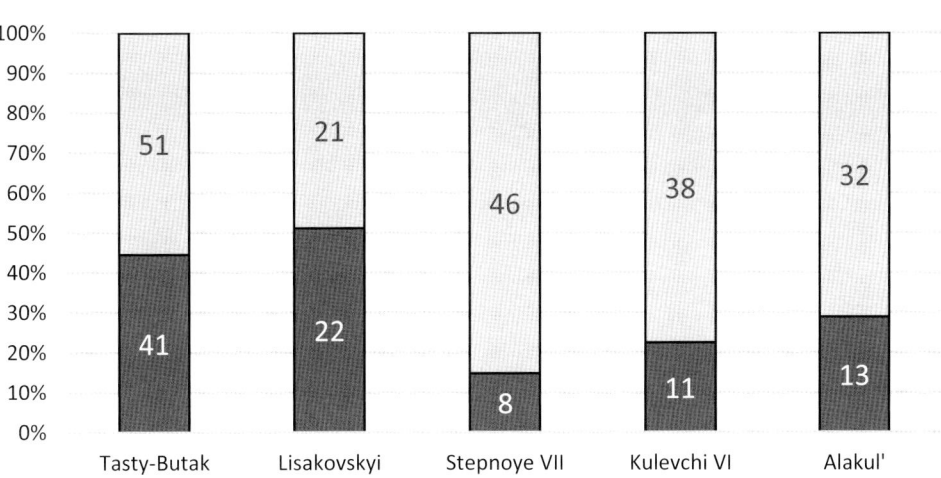

Figure 9.9. The proportions of adults and subadults at key Alakul' sites.

Table 9.5. Correlation of grave goods and age-at-death of subadults (Alakul' Culture).

Grave Goods	Number of subadults	Age-at-death of subadults (years)
Shell	8	6–10
Beads	70	0–15
Astragali	15	1.5–10
Bronze bracelet	37	0–15
Bronze finger-ring	2	6–8
Hair decoration	3	6–12
Drilled fang	10	1–13
Bone knitting needles	1	0–3 months

been afforded alternative ways of interment. The Alakul' model of funerary sites seems to have been horizontally organised and based on age/family relationships. It is probable that Alakul' burial grounds were 'family cemeteries' or were intentionally arranged to look like them.

The burial ritual of Alakul' infants is similar to that for Petrovka and involved interment in small rectangular pits with wooden ceilings. The deceased were laid in a flexed position on the left side. The main type of grave good comprised pottery, with the second most prevalent category consisting of ornaments which were associated with up to 70% of all buried infants (Table 9.5). It is interesting to note that the total number of ornaments and the complexity of their combinations greatly increases in contrast to Petrovka graves (Fig. 9.10). Astragali were also often found in infant pits and in all of these cultures they appear to have been associated with child burial.

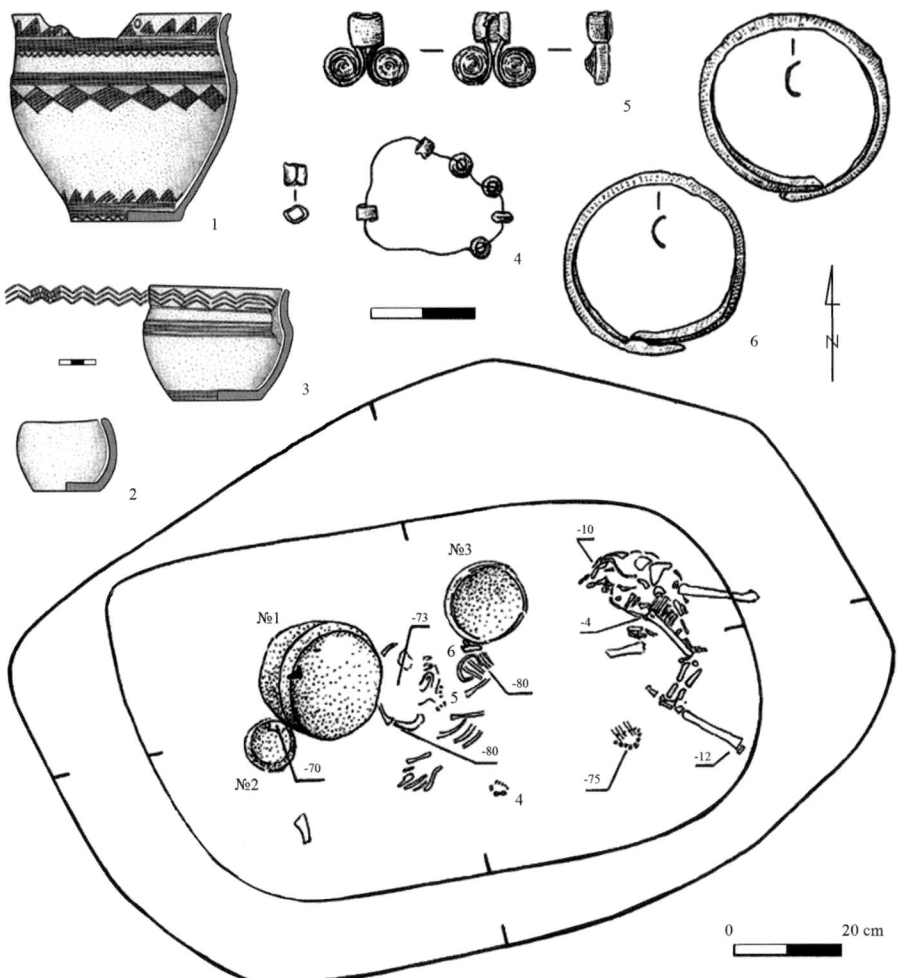

Figure 9.10. Stepnoe VII cemetery (Alakul' Culture). Burial 63. Alakul' child burial (9–12 months old). 1–3 – ceramic vessels; 4 – beads; 5 – bronze pendants; 6 – bronze bracelets (after Kupriyanova and Zdanovich 2015).

In the Sintashta and Petrovka graves astragali were mainly associated with older children and pre-teens, but 90% of astragali were discovered in infant graves in Alakul' mortuary sites. This presumably is related to the fact that the burial populations of the latter group overwhelming comprised infants (see Figs. 9.7 and 9.9). Other types of artefacts are rare, with only one case present which consisted of five bone 'knitting needles' recovered from an infant (0–3 months) burial (Grave 59 in the Stepnoye VII cemetery). Infants were sometimes accompanied by sacrificial deposits represented by the skulls and extremities of sheep. Usually these sacrifices were found in communal children's burials but, in 10–20% of cases, they were recovered from individual graves.

The burials of children generally mirrored those of infants with an absence of tools, although the number of ornaments increased slightly. Children's grave goods predominantly included ornaments related to clothing attachments as well as astragali, amulets and shells. Unfortunately, the graves of pre-teens and adolescents are relatively rare in Alakul' burial grounds which makes it difficult to draw comparisons between them and the other age groups. Pottery was an obligatory element in the Alakul' burial rite and vessels accompanied all adults and subadults regardless of their age. No artefacts (except for pottery) were present in almost half (46.6%) of all subadult individual graves.

Socialisation

It is possible to draw some conclusions concerning the socialisation of children based on the burial data. This is particularly the case for Sintashta sites since tools were included as grave goods and it is therefore possible to estimate the age at which children became first involved in work-related activities. Indeed, the relatively high number of grave goods in general in the Sintashta child burials enables important stages of the life course and gender differences to be identified. Certain features of the burial contexts, however, which cause difficulties for the interpretation should first be noted – namely the collective nature of the graves (55%) in addition to the fact that at least of half of the burials have been disturbed by animals and robbers (perhaps in antiquity). As such, only burials where it is possible to associate artefacts with specific individuals are used in the artefact analysis.

Gender Socialisation

In many traditional societies, gender-differentiated socialisation starts with the birth of a child or very soon after. This is a pan-cultural phenomenon and in all societies boys and girls are socialised in different ways. Children adopt cultural norms of behaviour, gender appearance, values and so on. Gender socialisation by the age of three years is considered normal in contemporary sociology. At this age, the child begins to attribute herself or himself to a certain gender with confidence (so-called *gender identification*) (Bern 2007, 48).

Gender Markers

In many societies, gender distinctions are marked in burials by artefacts that can be traced archaeologically (Binford 1972, 233; Carr 1995, 169–70; Nelson 1997, 58–64; Sorensen 2000 etc.). As a rule, these artefacts are associated with 'femaleness' and 'maleness', and may reflect social roles performed by individuals of different sexes. In burials, especially of children, gender and age symbols can 'cross-cut' each other. For our purpose, it is necessary to distinguish between possible age and gender markers.

Only artefacts that are associated with all age groups of the same sex can be considered to be firm markers of gender. It is logical that it is necessary to determine so-called 'sex markers', for anthropologically-sexed adults first. As mentioned above, clear gender differentiation related to burial location in a barrow has not been observed and both male and female graves were organised along the same principle. There are no systematic differences in the volume and depths of the grave pits of adult males and females.

In the Sintashta Culture, the correlation of the biological sex of the adult dead and their accompanying artefacts allows the following conclusions to be made. Artefacts that were exclusively found in male graves comprise rare categories (e.g. chariots, bronze axes, spearheads), with the exception of bronze adzes and bronze, stone, bone arrowheads which were quite common. Despite the fact that weaponry is typically found in male burials only, arrowheads, parts of bows and cheek-pieces are rarely found in children's tombs (see Table 9.3). There is no doubt that bronze hair ornaments are female attributes. However, some types of ornaments – for example, finger rings or single beads and animal fangs – sometimes belong to male burials. Awls are represented in male, female and child grave goods, whereas sewing needles are usually found in female pits. The arrowhead and the bronze adze can be considered a firm male marker.

Sintashta children received gender-distinctive grave goods beginning at the age of three years. For male children, these gender-linked objects included bronze adzes, arrowheads, the horn parts of a bow and stone tools (abrasives, pestles and so on). For female children, gender-linked objects could be bronze ornaments (bracelets, pendants, hair decorations and beads) but, in general, they do not seem to have been given these objects until the age of 10–11 years old. The girls also received certain types of tools – needles and bone spindle-whorls – after the age of three years, and 16% of child burials contained these items. We can, therefore, assume that the age of approximately three years was the normal age for the start of gender-differentiated socialisation in Sintasha society (see Table 9.3).

One may propose that the Sintasha society did not accentuate the distinctions between boys and girls before they reached the age of three years, at least in the funerary sphere. A child that died prior to three years of age therefore necessarily received gender-neutral grave goods. It should be noted that among the children aged three years and older a large number of burials also included gender-neutral artefacts. As such, it appears to be the case that the adult mourners treated only a small proportion of children as (future) males and females. It is difficult to explain why this attitude was only shown towards certain children and not all and perhaps they came from important families where gender identification was important from an early age.

In Sintashta mortuary practice the *age* of subadults was clearly marked. The graves of children of all ages (and, obviously, both boys and girls) include items rarely associated with adults – astragali (mostly sheep astragali; goat, cattle, horse and pig astragali are represented to a lesser extent), the drilled fangs of small predators, mollus shells and ornaments, such as faience and bronze beads. Astragali are frequently found in child and female burials of the Bronze and Iron Ages in the steppes of Eurasia (Berseneva 2006; Yudin 2009). It is possible that the astragali and mollusk shells were used as toys or play things, while the drilled fangs were amulets or ornaments.

In the Sintashta period the male gender was mainly emphasised in the child burials. 'Adult' type clothes were usually found in female (?) graves of individuals aged 10–11 years of age and older and the graves rarely included large metal ornaments. In the Petrovka and Alakul' child burials the female gender was similarly reflected. As mentioned above, no tools or weaponry were present in the graves and firm male clothing attributes have also not been found. At the Petrovka and Alakul' sites the most complete set of dress decorations (bronze hair decorations, bracelets, finger rings and bronze beads as shoe ornaments) can be found with eight- to nine-year-old Petrovka children, but they occur as early as 9–12 months of age for Alakul' subadults (bronze bracelets, beads and pendants, see Fig. 9.9 and Tables 9.4 and 9.5). Thus, from a very early age the clothing of Alakul' girls looked similar to that of adult females, with the exception of hair decorations and finger rings, the latter of which only appears at approximately four to five years of age. It is difficult to say if girls wore all these ornaments when they were alive but presumably they, and particularly the infants, did not wear the entire set of decorations in everyday life (Kupriyanova 2008, 186–90).

Almost nothing can be said in relation to the gender socialisation of Petrovka and Alakul' boys since they had very little in the way of personal grave goods and clothing ornaments. In this sample (up to 14–15 years) there are no burials with any male gender-linked markers or possessions. The girls received gender-linked accessories much earlier than boys but they, like boys, had no tools.

The tradition of the placement of some dead children in an adult-like 'embrace' position in the grave is most intriguing from a gender point of view. Does it mean that the Bronze Age population practised very early marriages? Biologically, this would have made no sense since there would have been a high risk of the death of one or other of the 'spouses' prior to their fertile ages because of the high rate of child mortality. Accordingly, later nomads of the Ural-Kazakhstan steppes (Kazakhs and Bashkirs) rarely practised very early (9–12 years) marriages. A girl was usually considered to be ready for marriage by 13–15 years of age, but the majority of young people entered into marriage at the age of 15–18 years during the 18th to early 20th centuries AD (Anon n.d.; Asfandiyarov 1989). In the case of the Bronze Age, the burial of children in the 'embrace' position appears to reflect a symbolic concept (about sacred marriage or fertility cult?) because the burials were evidently composed of persons of different sexes (a girl and a boy), and the girls were dressed in clothing with female ornaments.[5] It is a very complicated issue, however, which deserves more careful investigation that is beyond the scope of the current paper.

Labour Socialisation

Archaeologically, the inclusion of children in the economic activities of a community could be reflected in the mortuary record through the presence of tools in their graves. One may suggest that in Sintashta society labour socialisation started around the age of three years in tandem with the development of gender stereotyping. The tools recovered from child burials are mostly of common types and are simple and multifunctional. In the Sintashta *adult* burials many types of metal tools are present, including bronze knives, awls, adzes, needles, chisels and sickles. Stone tools used for ore grinding, polishing and sharpening are also numerous, while spindle whorls, various handles, and harpoon points represented the bone artefacts. The overwhelming majority of adults were buried with tools (73.5%), but *children* (from 3–15 years) were not. Only 29 graves (16.0%) contained metal tools (knives, awls, a sickle and fishhooks) and, in several cases, abrasive stones and bone spindle-whorls. In the adolescent burials (15–17 years) tools were often found (70%), however, in much the same proportion as was the case for adults. The adult tools are identical to those found with children and adolescents, although a greater range of stone artefacts are represented (e.g. a rubber, a pestle and abrasives).

These tools were associated with day-to-day activities and were not miniatures or imitations. One may suggest that the division of labour was organised along the lines of gender and age. In Sintashta child burials (3–15 years) some tools were present but a notable majority of the subadults, whose age exceeded 14 years, had tools as grave goods. The evidence is suggestive that, from the age of 14 years, subadults were fully included in adult work activities and may, in fact, have been considered to be adults. Returning to the younger age group (3–15 years), it should be reiterated that 80% of children did not receive tools as part of their grave gifts although they very probably would have worked during their lifetimes. The presence of tools in these graves can therefore be explained in relation to the value of metal tools, or, alternatively, ideological or sentimental factors.

In summary, the archaeological evidence is suggestive that three stages of child labour socialisation occurred in Sintashta society:

1. Infants and children up to three years – All burials are gender-neutral in terms of grave goods. Tools are completely absent.
2. Children 3–15 years – Burials contained tools and weapons to a small extent.
3. Adolescents of 15–17 years – Tools are represented in the majority of burials, but their variety is less numerous than in adult graves. Individuals in this age group may have been considered to be adults.

As such, it would appear that the subadult category in Sintashta society ended at approximately 14 years of age.

Cross-culturally, ethnographical and historical observations demonstrate that child labour was important in agricultural and pastoral societies (Kamp 2001, 15). In many cases, the contribution of children in economic activity is quite notable and they perform work essential to the economy as well as take part in religious and temporal rites (Rogoff *et al.* 1975, 355, 367). In general, the responsibilities of children tend to parallel those of adult gender divisions. The conclusions concerning labour socialisation in the Sintashta period can perhaps be extrapolated to Petrovka and Alakul' children since these groups followed a similar way of life and lived in the same landscape. Despite the changes in mortuary traditions that occurred after the Sintashta period, the labour socialisation may well have been the same because it was mostly dependent on the economic base of subsistence, i.e. stockbreeding, and a settled way of life.

Conclusions

Common traits throughout the Sintashta, Petrovka and Alakul' periods are the large quantity of subadult burials and, as a rule, their peripheral location in the burial sites. Intramural interments are found at the settlements of all of these cultures. Burial on the left side, with the hands near the face also remain constant throughout these groups. Other aspects of the burials, however, display obvious distinctions. The number of double and multiple child graves sharply decreased in the Petrovka period. Furthermore, the overall energy expenditure on the burial, including the provision of artefacts and animal sacrifices, was gradually reduced to a minimum as well. The grave goods given to children became totally represented by ornaments, astragali and clothing details during the post-Sintashta period.

The nature of child burials appears to have been progressively simplified along a Sintashta–Petrovka–Alakul' continuum, although this is not the case for the amount of ornaments found within the graves of each culture. Nevertheless, the same trend can generally be observed for adult graves. It is thus very important to note that the post-mortem treatment of children, even neonates, was largely the same as for adults.

It would be incorrect to characterise the status of either children or adults in the Ural Bronze Age in terms of 'high/low'. Children, from birth onwards, were incorporated into the social structure and the degree of adult care invested in them was probably very high. In spite of a gradual simplification and transformation of the funerary rites, we may suppose, that the attention given to children remained quite constant within these societies judging from the nature of the mortuary rituals. Children were well-represented in the common cemeteries during all periods and their mortuary treatment was generally the same as that for adults. Children, as well as adults, buried in the 'embrace' position may well have participated (possibly, post-mortem) in very important cult activities of the society. It could even be the case that children took on special significance and value in the Late Bronze Age because the number of child, and especially infant, burials increased in comparison with the Sintashta period.

It is possible to suppose that labour socialisation in Sintashta society started around the age of three years in tandem with the development of gender stereotyping. There is no firm evidence about labour socialisation for the Petrovka and Alakul' cultures. Gender socialisation was reflected better in the Petrovka and Alakul' mortuary sites, especially in the Alakul' society where female gender was marked as early as 9–12 months of age.

It is quite possible that the mortuary ritual of the Sintashta, Petrovka and Alakul' cultures were modelled on different types of social relationships. The Sintashta rite could primarily reflect vertical stratification, whereas the Alakul' and Petrovka rites appear to have been based on family/kinship relations. The increasing number of child burials, and a notable rise in the number of infants among them, would tend to support this interpretation.

Acknowledgements

I would like to thank the editors, Eileen Murphy and Mélie Le Roy, for the invitation to contribute to the book and for editing my text. My special gratitude goes to Karlene Jones-Bley (University of California, Los Angeles) for her help with my English and kind attention to my work. South Ural State University is grateful for the financial support of the Ministry of Education and Science of the Russian Federation (Grant No. 33.5494.2017/BP).

Notes

1. Institute of History and Archaeology, Ural Branch of Russian Academy of Sciences, 16 S. Kovalevskaya St., 620990, Ekaterinburg, Russia. Also South-Ural State University, Department of Eurasian Investigations, 76 Lenina Av., 454000, Chelyabinsk, Russia. Email: bersnatasha@mail.ru.
2. For more details about the Ural Bronze Age see Koryakova and Epimakhov (2007).
3. The Srubnaya (Timber grave) sites occupied a huge area of East Europe. The South Trans-Urals was only their very eastern periphery. I will not consider the Srubnaya Culture here due to word restrictions.
4. Cheek-pieces are not exactly weapons but they are part of so-called 'chariot-complex' discovered in the Sintashta and Petrovka graves. The 'chariot' complex included traces of the chariot (presumably imprints of wheels), horn cheek-pieces and arrowheads and sometimes other weapons. Usually, such a burial was accompanied by the complete bodies of two or more sacrificed horses. It is supposed that the use of these chariots was related to both warfaring and ritual activities. Cheek-pieces are an extremely unusual find for a child burial. For more details see Anthony and Vinogradov (1995); Koryakova and Epimakhov (2007).
5. In all adult anthropologically-identified cases (n=42) investigated to date, such burials belonged to a male (on the left side) and a female (on the right side). The female was usually associated with a large quantity of ornaments (Rafikova 2014).

References

Anon., n.d. *History of Kazakhstan*. Available at: <https://www.nur.kz/859908-tradiciya-rannikh-brakov-u-kazakhov-foto.html>.

Anthony, D. W. and Vinogradov, N. B. 1995. Birth of the chariot. *Archaeology* 48, 36–41.

Asfandiyarov, A. Z. 1989. *Brak i Razvod u Bashkir v XVIII–Pervoi Polovine XIX Veka*. Ufa: Bashkirskyi Gocudarstvennyi Universitet.
Bern, S. M. 2007. *The Social Psychology of Gender*. St. Petersburg: Prime-Evroznak.
Berseneva, N. 2006. Archaeology of children: Sub-adult burials during the Iron Age in the Trans-Urals and Western Siberia, in Gallou, C. and Georgiadis, M. (eds.), *The Archaeology of Cult and Death. Proceedings of the session held in the 9th Annual Meeting of the European Association of Archaeologists, St Petersburg, Russia, 10–14 September 2003*, 179–192. Budapest: Archaeolingua.
Berseneva, N. 2010. Child burials during the Middle Bronze Age of South Urals (Sintashta Culture), in Dommasnes, L. H., Hjorungdal, T., Monton-Subias, S., Sanchez Romero, M. and Wicker, N. L. (eds.), *Situating Gender in the European Archaeologies*, 161–180. Budapest: Archaeolingua.
Binford, L. R. 1972. *An Archaeological Perspective*. New York: Seminar Press.
Carr, C. 1995. Mortuary practices: their social, philosophical-religious, circumstantial, and physical determinants. *Journal of Archaeological Method and Theory* 2, 105–200.
Epimakhov, A. V. 2005. *Rannie Kompleksnye Obshtsva Severa Zentralnoi Evrasii (po Materialam Mogil'nika Kamennyi Ambr-5)*. Chelyabinsk: Chelyabinskii Dom Pechati.
Jones-Bley, K. 1994. Juvenile grave goods in Catacomb Graves from the South Russian steppe during the Eneolithic and Early Bronze Age. *The Mankind Quarterly* 34, 323–335.
Kamp, K. A. 2001. Where have all the children gone? The archaeology of childhood. *Journal of Archaeological Method and Theory* 8, 1–33.
Koryakova, L. N. and Epimakhov, A. V. 2007. *The Urals and Western Siberia in the Bronze and Iron Ages*. Cambridge: Cambridge University Press.
Kostyukov, V. P. and Epimakhov, A. V. 1999. Predvaritelnye itogi issledovaniya mogilnika bronzovogo veka Troizk-7, in Kovaleva, V. T. (ed.), *120 Let Arkheologii Vostochnogo Sklona Urala*, 66–70. Ekaterinburg: Ural'skii Gosudarstvennyi Universitet.
Kupriyanova, E. V. 2008. *Ten Zhenshiny: Zhenskyi Kostyum Bronzovogo Veka kak 'Tekst' (po Materialam Nekropoley Yuzhnogo Zaural'ya i Kazakhstana)*. Chelyabinsk: Avto Graf.
Kupriyanova, E. V. 2016. *Pogrebalnye Praktiki Epokhi Bronzy Yuzhnogo Zaural'ya: Mogilnik Stepnoye-1*. Chelyabinsk: Enziklopediya.
Kupriyanova, E. V. and Zdanovich, D. G. 2015. *Drevnosti Lesostepnogo Zaural'ya: Mogil'nik Stepnoye VII*. Chelyabinsk: Enziklopediya.
Lewis, M. 2011. The osteology of infancy and childhood: Misconceptions and potential, in Lally, M. and Moore, A. (eds.), *(Re)Thinking the Little Ancestor: New Perspectives on the Archaeology of Infancy and Childhood* (BAR International Series 227), 1–13. Oxford: Archaeopress.
Lindstrom, R. W. 2002 Anthropological characteristics of the population of the Bolshekaraganskyi Cemetery, Kurgan 25, in Zdanovich, D. G. (ed.), *Arkaim – Nekropol*, 159–164. Chelyabinsk: Yuzhno-Uralskoye Knizhnoye Izdatelstvo.
Nelson, S. M. 1997. *Gender in Archaeology. Analyzing Power and Prestige*. Walnut Creek, CA: AltaMira Press.
Sorensen, M. L. S. 2000. *Gender Archaeology*. Cambridge: Polity Press.
Rafikova, Ya. V. 2014. Parnye pogrebeniya alakul'skoi kul'tury na Yuzhnom Urale, in Molodin V. I. and Epimakhov A. V. (eds.), *The Aryans in the Eurasian Steppes: The Bronze and Early Iron Ages in the Steppes of Eurasia and Contiguous Territories. Elena Kuz'mina Memorial Volume*, 228–243. Barnaul: Altaiskii Gosudarstvennyi Universitet.
Rogoff, B., Sellers, M., Pirrotta, S., Fox, N. and White, S. 1975. Age of assignment of roles and responsibilities to children: A cross-cultural survey. *Human Development* 18, 353–369.
Rykushina, G. V. 2003. Antropologicheskaya charakteristika naseleniya epokhi bronzy Yuzhnogo Urala po materialam mogil'nika Krivoe Ozero, in Vingradov, N. (ed.), *Mogil'nik Bronzovogo Veka: Krivoe Ozero v Yuzhnom Zauralye*, 345–360. Chelyabinsk: Yuzhno-Uralskoye Knizhnoye Izdatelstvo.
Sal'nikov, K. V. 1952. Kurgany na ozere Alakul'. *Materialy i Issledovaniya po Arkheologii SSSR* 24, 51–71.

Shevnina, I. V. and Logvin, A. V. 2015. *Bronze Age Burial Ground Halvay III in North Kazakhstan* (Materials and Researches on Archaeology of Kazakhstan VII). Astana: Institute of Archaeology.

Shevnina, I. V. and Voroshilova, S. A. 2009. Detskie pogrebeniya epochi razvitoi bronzy (po materialam mogil'nika Bestamak), in Tairov, A. D. and Ivanova, N. O (eds.), *Etnicheskie Vzaimodeistvia na Yuzhnom Urale*, 59–63. Chelyabinsk: Yuzhno-Ural'skii Gosudarstvennyi Universitet.

Sorokin, V. S. 1962. Mogilnik bronzovoi epokhi Tasty-Butak I v Zapadnom Kazakhstane. *Materialy i Issledovaniya po Arkheologii SSSR* 120, 1–207.

Usmanova, E. R. 2005. *Mogilnik Lisakovskii I: Fakty i Paralleli*. Karaganda-Lisakovsk: Karagandinskii Gosudarstvennyi Universitet.

Ventresca Miller, A., Usmanova, E., Logvin, V., Kalieva, S., Shevnina, I., Logvin, A., Kolbina, A. and Suslov, A. 2014. Dental health, diet, and social transformations in the Bronze Age: Comparative analysis of pastoral populations in northern Kazakhstan. *Quaternary International* 348, 130–146.

Vinogradov, N. B. 1984. Kulevch VI – novyi alakul'skii mogilnik v lesostepi Yuzhnogo Zaural'ya. *Sovetskaya arkheologiya* 3, 135–153.

Vinogradov, N. B. 1998. Novye materialy dlya rekonstrukzii oblika odezhdy alakul'skikh zhentshin (po resul'tatam isucheniya mogil'nika Kulevchi VI). *Problemy Istorii, Filologii, Kul'tury* 6, 186–202.

Vinogradov, N. B. 2003. *Mogil'nik Bronzovogo Veka Krivoe Ozero v Yuzhnom Zaural'e*. Chelyabinsk: Yuzhno-Ural'skoe Knizhnoe Izdatel'stvo.

Yudin, A. I. 2009. Pogrebeniya s astragalami iz Novopokrovki II: sluzhiteli kul'ta ili 'igroki'? *Arkheologia Vostochno-Evropeiskoi Stepi* 7, 146–170.

Zdanovich, D. G. 2002. Arkheologiya kurgana 25 Bolshekaraganskogo mogilnika, in Zdanovich, D. G. (ed.), *Arkaim – Nekropol*, 17–110. Chelyabinsk: Yuzhno-Ural'skoe Knizhnoe Izdatel'stvo.

Chapter 10

Juvenile Burial and Age as a Social Category in Funerary Contexts of Pre- and Protopalatial Crete

Nathalja Calliauw[1]

Abstract: Attention for juvenile burials has risen considerably in archaeological research, but they remain remarkably invisible in the archaeology of Bronze Age Crete. This finding can be explained as a consequence of a traditional focus by archaeologists on socio-political organisation, leaving little room for juveniles. In addition, many older excavations of funerary contexts had little regard for human remains, which were neither fully excavated nor analysed by osteoarchaeologists. This paper reviews the characteristics of juvenile burials in Pre- and Protopalatial Crete, and compares their features to those of adult burials. The substantial variability in funerary practice throughout the island, related to juvenile burial, will be discussed. It will explore if age was expressed as a social category in death, which may hint at specific attitudes and social roles connected to young age.

Keywords: Crete, Early and Middle Bronze Age, juvenile burial, social age

Introduction

Attention for juveniles in the archaeological record and their potential roles in society has increased considerably in archaeological studies since the turn of the century, exemplified by numerous conference sessions and publications (e.g. Sofaer Derevenski 2000; Kamp 2001; Baxter 2005; Pomadère 2007; Lebegyev 2009; Georgiadis 2011; Lally and Moore 2011; Smith 2011; Coskunsu 2015; this volume). This increasing attention has, however, had a limited impact on the archaeology of Bronze Age Crete (exceptions being Pomadère 2007; McGeorge 2012). While juveniles do feature in iconographical studies,[2] centring on a number of famous depictions such as the ivory figurines of Palaikastro (Rutter 2003, 37–8), the Taureador fresco of Knossos (Rutter 2003, 42–3) and several frescoes of Akrotiri (Davis 1986; Doumas 2000; Rutter 2003, 44),[3] they are conspicuously under-represented in studies of archaeological contexts. Depictions

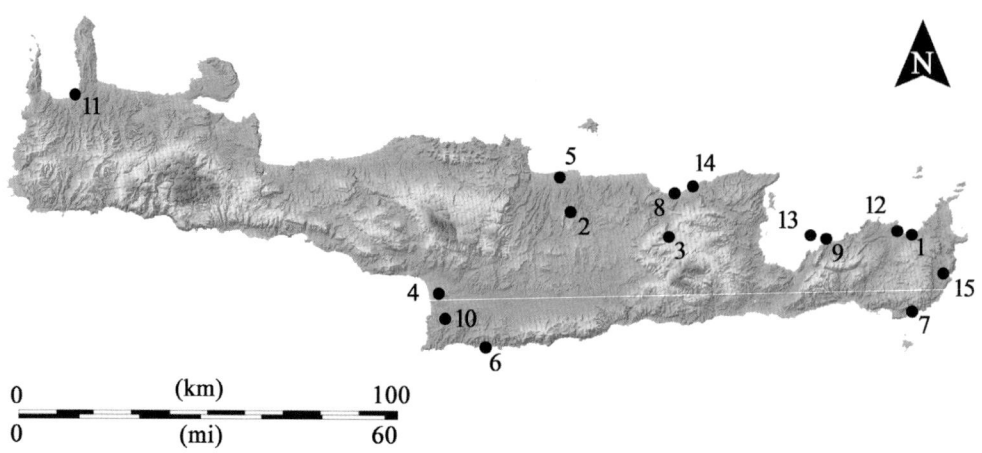

Figure 10.1. Map of Crete with the location of the sites mentioned in the text: 1. Agia Photia, 2. Archanes, Phourni, 3. Hagios Charalambos, 4. Kamilari, 5. Knossos, 6. Lebena, 7. Livari Skiadi, 8. Mallia, 9. Mochlos, 10. Moni Odigitria, 11. Nopigeia, 12. Petras, 13. Pseira, 14. Sissi, 15. Zakros.

may give some impressions of age as an element of social organisation and identity, but they are an idealised source of information with a strong focus on ritual scenes, thereby omitting information about daily life. Iconographical sources thus only provide a limited window on 'age', which may be broadened by further archaeological inquiry. This paper will focus on the available data from Pre- and Protopalatial (i.e. Early Minoan I–Middle Minoan II, c. 3100–1750/1700 BC[4]) funerary contexts (Fig. 10.1 and Table 10.1) to explore potential attitudes towards young age and the expression of age as a social category in death.

Juveniles in Cretan Bronze Age Archaeology

Until quite recently juveniles were largely absent from studies of Cretan archaeological contexts. The main 'excuse' provided to explain this omission is the rather poor state of preservation of many funerary assemblages – as a result of secondary funerary practices, taphonomic processes and looting – and the resultant low visibility of juvenile remains. It is likely that the traditional focus of Minoan archaeology on political and socioeconomic organisation, however, is equally, or even more, responsible for their under-representation. This focus has, furthermore, resulted in a lack of interest in relation to human remains which were often excavated without much care and recording, and published without detail (Pomadère 2007, 295; Triantaphyllou 2009, 19; Schoep in press; Triantaphyllou in press). These factors resulted in a perceived absence of juveniles from the Bronze Age Cretan record, or at least from the record created by archaeologists.

More recent excavations, however, have shown that juveniles can sometimes comprise a considerable part of the funerary record. An example is the cemetery of

Table 10.1. Selected burial contexts used in the paper: some excavation reports do not offer an exact Minimum Number of Individuals (MNI) or number of juvenile burials; the burial mode is sometimes unclear because of heavy disturbance or old excavation practices; the majority of the publications use 18 years as the upper limit for the juvenile age group (in most other cases – especially in older publications – age limits are not specified). Where possible, details of more precise age estimates will be included in the text.

Burial context	Date	MNI	Juveniles	Burial mode of juveniles
Archanes, Phourni, Burial Building 19, layer 1	EM II–MM IA	105	2	primary inhumation
Archanes, Phourni, Burial Building 19, layer 2	MM IB–MM II	76	12	primary inhumation, some in containers
Archanes, Phourni, *Tholos* Gamma	EM III–MM I	30	4	closely associated with *larnakes* and *pithos*
Hagios Charalambos Cave	EM III–MM IIB	c. 400	unclear	secondary
Kamilari, *Tholos* A	MM IB–LM III	134	30	unclear
Lebena, Tomb 1	MM I	50	1	secondary
Livari Skiadi, *tholos*	EM IB–EM III	84	7	secondary
Mallia, Pières Meulières	MM I	unclear	unclear	'several' juveniles in containers
Moni Odigitria, *Tholos* A	EM I–EM II	133	34	primary/secondary
Moni Odigitria, *Tholos* B	EM III–MM IB	64	4	secondary
Nopigeia, settlement	EM IIA	1	1	container
Petras Kephala rock shelter	EM I–MM IB	165	36	secondary
Pseira, Tomb 2	MM II (?)	15	2	unclear
Pseira, Tomb 8	MM II	unclear	2	unclear
Pseira, Tomb 9	EM I–MM II	unclear	-	-
Pseira, Tomb 10	MM I–MM II	unclear	2	unclear
Sissi, Room 1.7	MM IB–MM II	7	3	primary inhumation
Sissi, Room 1.9	-	13	3	secondary
Sissi, Room 1.11–12	EM IIA	11	6	container
Sissi, Room 1.18	MM IB–MM IIA	2	1	container
Sissi, Room 1.30–31	MM IIA	16	-	-
Sissi, Room 9.1, upper layer	MM II	7	-	-
Zakros, Pezoules Kephala, Enclosure A	MM I	74–81	1	unclear

Sissi, on the northern coast of the island, which was used between EM II and MM II (c. 2650–1750/1700 BC; Schoep 2009; Schoep et al. 2011; Schoep et al. 2013). The Sissi cemetery was excavated according to the rules of 'archaeothanatology' or 'field anthropology' (Duday 2009; Crevecoeur et al. 2015), a method which results in the recording of detailed information on the human remains and their context, and which has proven to be particularly beneficial for the study of juvenile burials. The first advantage is the presence of osteoarchaeologists during the excavation. While this may seem self-evident, their presence in the field was, until recently, rare in Cretan Bronze Age archaeology and is still not part of standard practice. As the remains of juveniles are often small and fragile, the presence of 'trained eyes and hands' is thus of indisputable importance, especially in burial environments such as those on Crete where the state of preservation is often poor. It is thus conceivable that much information on juvenile (and other) burials has been lost in the past. Another advantage of archaeothanatology is that the first analyses occur on site when the remains are still *in situ*. This partly remedies problems related to poor preservation, since bones often fall apart when they are being lifted from the soil, thereby complicating further analysis. A final advantage is the detailed nature of the associated spatial recording which results in the production of high quality contextual information. Several other recent excavations, including those at Moni Odigitria (Triantaphyllou 2010; Vasilakis and Branigan 2010), Petras (Triantaphyllou 2009; Tsipopoulou 2012; 2017) and Livari Skiadi (Papadatos and Sofianou 2015), characterised by the careful analysis of their human skeletal assemblages, have also provided new information concerning the presence and burial treatment of juveniles. These recently acquired data enable the study of age as a potential element of social organisation.

Juvenile Burial in Pre- and Protopalatial Crete

The funerary landscape of Pre- and Protopalatial Crete is marked by great variety, and includes large circular tombs or *tholoi*, rectangular tombs or 'house-tombs', as well as caves and rock shelters (Soles 1992; Branigan 1993; Legarra Herrero 2009; 2014; Schoep in press). These make up the majority of the funerary contexts in this period and were used for collective burial, i.e. the deposition of multiple individuals in the same space over successive episodes. A few Early Minoan sites along the northern coast, such as Agia Photia (Davaras and Betancourt 2004) and Pseira (Betancourt and Davaras 2003), show a different type of burial, comprising one or a few deposits in cist graves or small rock-cut tombs, which seem to have resulted from the strong Cycladic influence that occurred in this region during the Early Bronze Age (e.g. Davaras and Betancourt 2004, 239; Legarra Herrero 2009, 38–9). Such variation in burial types and practices is likely to be a manifestation of differing attitudes towards death, and may also be reflected in the treatment of juveniles.

Location
The funerary landscape of Pre- and Protopalatial Crete, with its strong collective character, shows no clear age-based spatial segregation. In general, juveniles are buried in close association with adults, i.e. in the same structure, room or container, or even mixed with adult bones in larger collective contexts with a high degree of

disturbance as well as in secondary burial deposits. In Sissi, for example, the vast majority of the juveniles were deposited in the proximity of adults, such as in Room 1.7, where two juveniles of three to five years old were found in direct association with an adult (Crevecoeur and Schmitt 2009, 63–7; Crevecoeur *et al.* 2015, 291–2). The limited data currently available for smaller burial types indicates the same lack of age-based spatial segregation. In Pseira, for example, several fragmentary juvenile bones were found in association with adult bones (Betancourt and Davaras 2003, 155–61). One exceptional juvenile burial (*c.* three years old) has been found in Nopigeia (EM IIA, *c.* 2650–2450/2400 BC), inside a settlement context (Pomadère 2007, 282–4).[5] While this practice is known on Crete during the Neolithic (with a few examples at Knossos) and reappears in the Late Minoan period, with examples in Petras (LM IA, *c.* 1700/1675–1625/1600 BC) and Knossos (LM IA–IIIC, *c.* 1700/1675–1075/1050 BC) (McGeorge 2012), it does not seem to have been a common practice during the Pre- and Protopalatial period.

Secondary Burial

Secondary funerary practices, which include the clearance, relocation and reorganisation of bones, were rather prevalent in all types of Pre- and Protopalatial burial contexts (Branigan 1993; Papadatos 2005; Crevecoeur *et al.* 2015; Triantaphyllou in press). Combined with the collective character of most contexts, this has often resulted in mixed masses of bone, which not only impedes the identification of juvenile burials, but also erases our knowledge of any potential differentiation that may have occurred between burials in the primary stage. Burial contexts with a less collective character such as those at Agia Photia and Pseira could help in this respect but, unfortunately, their human remains have been heavily disturbed or lost as a result of old excavation practices (as in the case of Pseira – Betancourt and Davaras 2003, 153) or still await publication (Agia Photia). In some of the more recent excavations of secondary deposits, however, individuals of all ages have been identified. In the cemetery of Sissi, for example, the secondary deposit found in Room 1.9 contained the remains of at least 13 individuals, including three juveniles (one neonate and two of approximately two to five years old – Crevecoeur and Schmitt 2009, 77–86; Schoep 2009, 53; Crevecoeur *et al.* 2015, 292–4). Juveniles were also found in the large secondary deposits discovered in the Hagios Charalambos Cave (EM III–MM II, *c.* 2200–1750/1700 BC; Betancourt *et al.* 2008) and the rock shelter of Kephala Petras (EM I–MM IB, *c.* 3100–1875/1850 BC; Triantaphyllou 2012). There is still much debate as to the meaning of secondary burial practices, but several scholars see it as a way to erase individual identity and pass over to the collective ancestors (e.g. Soles 1992, 246–8; Branigan 1993, 120–1; Legarra Herrero 2012, 348). If this were the case, the examples provided here are indicative that age was not a criterion in this passage. Some secondary deposits show an under-representation of juveniles, however, such as the deposit of the Livari Skiadi *tholos* (EM IB–EM III, *c.* 2900–2100/2050 BC; Triantaphyllou 2009) and *Tholos* B at Moni Odigitria (EM III–MM IB, c. 2200–1875/1850 BC; Triantaphyllou 2010). This under-representation could be due to poor preservation of these collective secondary deposits, however, in addition to the small size of juvenile bones, making them less prone to be selected for re-deposition. On the other hand, it is possible that the lack of juvenile bones was in some cases the result of a deliberate choice, but whether this was related to the secondary practices is unclear (the possibility of deliberate exclusion will be discussed further below).

Burial in a Container

From the late Early Bronze Age onwards we see the appearance of ceramic burial containers in Cretan funerary contexts (Soles 1992; Branigan 1993; Legarra Herrero 2014). Both juveniles and older individuals were occasionally deposited in containers, and sometimes even in the same vessel. There are, however, a few differences to be noticed. While adults are only found in the most recurrent types, i.e. *pithoi* (large storage vases) and *larnakes* (rectangular or oval 'tubs'), juveniles were also deposited in other types of vessels such as jars, cooking pots and smaller storage vases. The

Figure 10.2. Sissi cemetery, Room 1.11–12: left: two container burials of neonates (top) and the primary inhumation of three adults and one adolescent; right: four more vessels containing the remains of neonates, two of which are partly covered by stones (bottom). The fourth adult was already removed at the time of the photo (published in Schoep 2009, 53; © Sarpedon - Sissi Archaeological Project).

majority of these juveniles were younger than three years old, and the use of smaller types of container may simply be linked to their smaller stature. The use of different vessel types may also have a chronological explanation, as the appearance of larger burial containers – and adult container burials – is dated to EM III (2200–2100/2050 BC). For most of the Early Minoan period, burial in a ceramic container thus seems to be reserved for juveniles.[6] In the only Early Minoan burial structure in Sissi identified to date, i.e. Room 1.11–12 (Fig. 10.2; EM IIA, *c.* 2650–2450/2400 BC), for example, six jars were found which contained the remains of neonates, alongside the primary deposit of four adults and one older juvenile ('less than 20 years old'), none of which were associated with a ceramic container (Crevecoeur and Schmitt 2009, 86; Schoep 2009, 53–4; Schoep *et al.* 2011, 44–7; Crevecoeur *et al.* 2015, 288–9). It is possible, however, that older (and bigger) individuals were wrapped in perishable materials during this early period. During the late Early and early Middle Minoan period, larger ceramic burial containers appear in funerary contexts and were used for individuals of all ages. It would thus appear to be the case that neither the type of vessel nor the actual practice of burial in a container was an important criterion of differentiation – in this period, at least. As the Middle Minoan period progresses, however, a decline in the number of juvenile container burials is apparent and this phenomenon will be returned to in the discussion.

Burial Gifts

The high prevalence of secondary burial practices – along with the fact that most contexts contained multiple individuals – makes the association between specific burials and material culture very difficult, if not impossible. Overall, there is no clear age-based differentiation (Pomadère 2007, 354–5). Most juvenile burials did not contain any objects that could be linked to them with certainty, but the same can also be said about the majority of adult burials. It is interesting, though, that no neonate burials have been discovered to date with which an object was associated with certainty. As the majority of neonates have been found inside a container, the identification of associated objects should have been straightforward. The fact that no such associations have been published may therefore be meaningful, although the lack of attention shown towards juvenile remains in older excavations and publications warrants caution. The majority of recent projects, furthermore, concern large collective and secondary deposits – where no associations are possible – and it may therefore be too early at this point to truly identify any differentiation in burial practice.[7] When associated objects have been identified there has been no clear distinction in type. The only possible association with young age is the presence of shells in juvenile burials, an association that is also known from other contexts in the Aegean (Pomadère 2007, 251–3, 359–60). On Crete, several juvenile burials were found with a shell, or shell necklace, directly associated with them, such as one of the juveniles recovered from Room 1.7 at Sissi (Crevecoeur and Schmitt 2009, 66–7; Schoep 2009, 51) and several in Archanes, Phourni (Sakellarakis and Sapouna-Sakellaraki 1997, 194, 212–3).

Representation

Although recent excavations have gone some way towards countering the perceived absence of juvenile burials, they nevertheless still seem to be under-represented in some cases. The reasons for this under-representation are not always easy to assess since it could be related to the fragility of their bones, taphonomic processes, a lack of care and attention shown towards human remains retrieved during older excavations or actual exclusion. The latter is suggested by some more recently excavated contexts such as the cemetery of Moni Odigitria (Vasilakis and Branigan 2010). In *Tholos* A (EM I–II, *c.* 3100–2200 BC) juveniles are well represented (n=34/133), but only four juveniles (one neonate, one of 1–6 years and two of 6–12 years) were found in *Tholos* B (EM III–MM IB, *c.* 2200–1875/1850 BC) among 60 adults (Triantaphyllou 2010; in press). Poor preservation does not suffice as an explanation in this case, as the material in *Tholos* B was better preserved according to the excavators (Triantaphyllou 2010). The possibility of exclusion from, or limited access to, the tomb for younger individuals thus becomes more plausible. A similar under-representation (seven juveniles and 77 adults) has also been noted for the *tholos* of Livari Skiadi (EM IB–EM III, *c.* 2900–2100/2050 BC; Triantaphyllou 2009; in press). In Burial Building 19 of Archanes, Phourni, two burial layers were identified (Sakellarakis and Sapouna-Sakellaraki 1997, 218–20; Maggidis 1998). The oldest layer (EM II–MM IA, *c.* 2650–1925/1900 BC) showed a clear under-representation of juveniles (n=2/105), while they were better represented (n=12/76) in the upper layer (MM IB–II, *c.* 1925/1900–1750/1700 BC). Finally, while the cemetery of Sissi shows a general balance between the number of juveniles and older individuals, some contexts dated to MM II (*c.* 1875/1850–1750/1700 BC) stand out through their distinct under-representation or even lack of juveniles. A large pottery deposit from Rooms 1.30–1, for example, contained the selected (secondary) remains of at least 16 adults, mixed up with pottery (Schoep *et al.* 2011, 59–61). The absence of juveniles may be explained by the secondary nature of this deposit, but the presence of several juveniles in the secondary deposit of Room 1.9 (Crevecoeur and Schmitt 2009, 77–86; Schoep 2009, 53) is suggestive that their absence from Rooms 1.30–1 may have a different explanation. Another example is Room 9.1, which was used in MM II for the primary inhumation of seven adults and no juveniles (Schoep *et al.* 2011, 64; Schoep *et al.* 2013, 40–1). We will return to this phenomenon of under-representation in the next section.

Discussion

The study of age as an element of social organisation has often focused on the identification of 'children' in burial contexts (e.g. Crawford 2000; Houby-Nielsen 2000; Baxter 2005, 97–8; Gowland 2006). At first glance the data from Pre- and Protopalatial Crete often seem to defy such strict categorisation, however, as a cursory overview does not show a lot of age-based differentiation and sometimes leaves us with the question of whether age groups can be identified at all. But the regional, or even local, character of burial practices in this period calls for an equally local focus in the study of juvenile burials and the potential role of age as a social category. Sadly, at this point in time, only a few sites have been excavated with sufficient care to enable a detailed analysis to be undertaken of the human remains, while the publication of some burial sites is still pending. As a result, information on juvenile burials of all burial types, or for all regions of the island, is currently unavailable.

Data for EM I (*c.* 3100–2650 BC) juvenile burials are, unfortunately, almost non-existent. Most EM I contexts were either excavated in the early 20th century, or their material has been heavily disturbed by later burial activity and secondary practices. Many EM II (*c.* 2650–2200 BC) contexts have been similarly affected. The EM II layer of *Tholos* Gamma at Archanes, Phourni, was, for example, removed by later burial (Papadatos 2005, 58), while the human remains of Pseira and Mochlos were poorly documented or even lost after their initial excavation. The few EM I–II contexts with documented juvenile burials clearly demonstrate the rich variety that is typical for this period. In Sissi on the northern coast, six neonates were found in an EM II house-tomb, all in a container, while four adults and one older juvenile were placed on the floor (see above and Fig. 10.2; Crevecoeur and Schmitt 2009, 86; Schoep 2009, 53–4; Schoep *et al.* 2011, 44–7; Crevecoeur *et al.* 2015, 288–9). In Nopigeia, also on the northern coast (west Crete), a three-year-old juvenile was found in a container, albeit in the settlement. This is a rather exceptional case and might be explained by the close proximity to the Greek mainland where this practice was more prevalent (McGeorge 2012, 295). In the south, juveniles were well represented in *Tholos* A of Moni Odigitria (Triantaphyllou 2010), where they are indistinguishable from adults, while they are under-represented in the *tholos* of Livari Skiadi (south-east; Triantaphyllou 2009). These four sites thus show four distinct burial treatments for juveniles. At Moni Odigitria and Sissi they are well represented but, while no age-based differentiation can be discerned in Moni Odigitria A, the neonates at Sissi are clearly distinguished by their burial in a container. While the exclusive use of containers for infant burials may establish them as a separate age group, their inclusion in the burial context implies that they were an active part of the group and burial community. This might indicate that inclusion in the burial context was primarily defined by kinship, which has been suggested elsewhere (e.g. Soles 1992, 254; Maggidis 1998, 95). The different burial treatment may indicate a specific attitude related to their young age, which did not affect their inclusion. Containers call up the image of cover and protection, and could thus potentially be linked with a caring attitude. This explanation seems to fit well with the interpretation of burial contexts as family tombs. The *tholos* of Livari Skiadi tells a very different story, however, since an under-representation of juveniles was noted (Triantaphyllou 2009), implying that some age-based selection (see above) may have been practised in this Early Minoan context.

EM III contexts are often hard to distinguish from MM I contexts, and are hence considered together (*c.* 2200–1875/1850 BC). In north-central and eastern Crete, juveniles are well represented in the secondary deposits of the Kephala

Petras rock shelter (Triantaphyllou 2012, 164) and the Hagios Charalambos Cave (Betancourt *et al.* 2008, 578), as well as in the house-tombs of Sissi (Schoep *et al.* 2011; Schoep *et al.* 2013; Crevecoeur *et al.* 2015). Following the reasoning from the previous paragraph, this could indicate that kinship was still the main relationship expressed in death. This hypothesis seems to be strengthened by several close associations between adults and juveniles in the Sissi cemetery, as well as by the two juveniles of Room 1.7, which show several similar morphometric traits hinting at a

Figure 10.3. Sissi cemetery, Burial Room 1.18, showing the only juvenile container burial of the MM IB-MM II period found thus far (published in Schoep et al. 2013, 34; © Sarpedon - Sissi Archaeological Project).

possible genetic relationship (Crevecoeur et al. 2015, 291). The clear differentiation of neonates – noted in the EM II context of this site – has disappeared, and the use of containers for juveniles has declined, while they are frequently used for adults.[8] Only one juvenile (under nine years old) of the 30 known for the EM III – MM II period (c. 2200–1750/1700 BC) was found inside a container (a small storage vase, Room 1.18 (Fig. 10.3); Schoep et al. 2013, 34) and one (four to eight years) in close association with a *pithos* (probably removed for a later burial; Schoep et al. 2011, 51–2). This remarkable scarcity of juvenile container burials compared with the number of adult container burials could possibly hint at a shifting age differentiation, which no longer differentiates among juveniles, but rather between juveniles and older individuals. This potential age-based differentiation in burial mode is less apparent in other contexts in the vicinity where juvenile container burials have been reported as, for example, in Mallia (van Effentere 1963, 92–5) and Phourni, Archanes (Sakellarakis and Sapouna-Sakellaraki 1997). On the other hand, two of the burial structures in Archanes – *Tholos* Gamma and Burial Building 19 – reveal a distinct under-representation of juveniles in this period (Maggidis 1998; Triantaphyllou 2005), while for the other contexts no exact numbers were provided in the publication. Clear under-representation of juveniles can also be noted in the secondary deposits of *Tholos* B at Moni Odigitria (Triantaphyllou 2010) and of Tomb 1 in Lebena (Alexiou and Warren 2004, 12) in the south, as well as in Enclosure A of Pezoules Kephala, Zakros (Becker 1975, 274), in the east. The data currently available thus seem to imply that under-representation of juveniles becomes more frequent all over the island and in different burial types.

Under-representation may sometimes be explained by poor excavation and/or preservation, the latter being quite typical for the larger collective deposits with a secondary character, but most of the examples included here suggest that age may have been a distinguishing factor, leading to the deliberate exclusion of a younger age group from the burial context or from secondary practices. This could imply that inclusion in the burial space was somehow connected to a social position acquired with older age, and that age was a factor of social organisation in these communities. Unfortunately, with the data currently available, we cannot assess at what point in the life course this potential transition may have occurred, as published ages tend to be rather vague – often limited to terms such as 'infant' or 'child'[9] – and the analysis of the human remains from several other contexts still awaits publication. We also currently lack additional biological data which may help us assess whether this potential transition coincided with a developmental change, such as the transition from crawling to walking and the process of weaning. As chronological age did not necessarily have meaning in the Bronze Age, such phases in the juvenile's development were more likely to effect a transition in social age (*cf.* Perry 2006, 92; Halcrow and Tayles 2011, 348–9). Exclusion of a younger age group also seems to imply that kinship or family ties were not the only, or most important, relationship materialised in these funerary contexts.

Few juvenile burials from MM II (c. 1875/1850–1750/1700 BC) are known, which is mainly explained by the fact that many funerary contexts were abandoned during this period, resulting in a very limited overall dataset. Some contexts with a recorded MM II occupation, such as Sissi, again show an under-representation of juveniles that may be explained by deliberate exclusion. This may also have been the case at Pseira, but here the human remains are too fragmentary to enable any conclusions to be made. It would thus seem that the trend initiated in the early Middle Minoan period saw some continuation. On the other hand, juveniles are well represented at *Tholos* A at Kamilari in the Mesara (Triantaphyllou and Girella in press), demonstrating that local/regional variation in the treatment of juveniles remained.

Conclusions

Through a description of juvenile burials in Pre- and Protopalatial contexts, this paper aimed to explore the attitudes shown towards young age in death and the potential role of age as an element of social organisation. The rich variety of the funerary landscape in this period found its reflection in the treatment of juveniles, which differs between sites and changed through time. In some contexts, juveniles can hardly be distinguished from older individuals in terms of burial treatment. This apparent absence of age-based differentiation does not signify that age was not an element of social organisation, but rather that other social roles or identities were chosen to be materialised in death, such as kinship, which seems plausible in several occasions. Future studies, including DNA analysis, will hopefully help to either confirm or refute this hypothesis. In other cases, age could be recognised as a differentiating factor in the burial context, either in the form of variation in the burial treatment or in the under-representation or absence of a younger age group. While the former may indicate that a different attitude was shown towards a certain age group, such as extra care for infants in the Early Minoan period, the latter seems to imply a different social role. Whatever the rules for inclusion in the burial context, in some cases these seem to have excluded young people.

The current data are rather restricted in terms of age assessment and additional biological information, because of poor preservation and/or old excavation on the one hand, and the ongoing analysis of human remains on the other. It is therefore impossible, at the present time, to establish what point in the life course and in a juvenile's biological development may have been linked to different burial treatment or inclusion. There are, furthermore, still some regional and chronological gaps in our current understanding of juvenile burial, which will hopefully be filled by additional data in the future. In the meantime, we can conclude that age was expressed as a differentiating factor in several Pre- and Protopalatial burial contexts, possibly hinting at different attitudes and social roles connected to young age.

Acknowledgements

I would like to thank Dr Eileen Murphy and Dr Mélie Le Roy for the organisation of the EAA session, as well as the anonymous reviewer who offered several helpful suggestions on the first draft of this paper. This research follows from my MA thesis and is part of my doctoral research at the University of Leuven, Belgium, funded by the Research Foundation Flanders. I am grateful to my supervisors Professor Ilse Schoep and Dr Peter Tomkins for their comments and suggestions, and to Dr Frank Carpentier for his assistance with the illustrations. The 2007–11 Sissi excavations were conducted under the auspices of the Belgian School at Athens, under the direction of J. Driessen (UCL) and in collaboration with the KU Leuven. The human remains were excavated and analysed under the direction of I. Crevecoeur (Université de Bordeaux/CNRS) and A. Schmitt (Aix-Marseille Université/CNRS/EFS).

Notes

1. KU Leuven/Research Foundation-Flanders, Blijde Inkomststraat 21, bus 3313, 3000 Leuven, Belgium. Email: nathalja.calliauw@kuleuven.be.
2. Young people have been identified on the basis of hairstyle, which has been used to discern several age categories. The majority of studies are concerned with the kind of activities they are involved in, most of which seem to be ritual in nature (Marinatos 1984, 78–84; Davis 1986; Koehl 1986; Doumas 2000; Chapin 2007; Rehak 2007).
3. Though this site is located on Thera, to the north of Crete, the style and technique of the frescoes have been labelled 'Minoan', and therefore the depictions are also referred to when addressing Minoan age and identity.
4. The dates used in the paper were taken from Manning (2010, 23, tab. 2.2); in the remainder of this paper, the periods will be abbreviated as EM (Early Minoan), MM (Middle Minoan) and LM (Late Minoan).
5. This practice of intramural juvenile burial is quite common in the Near East, Anatolia, the Greek mainland and the Cyclades from the Neolithic onwards (McGeorge 2012; several papers in this volume).
6. The practice of juvenile container burial goes back to the Neolithic in Crete and the wider Aegean (see Pomadère 2007; several papers in this volume).
7. The publication of the human remains from Agia Photia and their associated objects may provide additional data, although the Cycladic character of this context sets it apart from others.
8. The increased use of containers for (adult) burial has been interpreted as an indication for rising individualism, and hence an increasing importance of personal status in the burial context, linked to a diminishing importance of kinship ties (e.g. Branigan 1993, 65–6, 141). This interpretation of container burial has, however, met with several criticisms (e.g. Hamilakis 2013, 149; Legarra Herrero 2014, 162–3).
9. For many burial contexts, the precise assessment of age is almost impossible because of the disorderly state of the remains.

References

Alexiou, S. and Warren, P. 2004. *The Early Minoan Tombs of Lebena, Southern Crete*. Sävedalen: Paul Åströms Förlag.

Baxter, J. E. 2005. *The Archaeology of Childhood: Children, Gender and Material Culture*. Walnut Creek, CA: Altamira Press.

Becker, M. J. 1975. Human skeletal remains from Kato Zakro. *American Journal of Archaeology* 79, 271–276.

Betancourt, P. P. and Davaras, C. 2003. *Pseira VII. The Pseira Cemetery 2. Excavations of the Tombs.* Philadelphia, PA: INSTAP Academic Press.

Betancourt, P. P., Davaras, C., Stravopodi, E., Panagiotis, K., Langford-Verstegen, L., Muhly, J. D., Hickman, J., Dierckx, H., Ferrence, S. C., Reese, D. S. and McGeorge, P. J. P. 2008. Excavations in the Hagios Charalambos Cave. A preliminary report. *Hesperia* 77, 539–605.

Branigan, K. 1993. *Dancing with Death: Life and Death in Southern Crete, c. 3000-2000 B.C.* Amsterdam: Adolf M. Hakkert.

Chapin, A. P. 2007. Boys will be boys: youth and gender identity in the Theran frescoes, in Cohen, A. and Rutter, J. (eds.), *Constructions of Childhood in Ancient Greece and Italy* (Hesperia Supplement 41), 229–255. Athens: The American School of Classical Studies at Athens.

Coskunsu, G. 2015. *The Archaeology of Childhood: Interdisciplinary Perspectives on an Archaeological Enigma.* New York: State University of New York Press.

Crawford, S. 2000. Children, grave goods and social status in Anglo-Saxon England, in Sofaer Derevenski, J. (ed.), *Children and Material Culture*, 169–179. New York: Routledge.

Crevecoeur, I. and Schmitt, A. 2009. Etude archéo-anthropologique de la nécropole (zone 1), in Driessen, J., Schoep, I., Carpentier, F., Crevecoeur, I., Devolder, M., Gaignerot-Driessen, F., Hacigüzeller, P., Jusseret, S., Langohr, C., Letesson, Q. and Schmitt, A., *Excavations at Sissi. Preliminary Report on the 2007-8 Campaigns*, 57–94. Louvain-la-Neuve: Presses Universitaires de Louvain.

Crevecoeur, I., Schmitt, A. and Schoep, I. 2015. An archaeothanatological approach to the study of Minoan funerary practices. Case-studies from the Early and Middle Minoan cemetery at Sissi. *Journal of Field Archaeology* 40, 283–299.

Davaras, C. and Betancourt, P. P. 2004. *The Hagia Photia Cemetery I: The Tomb Groups and Architecture* (Prehistory Monographs 14). Philadelphia, PA: INSTAP Academic Press.

Davis, E. 1986. Youth and age in the Thera frescoes. *American Journal of Archaeology* 90, 399–406.

Doumas, C. 2000. Age and gender in the Theran wall paintings, in Sherratt, S. (ed.), *The Wall Paintings of Thera, Vol. 2*, 971–980. Piraeus: Thera Foundation.

Duday, H. 2009. *The Archaeology of the Dead: Lectures in Archaeothanatology.* Oxford: Oxbow Books.

Georgiadis, M. 2011. Child burials in Mesolithic and Neolithic Southern Greece: a synthesis. *Childhood in the Past* 4, 31–45.

Gowland, R. 2006. Ageing the past: examining age identity from funerary evidence, in Gowland, R. and Knüsel, C. (eds.), *Social Archaeology of Funerary Remains*, 143–154. Oxford: Oxbow Books.

Halcrow, S. and Tayles, N. 2011. The bioarchaeological investigation of children and childhood, in Agarwal, S. and Glencross, B. (eds.), *Social Bioarchaeology*, 333–360. Oxford: Wiley-Blackwell.

Hamilakis, Y. 2013. *Archaeology and the Senses: Human Experience, Memory, and Affect.* Cambridge: Cambridge University Press.

Houby-Nielsen, S. 2000. Child burials in ancient Athens, in Sofaer Derevenski, J. (ed.), *Children and Material Culture*, 151–166. London: Routledge.

Kamp, K. A. 2001. Where have all the children gone? The archaeology of childhood. *Journal of Archaeological Method and Theory* 8, 1–34.

Koehl, R. B. 1986. The Chieftain Cup and a Minoan rite of passage. *Journal of Hellenic Studies* 106, 99–110.

Lally, M. and Moore, A. (eds.). 2011. *(Re)Thinking the Little Ancestor: New Perspectives on the Archaeology of Infancy and Childhood* (BAR International Series 2271). Oxford: Archaeopress.

Lebegyev, J. 2009. Phases of childhood in Early Mycenaean Greece. *Childhood in the Past* 2, 15–32.

Legarra Herrero, B. 2009. The Minoan fallacy: cultural diversity and mortuary behaviour on Crete at the beginning of the Bronze Age. *Oxford Journal of Archaeology* 28, 29–57.

Legarra Herrero, B. 2012. The construction, deconstruction and non-construction of hierarchies in the funerary record of Prepalatial Crete, in Schoep, I., Tomkins, P. and Driessen, J. (eds.), *Back to the Beginning. Reassessing Social and Political Complexity on Crete during the Early and Middle Bronze Age*, 325–357. Oxford: Oxbow Books.

Legarra Herrero, B. 2014. *Mortuary Behavior and Social Trajectories in Pre- and Protopalatial Crete*. Philadelphia, PA: INSTAP Academic Press.

Maggidis, C. 1998. From polis to necropolis: social ranking from architectural and mortuary evidence in the Minoan cemetery in Phourni, Archanes, in Branigan, K. (ed.), *Cemetery and Society in the Aegean Bronze Age Societies*, 87–102. Sheffield: Sheffield Academic Press.

Manning, S. W. 2010. Chronology and terminology, in Cline, E. H. (ed.), *The Oxford Handbook of the Bronze Age Aegean*, 11–28. Oxford: Oxford University Press.

Marinatos, N. 1984. *Art and Religion in Thera*. Athens: D. & I. Mathioulakis.

McGeorge, P. J. P. 2012. The Petras intramural infant jar burial: context, symbolism, eschatology, in Tsipopoulou, M. (ed.), *Petras, Siteia - 25 Years of Excavations and Studies*, 291–304. Århus: Aarhus University Press.

Papadatos, Y. 2005. Mortuary practices, in Papadatos, Y. (ed.), *Tholos Tomb Gamma: A Prepalatial Tholos Tomb at Phourni, Archanes* (Prehistory Monographs 17), 55–61. Philadelphia, PA: INSTAP Academic Press.

Papadatos, Y. and Sofianou, C. 2015. *Livari Skiadi, A Minoan Cemetery in Lefki, Southeast Crete. Volume I: Excavation and Finds*. Philadelphia, PA: INSTAP Academic Press.

Perry, M. 2006. Redefining childhood through bioarchaeology: toward an archaeological and biological understanding of children in antiquity, in Baxter, J. E. (ed.), *Children in Action. Perspectives on the Archaeology of Childhood* (Archaeological Papers of the American Anthropological Association 15), 89–111. Berkeley, CA: University of California Press.

Pomadère, M. 2007. Les Enfants Dans Le Monde Égéen, Du Néolithique Au Début de l'Âge Du Fer. Unpublished Ph.D. thesis, Université de Paris I – Panthéon Sorbonne.

Rehak, P. 2007. Children's work: girls as acolytes in Aegean ritual and cult, in Cohen, A. and Rutter, J. (eds.), *Constructions of Childhood in Ancient Greece and Italy* (Hesperia Supplement 41), 205–225. Athens: The American School of Classical Studies at Athens.

Rutter, J. 2003. Children in Aegean Prehistory, in Neils, J. and Oakley, J. H. (eds.), *Coming of Age in Ancient Greece. Images of Childhood from the Classical Past*, 31–57. London: Yale University Press.

Sakellarakis, Y. and Sapouna-Sakellaraki, E. 1997. *Archanes. Minoan Crete in a New Light*. Athens: Ammos Publications.

Schoep, I. 2009. The excavation of the cemetery (zone 1), in Driessen, J., Schoep, I., Carpentier, F., Crevecoeur, I., Devolder, M., Gaignerot-Driessen, F., Hacigüzeller, P., Jusseret, S., Langohr, C., Letesson, Q. and Schmitt, A., *Excavations at Sissi. Preliminary Report on the 2007-8 Campaigns*, 45–56. Louvain-la-Neuve: Presses Universitaires de Louvain.

Schoep, I. in press. The cemetery of Sissi in context: assessing mortuary behaviour in Northeast Crete, in Relaki, M. and Papadatos, Y. (eds.), *From the Foundations to the Legacy of Minoan Society. Sheffield Round Table in Honour of Professor Keith Branigan*. Oxford: Oxbow Books.

Schoep, I., Schmitt, A. and Crevecoeur, I. 2011. The cemetery at Sissi. Report of the 2009 and 2010 campaigns, in Driessen, J., Schoep, I., Carpentier, F., Crevecoeur, I., Devolder, M., Gaignerot-Driessen, F., Hacigüzeller, P., Isaakidou, V., Jusseret, S., Langohr, C., Letesson, Q. and Schmitt, A., *Excavations at Sissi, II. Preliminary Report on the 2009-10 Campaigns*, 41–68. Louvain-la-Neuve: Presses Universitaires de Louvain.

Schoep, I., Schmitt, A., Crevecoeur, I. and Déderix, S. 2013. The cemetery at Sissi. Report on the 2011 campaign, in Driessen, J., Schoep, I., Anastasiadou, M., Carpentier, F., Crevecoeur, I., Déderix, S., Devolder, M., Gaignerot-Driessen, F., Jusseret, S., Langohr, C., Letesson, Q., Liard, F., Schmitt, A., Tsoraki, C. and Veropoulidou, R., *Excavations at Sissi III. Preliminary Report on the 2011 Campaign*, 27–50. Louvain-la-Neuve: Presses Universitaires de Louvain.

Smith, D. M. 2011. Reconciling identities in life and death: the social child in the Early Helladic Peloponnese. *Childhood in the Past* 4, 46–62.

Sofaer Derevenski, J. (ed.). 2000. *Children and Material Culture*. London: Routledge.

Soles, J. 1992. *Prepalatial Cemeteries at Mochlos and Gournia and the House Tombs of Bronze Age Crete*. Princeton, CA: American School of Classical Studies at Athens.

Triantaphyllou, S. 2005. Appendix: the human remains, in Papadatos, Y. (ed.), *Tholos Tomb Gamma: A Prepalatial Tholos Tomb at Phourni, Archanes* (Prehistory Monographs 17), 67–76. Philadelphia, PA: INSTAP Academic Press.

Triantaphyllou, S. 2009. EM/MM human skeletal remains from East Crete: the Kephala Petras rockshelter, Siteia, and the Livari Tholos Tomb, Skiadi. *Kentro, The Newsletter of the INSTAP Study Center for East Crete* 12, 19–23.

Triantaphyllou, S. 2010. The human remains, in Vasilakis, A. and Branigan, K. (eds.), *Moni Odigitria. A Prepalatial Cemetery and Its Environs in the Asterousia, Southern Crete*, 229–248. Philadelphia, PA: INSTAP Academic Press.

Triantaphyllou, S. 2012. Kephala Petras: the human remains and the burial practices in the rock shelter, in Tsipopoulou, M. (ed.), *Petras, Siteia – 25 Years of Excavations and Studies*, 161–170. Århus: Aarhus University Press.

Triantaphyllou, S. in press. Managing with death in Prepalatial Crete: the evidence of the human remains, in Relaki, M. and Papadatos, Y. (eds.), *From the Foundations to the Legacy of Minoan Society. Sheffield Round Table in Honour of Professor Keith Branigan*. Oxford: Oxbow Books.

Triantaphyllou, S. and Girella, L. in press. I Symvolí Tis Melétis Ton Anthrópinon Ostón Stin Ermineía Tou Tholotoú Táphou Sto Kamilári Stin Pediáda Tis Mesarás, N. Irakleíou, in Papadogiannakis, N.E. and Gryndakis, I. (eds.), *Proceedings of the 11th International Cretological Congress, Rethymnon, 21-7 October 2011*. Rethymnon: Association on Historical & Folklore Studies in Rethymnon.

Tsipopoulou, M. 2012. *Petras, Siteia. 25 Years of Excavations and Studies* (Monographs of the Danish Institute at Athens 16). Athens: Danish Institute at Athens.

Tsipopoulou, M. (ed.) 2017. *Petras, Siteia. The Pre- and Proto-Palatial Cemetery in Context. Acts of a Two-Day Conference Held at the Danish Institute at Athens, 14-15 February 2015*. Århus: Aarhus University Press.

van Effentere, H. 1963. *Fouilles Exécutées À Mallia, Etude Du Site (1956-57), Exploration Des Nécropoles (1915-28)* (Etudes Crétoises 13). Paris: Librairie Orientaliste Paul Geuthner.

Vasilakis, A. and Branigan, K. (eds.) 2010. *Moni Odigitria. A Prepalatial Cemetery and Its Environs in the Asterousia, Southern Crete*. Philadelphia, PA: INSTAP Academic Press.

Chapter 11

Geto-Dacian Child Burials in the Second Iron Age

Valeriu Sîrbu and Diana-Crina Dăvîncă[1]

Abstract: In this paper Geto-Dacian child burials from the fifth century BC to the first century AD will be reviewed based on the type of site where they were found (necropolises, fortresses, settlements, cult sites or isolated pits), for the purposes of ascertaining their similarities and differences. Based on the various sources of information available, namely the position and condition of the skeleton, the associated funerary inventory and evidence for offerings, as well as the eventual post-mortem actions performed on the dead, an attempt will be made to identify the rituals that may have taken place after death. In the absence of written sources or iconographic representations, and given the small number of anthropological analyses undertaken to date, however, this is all limited to hypotheses that remain to be confirmed. Fifteen necropolises contained the burials of 129 children, both cremated and inhumed; the number of children is therefore very small compared to that of adults (over 2,300 individuals). Twenty-nine non-funerary contexts (settlements, pit fields, isolated pit) contained 89 children, of whom 88 were inhumed and only one was cremated. Fourteen pit fields – out of the 19 we know about – contained skeletons or parts of human skeletons, from about 103 individuals; 40 of these skeletons were those of children and were recovered from six pit fields.

Keywords: Dacians, children, Second Iron Age, inhumation, cremation

Introduction

The Getae appear in written sources towards the end of the sixth century BC, on the occasion of the expedition of Darius against the Scythians (Herodotus, *Histories*, IV, 93), and the first definite mention of the Dacians dates to the middle of the first century BC (Caesar, *De Bello Gallico*, VI, 25), or perhaps from the end of the second century BC (Frontinus, *Strategemata*, II, 4, 3). The term 'Geto-Dacians' is a creation

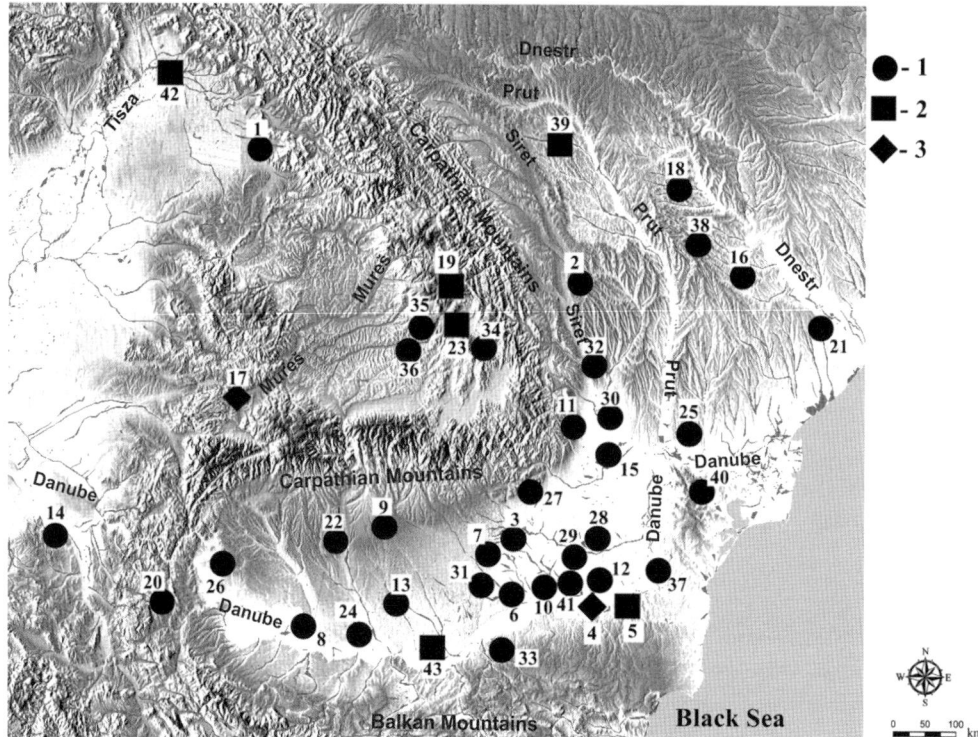

Figure 11.1. General map of discoveries according to the rite – 1. inhumation, 2. cremation, 3. bi-ritual. List of localities – 1. Berea, 2. Brad, 3. București, 4. Bugeac, 5. Canlia, 6. Căscioarele, 7. Cățelu Nou, 8. Celei, 9. Cetățeni, 10. Chirnogi, 11. Cîndești, 12. Coslogeni, 13. Dulceanca, 14. Gomolava, 15. Grădiștea, 16. Hansca, 17. Hunedoara, 18. Mășcăuți, 19. Merești, 20. Mokranje, 21. Mologa II, 22. Ocnița, 23. Olteni, 24. Orlea, 25. Orlovka, 26. Ostrovul Șimian, 27. Pietroasele, 28. Piscu Crăsani, 29. Platonești, 30. Poiana, 31. Popești, 32. Răcătău, 33. Russe, 34. Sf. Gheorghe, 35. Sighișoara-Albești, 36. Sighișoara-Wietenberg, 37. Stelnica, 38. Stolniceni, 39. Strahotin, 40. Telița-Celic Dere, 41. Unirea, 42. Zemplin, 43. Zimnicea.

of modern historians, which is used in relation to the general issue of northern Thracians, since the Getae and the Dacians were the most important Thracian tribes. The geographical area in question includes the territories north of the Balkans, namely the region inhabited by the northern Thracians (Geto-Dacians). The predominant funerary rite used by the Geto-Dacians between the fifth century BC and the first century AD was cremation, but they also practised inhumation.

The known subadult remains are spread over almost the entire area inhabited by the Geto-Dacians and significant regional differences are evident. It is uncertain whether these are genuine, however, or rather an artefact of the research that has been undertaken to date. From the period between the fifth century BC and the first century AD, we have 43 finds that include skeletal remains derived from 218 children (Fig. 11.1). The remains of children – complete or fragmentary skeletons, cremated

11. Geto-Dacian Child Burials in the Second Iron Age

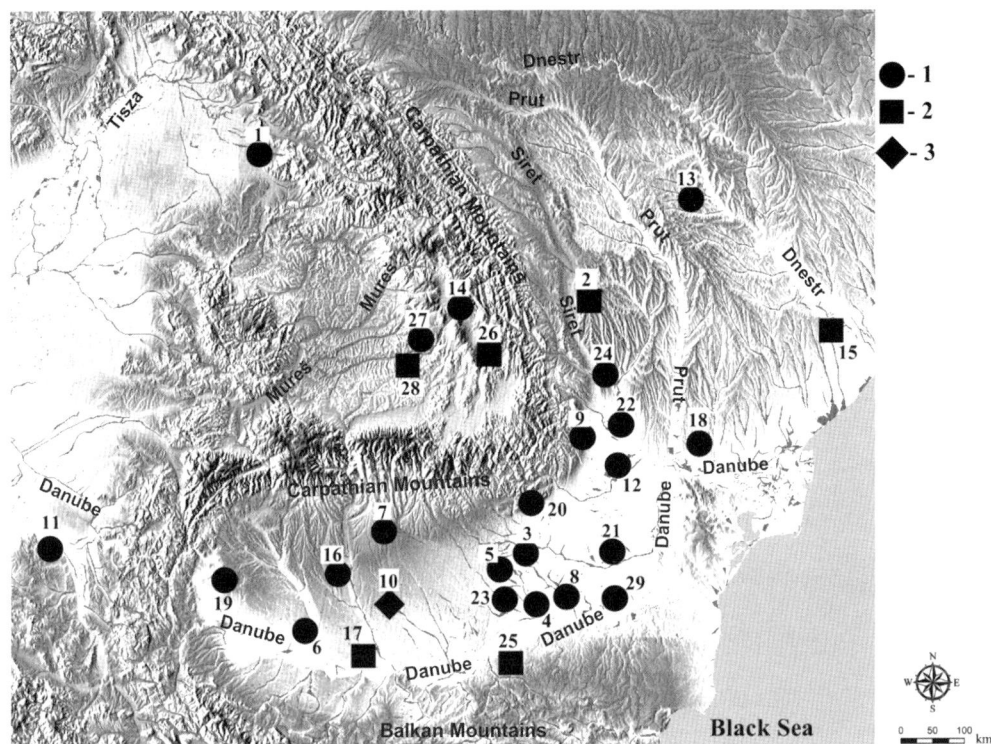

Figure 11.2. Map of discoveries of child remains from non-funerary contexts – 1. settlement, 2. 'field of pits', 3. isolated pit. List of localities – 1. Berea, 2. Brad, 3. București, 4. Căscioarele, 5. Cățelu Nou, 6. Celei, 7. Cetățeni, 8. Chirnogi, 9. Cîndești, 10. Dulceanca, 11. Gomolava, 12. Grădiștea, 13. Mășcăuți, 14. Merești, 15. Mologa II, 16. Ocnița, 17. Orlea, 18. Orlovka, 19. Ostrovul Șimian, 20. Pietroasele, 21. Piscu Crăsani, 22. Poiana, 23. Popești, 24. Răcătău, 25. Russe, 26. Sf. Gheorghe, 27. Sighișoara-Albești, 28. Sighișoara-Wietenberg, 29. Unirea.

or inhumed, were found in non-funerary settings (Fig. 11.2), as well as in funerary contexts (necropolises or isolated graves) (Fig. 11.3). On the basis of information derived from associated archaeological or anthropological analysis, the individuals were placed in three categories, namely *Infans* I (0–7 years) and *Infans* II (7–14 years), with the term 'children' having been used when details of age were not provided (Sirbu 2003, 34).

An obvious preference for inhumation is evident since the data reveals the presence of 184 inhumed and only 34 cremated individuals, but it is uncertain if this is a reflection of the genuine situation, given the small number of anthropological analyses that have been undertaken on cremated remains to date. The disparity is even larger in the case of *Infans I* (Fig. 11.4), for which 112 inhumed and only 15 cremated individuals have been identified. This situation has arisen because almost all of the non-funerary contexts (save for one case) have only inhumed children;

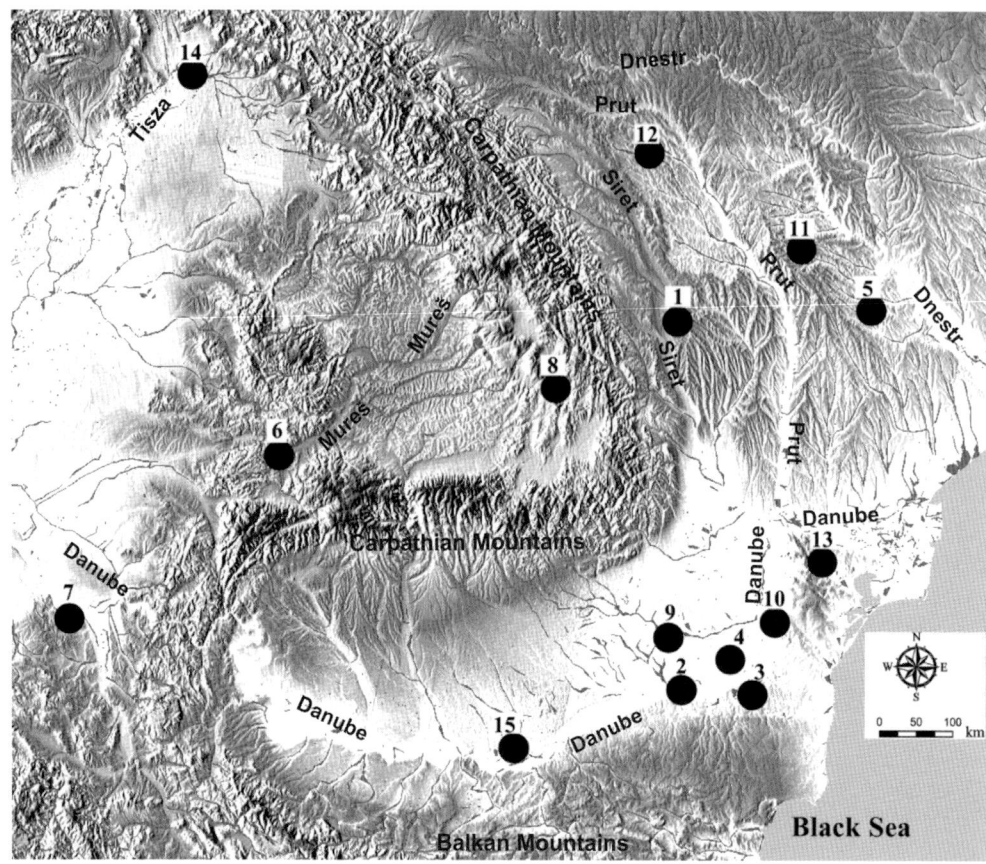

Figure 11.3. Map of discoveries of child remains from necropolises. List of localities – 1. Brad, 2. Bugeac, 3. Canlia, 4. Coslogeni, 5. Hansca, 6. Hunedoara, 7. Mokranje, 8. Olteni, 9. Platonești, 10. Stelnica, 11. Stolniceni, 12. Strahotin, 13. Telița-Celic Dere, 14. Zemplin, 15. Zimnicea.

funerary contexts for the last period (first century BC to first century AD) are very rare. The fact that inhumation appears to be the predominant rite has two explanations. Firstly, with the exception of the necropolis in Stelnica, most of the burials from the fifth to the third century BC are associated with cremations, and few anthropological analyses have been undertaken on these remains. Secondly, starting with the second century BC to first century AD, necropolises are no longer used for the burial of ordinary people and only clusters of tumuli graves are found, in addition to deposits of human remains (either complete burials or disarticulated or isolated bones) recovered from non-funerary contexts. The remains from both of these contexts are generally inhumed (Sîrbu 2006, 117–52). The following sections will provide a review of child burials by site type for the purposes of highlighting their specific features.

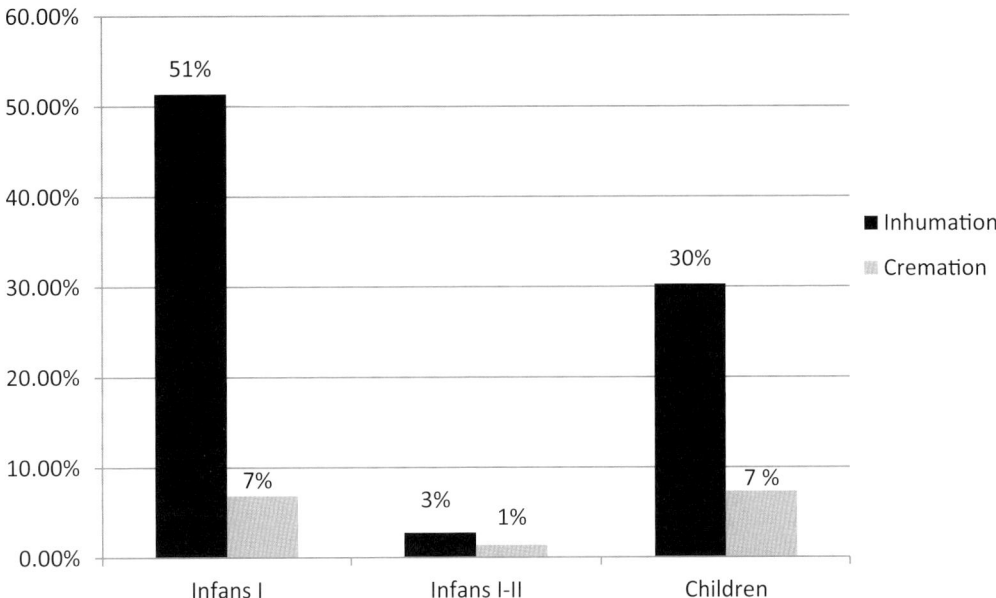

Figure 11.4. The ratio between inhumation and cremation across the different juvenile age groups, compared to the total number of discoveries.

Necropolises

In the 15 necropolises[2] from the Second Iron Age, the remains of 129 children have been recovered, 33 of which were cremated, while 96 were inhumed (Fig. 11.5). The inventory (personal belongings) of both cremation and inhumation graves is usually modest, consisting of adornments, dress items, rarely pottery and, in only two cases, weaponry. The offerings (food, liquids and animals) are also few and far between. The finds derived from two sites are particularly informative – the Stelnica necropolis contained the remains of a relatively high number of children, while evidence of the association of juveniles with specific rites and rituals was evident at the *Grădina Castelului* necropolis in Hunedoara.

The Stelnica necropolis dates from the fourth to the third century BC and contained the remains of 202 inhumed individuals as well as 214 cremation deposits. A total of 47 of the inhumations were those of children – thirty-three *Infans* I, four *Infans* I–II and ten children – which represents about 23% of the population (Conovici and Matei 1999, 102–5, 110; Conovici *et al.* 2005, 359–62; Babeș *et al.* 2012, 134–6; Zirra *et al.* 2014, 131–2; Dăvâncă 2015, 61). Unfortunately, the number of children represented in the 214 cremation deposits is unknown, which means it is impossible to estimate how many children were represented across the entire necropolis population which comprised a minimum of 416 individuals. The majority of inhumed individuals, both children and adults, were placed lying on either side, mostly with the head to the

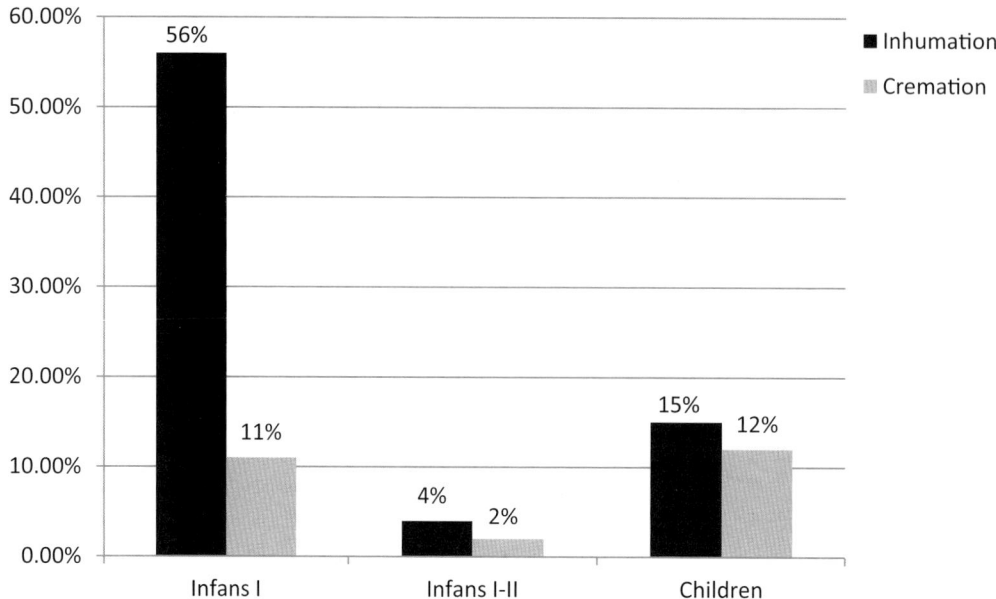

Figure 11.5. The ratio between inhumation and cremation across the different juvenile age groups, compared to the total number of discoveries from necropolises.

south. Their inventories, poor and unvaried, consisted mostly of adornments, while evidence for offerings was rare. Some juvenile skeletons were protected by pottery fragments. Double burials were also identified and these comprised either an adult and child or just children.

Grădina Castelului in Hunedoara is a necropolis where 'access' to the funerary space was restricted, at different chronological stages, to various categories of individuals. All in all, some 34 deposits of human bones have been discovered that derive from 57 individuals, nine of which were cremated while 48 were inhumed. Most of the inhumed were children; 38 were under seven years old, while 20 were less than one year of age (Fig. 11.6). All of the cremation graves are from the period between the end of the fourth century and the middle/end of the first century AD and, with the exception of one individual (Grave 32), they all contained adults (Sîrbu and Roman 2013, fig. 1). On the other hand, starting in the first century AD, the site became exclusively reserved for the inhumation of children. The relatively rich inventory associated with these burials mostly contained dress items and adornments (Sîrbu *et al.* 2007, 169–94). If we consider this to be a necropolis, the first stage of its development focused on cremation, and the dead may have been warriors since they were associated with various types of weaponry. This was followed by a transitional stage which involved both cremation and inhumation for adults, adolescents and children but, in the final stage of use, only inhumed small children were represented.

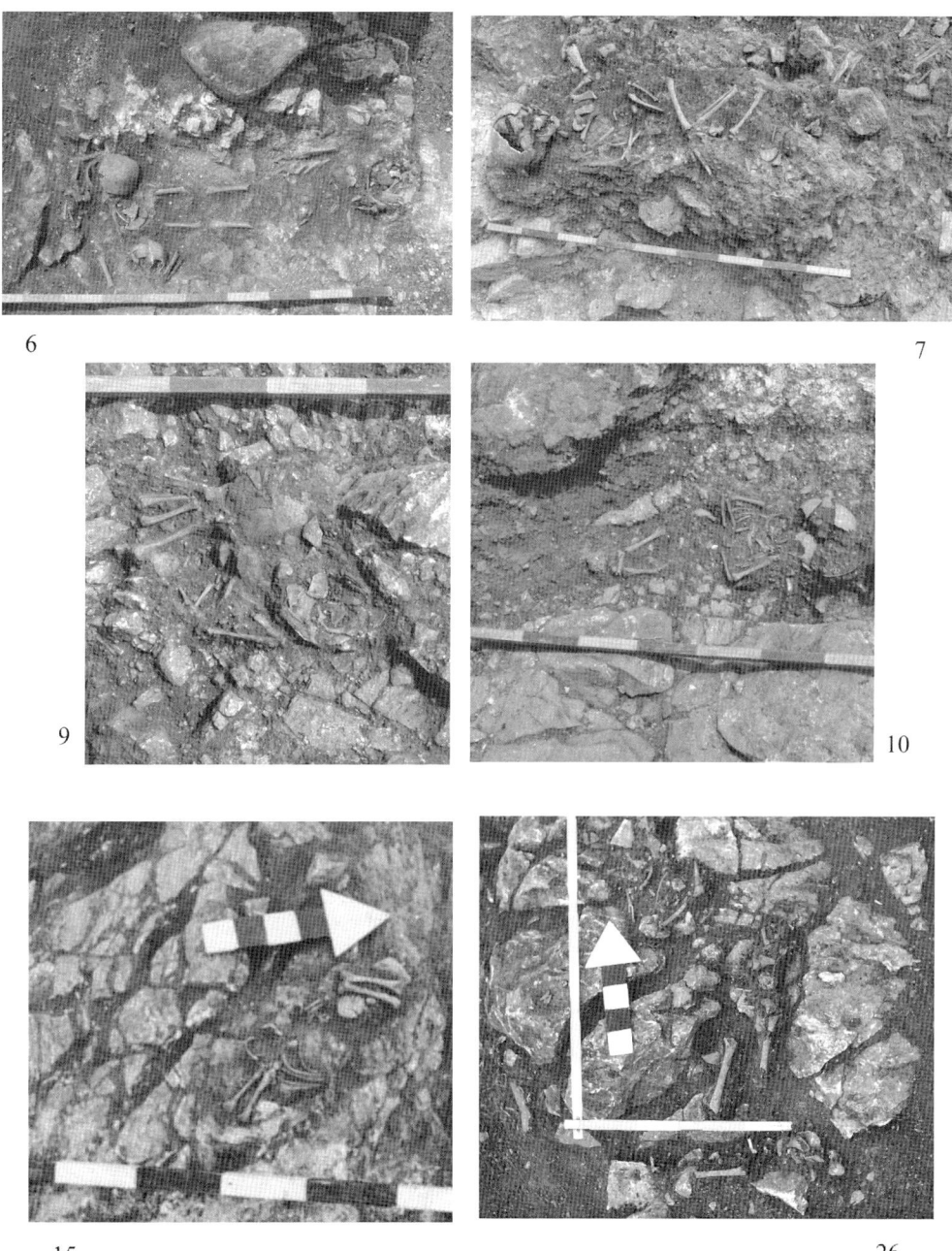

Figure 11.6. Grădina Castelului, in Hunedoara – children's burials from Structures 6, 7, 9, 10, 15 and 26 (photographs by Valeriu Sîrbu).

From the Second Iron Age, we have only two cases in which a child had definitively been buried with a weapon, and these derive from the necropolises of *Celic Dere* in Telița (Tumulus 44, Grave 2 – inhumation) and *Grădina Castelului* in Hunedoara (C73, Grave 32 – cremation) (Sîrbu and Roman 2013, fig. 1). Grave 2, Tumulus 44, from the tumuli and flat necropolis of *Celic Dere* in Telița (sixth to third century BC), dated to the fifth century BC and contained the inhumed remains of an 8–9-year-old child. Some bones were missing, and anthropological analysis indicated that the skull had been deliberately smashed and placed face down, while a round opening visible in the frontal bone was considered to be the result of a peri-mortem trepanation (Sîrbu et al. 2013, 354; fig. 13/h).[3] The funerary inventory contained the blade of a curved iron knife, five bronze arrowheads, an amber bead and fragments of a pottery bowl. Cremation Grave 32 from the *Grădina Castelului* necropolis in Hunedoara, was that of an *Infans* I, whose inventory contained a curved iron dagger with a scabbard, an iron arrowhead, a fragment from a silver earring, an iron belt buckle, a large iron clamp, a grinding stone and a fragment of bronze foil (Roman and Luca 2012, 75–6; Sîrbu and Roman 2013, 373–99). As such, even though we are dealing with children, in both situations we see their future status as warriors having been represented in their funerary inventory even though it is obvious, from their age-at-death that they could not yet have participated in battles. It should be noted that weapons were also found in a small number of double graves which contained the remains of an adult and a child but, in these cases, it seems more likely they were associated with the adult, rather than the child, since the weapons were positioned adjacent to the adult skeletons.

Non-funerary Contexts

The finds originate from 29 structures, pit fields, cult places and an isolated pit. In total, bones derived from 89 children have been discovered, 88 of which were inhumed, while only one was cremated (Sîrbu 2003, 21–2; Sîrbu and Dăvîncă 2014a, 363–4). It is important to remember, however, the difficulties of analysing burials from the first century BC to first century AD, since necropolises for the common people seem to be absent from the Geto-Dacian world at that time.

Settlements

The finds in settlements were recovered from pits that were identified in specific locations, either inside dwellings or in the spaces located between dwellings. The remains of 48 children were recovered from the pits, 47 of which were inhumed, while just one was cremated. The settlement of *Șuvița Hotarului* in Căscioarele (fourth to third century BC) contained a pit with the skeleton of a male aged 30–35 years who had been buried in an 'embracing', face-to-face, position with a woman aged 16–18 years, as well as a few bones from the skeleton of a 3–4-year-old child (Sîrbu 1993, 88). Several pits from the settlement of Grădiștea (second to first century BC) have yielded complete skeletons, as well as partial ones, from eight individuals, five of which were children. Some of the skeletons displayed possible evidence of dismemberment. For instance, Pit 116 contained the incomplete skeleton of a child aged 4–5 years, and fragments from over 20 clay vessels, as well as a part of a deer antler that was recovered from the associated pit fill (Sîrbu 1993, 89–91). The remains of a child aged 12–13 years were discovered beneath the foundation of Dwelling 2 in the Borduşani settlement (first century BC). The skeleton was in a crouched position and lying on the side (precise details are not known). It was incomplete, but in anatomical order and the absence of certain skeleton parts is suggestive that the child had either died a violent death or that the body had been exposed prior to burial. The upper part of the skeleton was covered with half a pottery pedestal cup, while a large stone was positioned beneath the hips (Sîrbu 1993, 86; Trohani 2005, 11–12, fig. 6–13).

The presence of human remains within or near settlements, some of which appear to have been exposed in open pits or other contexts as indicated by anthropological analyses, assumes a certain period of 'cohabitation' between the living and the dead.

Isolated Pit

In Dulceanca, the bottom of a pit with burnt and truncated walls, located near a settlement, contained the complete skeleton of a child. The skeleton was crouched vertically, face down, with the hands on the knees. Some pig bones and the bottom of a Dacian mug, of second to first century BC, were located beside the individual (Dolinescu-Ferche 1974, 50–1, fig. 34–5).

Pit Fields

The pit fields represent clusters of pits – sometimes hundreds – of various shapes, structures and arrangements, with diverse inventories, sometimes associated with human or animal skeletons, and in some instances which include fireplaces, shrines, platforms and deposits of various items. They are generally located outside or near settlements (Sîrbu 2006, 48; Sîrbu and Dăvîncă 2014b, 296). Pit fields are present throughout almost the entire area inhabited by the Geto-Dacians. On the basis of their various features, they can be regarded as cult areas, where certain rituals occurred that were followed by the placement of goods. Furthermore, they are not necropolises, nor are they associated with temples. Of the 19 sites classified as pit fields, only six contained human skeletons or skeleton parts (Sîrbu and Davîncă 2014b, 309). Out of a total of 1,000 pits, only 62 have been found to contain bones which derived from 103 individuals. These included 43 children, three adolescents and 28 adults, while the ages of the other individuals were not provided (Sîrbu and Dăvîncă 2014b, 309–10). In all cases, the individuals were inhumed, regardless of their age. Representative of this category of burial deposit is a site at Mologa II which, to date, is the most impressive cult site of this type to have been discovered (Maljukevich 2003, 105–11).

Its pits contain a notable diversity of structures, in addition to evidence for complex rituals and artefact depositions, all of which appear to have been of Getae origin (Maljukevich 2003, 105–11; Sîrbu and Dăvîncă 2014a, 369; 2014b, 301–3).

The spatial relationship between the pit fields and other types of sites is not always known due to research approaches which rarely include intensive field research and surveys in the surrounding areas. Some of the pit fields are located near fortresses or residential centres, such as Brad, Sighișoara-Wietenberg and Mologa II (Sîrbu and Dăvîncă 2014b, 313).

The complete skeletons recovered from such non-funerary contexts display various positions, including lying on the side, crouching, prone or even a variety of contorted positions. Some of the skeletons, or skeleton parts, were protected with pottery fragments. Skeletons with evidence of violence have been identified at Cetățeni (Sîrbu 1993, 89), Măscăuți (Zanoci 2004, 45), Căscioarele (Sîrbu 1993, 88) and Sighișoara-Wietenberg (Sîrbu 1993, 95), amongst others. The inventory associated with the dead, was neither rich nor diverse and included dress items, adornments, pottery that could sometimes be reconstituted or was fragmentary, animal bones and various other materials, such as ash, coal and fragments of clay (Sîrbu and Dăvîncă 2014a, 369). It should be noted that the inventory in non-funerary contexts was often fragmentary and could not always be associated with particular individuals. This was particularly the case for larger pits, in which objects were discovered at different depths to the human remains and may have arrived there by accident, as a result of the pit having been filled with household remains, and not as part of ritual activity.

A unique association between child bones and dog skeletons was identified at Mologa II. Some of the animals showed signs of sectioning (Maljukevich 2003, 105–11; Sîrbu and Dăvîncă 2014a, 369) and it can be assumed that this is evidence of ritual activities. Alternatively, their inclusion may simply have been symbolic of the animal's attachment to people, including in the 'afterworld' (Sîrbu 2006, 54). Analogies regarding the placement of whole animals together with children can be found with the southern Thracians at sites such as in Malko Tarnovo, where Pit A contained a child skeleton, the remains of a dog that had been cut into three pieces and lamb and calf bones (Tonkova 2010, 507).

Necropolises or cult sites?

Human remains, including those of children, have been recovered from two sites where it is difficult to decide, because of the presence of special features and prior to completion of the research, if the sites should be considered as necropolises or ritual sites (Sîrbu et al. 2007, 169–94; Sîrbu and Dăvîncă 2014a, 370). The *Grădina Castelului* site in Hunedoara is placed on a rocky plateau where 34 burials, seven deposits of items and five concentrations of animal bones have been identified (third century BC to the first century AD) (Sîrbu et al. 2007, 19–53). Some unusual features make it difficult to classify these finds as being representative of one type of site or another.

No habitation structure or vestiges of other activities associated with settlement could be identified nearby, no rules relating to the orientation or position of the dead were evident, and the number of individuals in a pit ranged from one to six, and consisted of complete skeletons, partial skeletons or just isolated bones, with some situations being suggestive of exposure/decomposition practices (Sîrbu *et al.* 2007, 162).[4] Was this just a 'normal' necropolis where the community applied, over four centuries, a very strict 'filter' regarding the age and social status of the dead and the rites and rituals applied to them? Or did this become in its final stages, namely the first century AD, an area reserved only for children ('children's necropolis'), whose remains first underwent a phase of exposure/decomposition? One must also wonder where the remainder of the dead, not eligible for burial at the site because of their age, were placed, and why age-related burial practices changed so much over time.

Research at a fortified settlement at Stolniceni in the Republic of Moldova (fourth to third century BC) which has two *valla* and two ditches, and differs to fortifications built for defense purposes, has resulted in the discovery of over 20 tumuli within the enclosure. Only four of these were investigated and they were found to contain human skeletons, complete or partial, with diverse orientations and positions. Only two of the tumuli included parts of children's skeletons. In Tumulus 2, in addition to seven incomplete human skeletons, the remains of four hearths, numerous pottery fragments, animal bones and two pits were identified at the bottom of the tumulus. Excavations undertaken in 2014 identified some habitation features, but the stratigraphical and chronological relationship between them and the tumuli is as yet unclear (Sîrbu and Arnăut 1995, 378–400; Arnăut and Ursu-Naniu 1996, 54–6; Sîrbu and Dăvîncă 2014a, 371). Was this a fortified necropolis, which would be unique in the Thracian world, or do the tumuli succeed the fortress after it has gone out of use? Alternatively, was this a sacred area delineated by earthen *valla*, where certain rituals were performed on particular categories of the dead? In any case, the state of the skeletons and the arrangements within the tumuli do not find direct analogies within contemporary necropolises throughout the entire Thracian world (Sîrbu 2006, 24).

The Southern Thracians

Complete skeletons, partial skeletons or isolated children's bones have also been discovered in Thracian non-funerary contexts, particularly pit fields, south of the Balkan Mountains in Bulgaria. Pits containing whole or partial child skeletons, sometimes in association with the remains of adolescents or adults, have been found at the sites of Yabalkovo (Tonkova 2010, 507–8), Koprivlen (Baralis 2009, 11–36), Gledacévo (Tonkova 1997, 594–605) and Malko Tarnovo (Tonkova 2010, 507). These finds date from the seventh to the fourth centuries BC. So far, no finds dating to between the second century BC and the first century AD have been identified, but this is because the area experienced particular historical developments from the middle of the fourth century BC when it came under the control of Macedonia. As such, it seems to be the case that children's remains have been discovered in non-funerary contexts throughout the entire area inhabited by the Thracians.

Greece

Only two finds in Greece will be discussed since these are considered representative of the situation for the Greek world. A terrace on the island of Astypaleia, in the Aegean Sea, has been found to contain thus far 2,754 amphorae burials of newborn infants and children less than three years of age. This area appears to have been a funerary space reserved for the burial of small children which dated from the Archaic to the Hellenistic period (eighth to first centuries BC) (ThesCRA VIII, 384). In the enclosure of the Plato Academy in Athens a pit dating from the second half of the eighth century to the beginning of the seventh century BC was discovered which contained approximately 40 amphorae. Each of the amphorae contained the remains of children, and they appear to have been placed within the pit in nine succeeding layers. Another pit, in Kerameikos and of fifth-century BC date, contained the skeletons of 150 children and adults; the children's skeletons had been protected by large pottery fragments (ThesCRA VIII, 378 and 383).

Conclusions

There are no known written sources or iconographic representations that provide any information about how the northern Thracians/Geto-Dacians dealt with the remains of deceased children. It is therefore necessary to turn to the archaeological record to gain insights about such practices. In necropolises, children were both inhumed (n=184 individuals) and cremated (n=34 individuals), but we cannot estimate the true percentage of children represented because of the small amount of anthropological analyses, particularly for cremation deposits, that has been undertaken to date. In fact, from the first century AD no normal necropolises have been identified in the area inhabited by the Geto-Dacians, which is suggestive of funerary practices that are undetectable by means of standard archaeological methods. The children inhumed in dwellings were perhaps connected to certain foundation sacrifices, and it may have been believed that new 'life' was given to a new building through the burial.

Both the necropolises and the non-funerary contexts contained a combination of complete and partial skeletons as well as isolated bones derived from various parts of the body (Fig. 11.7). Anthropological analyses suggest that, in some cases, bodies were temporarily exposed to the elements prior to burial and that some may have been exhumed and reburied for various reasons. These practices are suggestive of a period of 'cohabitation' between the dead and the living.

Some child skeletons from non-funerary contexts displayed traces of violence that had been made by both sharp and blunt instruments. Children were placed in

11. Geto-Dacian Child Burials in the Second Iron Age

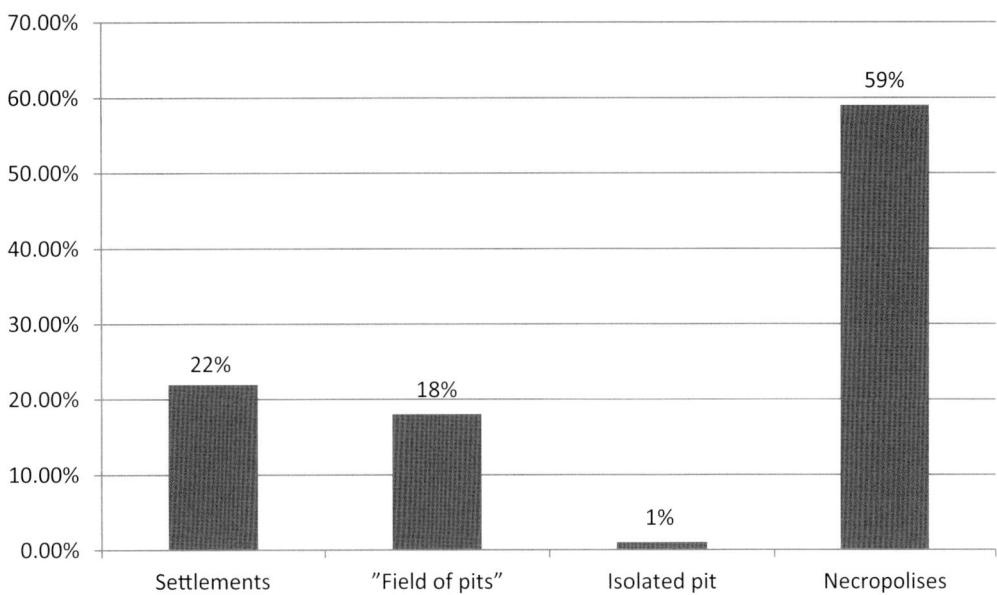

Figure 11.7. The distribution of child remains according to the type of site.

necropolises and non-funerary contexts, alone or in association with adolescents and adults, both men and women. The funerary inventory is usually poor and shows little diversity with notable exceptions comprising two graves from *Grădina Castelului*, in Hunedoara, and *Celic Dere*, in Telița, which both contained weapons and may be representative of the intended future status of these children as warriors. In non-funerary contexts (settlements, fortresses, pit fields and isolated pits), the children were all inhumed, with only one exception – Merești in Transylvania. It should be noted that no inhumed or cremated human or animal remains have been found in any of the dozens of temples discovered, that is in the impressive cult places of the Dacians (Sîrbu 2006, 21–39).

Almost all of the child skeletons subject to anthropological analyses have been found to be those of small children – *Infans* I (0–7 years), but the existence of a high number of cases which have just been classified as 'children' means that this trend cannot be considered to be definitive. The presence of a large number of small children in pits located in non-funerary contexts continues to be an important argument in favor of the existence of beliefs that required their separation from adults in death. In the absence of written sources, however, it is impossible to ascertain the selection criteria for this practice of separate burial.

It may be the case that skeletons (adults, adolescents and children) found in abnormal positions, sometimes with traces of violence, with missing parts or consisting of just parts of human skeletons in pits in non-funerary contexts are evidence of sacrifice. In the absence of anthropological analyses, however, the situation remains uncertain and it is possible that the remains of these individuals had been exposed in the open air prior to burial, resulting in a degree of dismemberment as part of this process.

It is only in the case of *Grădina Castelului*, in Hunedoara, where only inhumed children are found from the first century AD, that we can be certain of the existence of the practice of burying children under a certain age ('children necropolis') separate from the usual necropolises of a community. The fact that only 129 children, inhumed or cremated, are represented in the 2,500 or so burials known for the Geto-Dacian area, means that these burials are not an accurate reflection of the composition of society, even if the 140 children derived from non-funerary contexts are added to this figure. It should also be noted that non-funerary contexts also yielded a large number of adults.

In any case, with the exception of the Greek world, which had genuine necropolises reserved for children, particularly those born prematurely and newborns, it seems that the Balkan peoples had funerary practices for children that were not fundamentally different to those of other contemporary societies outside of the Greek and Roman worlds. Future research is needed to enable the nature of child burials in the northern Balkan area to be placed within its broader context. This research would enable further comparisons to be made with the practices of neighbouring peoples, thereby highlighting their similarities and differences and enabling the identification of any reciprocal influences that may have existed with regard to these funerary practices.

Acknowledgements

We thank Mrs Anca Ganciu for her kindness in providing information about recent research at the Stelnica necropolis.

Notes

1. Museum of Brăila 'Carol I', Piața Traian, No. 3, 810153 Brăila, Romania. Emails: valeriu_sirbu@yahoo.co.uk, ddavinca@yahoo.com.
2. This includes the finds from Hunedoara-*Grădina Castelului* and Stolniceni, although the nature of these sites has some particularities.
3. Analysis was performed by anthropologist, Andrei Soficaru.
4. Some features of the site have already been discussed so we will not dwell on them again.

References

Contemporary Sources

Caesar, *De Bello Gallico* (cited here from Vilan Unguriu, J. 1964. [Romanian translation], 224. Bucharest: Editura Științifică).

Frontinus, *Strategemata* (cited here from Iliescu, Vl. 1964. *Izvoare privind istoria României, I*, 431–433. 1964. Bucharest: Editura Academiei).

Herodotus, *Istoriai* (cited here from Vanț-Ștef, F. 1961. [Romanian translation]. Bucharest: Editura Științifică).

Publications

Arnăut, T. and Ursu-Naniu, R. 1996. Stolniceni – a new aspect of the funerary ritual of the Getic population, in Sîrbu, V. and Ștefănescu, R. (eds.), *Funerary Practices in Central and Eastern Europe (10th c. BC–3rd c. AD). Proceedings of the 10th International Colloquium of Funerary Archaeology*, 31–40. Brăila-Brașov: Editura C2 Design.

Babeș, M., Ganciu, A., Matei, Gh., Rențea, E. and Laquay, L. 2012. Stelnica, com. Stelnica, jud. Ialomița. Punct: Grădiștea Mare, in Angelescu, M. V. (ed.), *Cronica Cercetărilor Arheologice din România, Campania 2011*, 134–136. Bucharest: Institutul Național al Patrimoniului.

Baralis, A. 2009. Les champs de fosses rituelles en Thrace au Premier et Second Âge du Fer. Etat des lieux de la recherche, in Ailincăi, S. C., Micu, C. and Mihail, F. (eds.), *Omagiu lui Gavrilă Simion la a 80-a aniversare*, 140–153. Tulcea: Institutul de Cercetări Eco-Muzeale.

Conovici, N. and Matei, Gh. 1999. Necropola getică de la Stelnica – Grădiștea Mare (jud. Ialomița). Raport general pentru anii 1987–96. *Materiale și Cercetări Arheologice* I, New Series, 199–244.

Conovici, N., Ganciu, A. and Matei, Gh. 2005. Stelnica, com. Stelnica, jud. Ialomița. Punct: Grădiștea Mare, in Angelescu, M. V., Oberländer-Târnoveanu, I. and Vasilescu, F. (eds.), *Cronica Cercetărilor Arheologice din România, Campania 2004*, 359–362. Bucharest: cIMeC-Institutul de Memorie Culturală.

Dăvâncă, D. 2015. *Credințe și practici mortuare privitoare la copiii tracilor nordici (sec. XI a. Chr.–I p. Chr.)*. Cluj-Napoca: Academia Română. Centrul de Studii Transilvane.

Dolinescu-Ferche, S. 1974. *Așezări din sec. III și VI e. n. în sud-vestul Munteniei. Cercetările de la Dulceanca*. Bucharest: Editura Academiei.

Maljukevich, A. 2003. Žertvennye zahoronenija sobak v moložskom mogilnike, *Drevnee Pričernomor'e* 1, 105–111.

Roman, C. and Luca, S. A. 2012. Incinerated knights from Hunedoara-Grădina Castelului (Plateau) (Hunedoara County); Archaeological Campaigns from 2008 and 2009). *Brukenthal. Acta Musei* 7, 75–89.

Sîrbu, V. 1993. *Credințe și practici funerare, religioase și magice în lumea geto-dacilor*. Galați: Porto-Franco.

Sîrbu, V. 2003. *Arheologia funerară și sacrificiile: o terminologie unitară; Funerary Archaeology and Sacrifices: A Unifying Terminology*. Brăila: Editura Istros.

Sîrbu, V. 2006. *Oameni și zei în lumea geto-dacilor / Man and Gods in the Geto-Dacian World*. Brașov: C2 Design.

Sîrbu, V. and Arnăut T. 1995. Incinta fortificată de la Stolniceni, raionul Hâncești – Rep. Moldova. *Cercetări arheologice în aria nord-tracă* 1, 378–400.

Sîrbu, V. and Dăvâncă, D. 2014a. Schelete și părți de schelete/oase izolate de copii din epoca fierului descoperite în contexte nefunerare la tracii nordici, in Forțiu, S. and Cîntar, A. (eds.), *Arheovest III1 - In Honorem Gheorghe Lazarovici - Interdisciplinaritate în Arheologie*, 363–386. Szeged: JATEPress Kiadó.

Sîrbu, V. and Dăvâncă, D. 2014b. The 'fields of pits' in the Geto-Dacian Area (4th *c.* BC–1st *c.* AD). Sacred or profane spaces? in Sîrbu, V. and Matei, S. (eds.), *Mousaios XIX. Residential Centres (dava, emporium, oppidum, hillfort, polis) and Cult Places in the Second Iron Age of Europe. Proceedings of the International Colloquium Buzău (Romania)*, 295–342. Buzău: Buzău County Museum.

Sîrbu, V. and Roman, C. 2013. Cremation Graves and Item Deposits (4th–1st *c.* BC) in Hunedoara-Grădina Castelului, Hunedoara County (Romania), in Sîrbu, V. and Ștefănescu, R. (eds.), *The Thracians and their Neighbors in Bronze and Iron Ages. Necropolises, Cult places, Religion, Mythology, II, Proceedings of the 12th International Congress of Thracology*, 373–399. Brașov: Editura Istros.

Sîrbu, V., Luca, S. A., Roman, C., Purece, S., Diaconescu, D. and Cerișer, N. 2007. *The Dacian Vestigies in Hunedoara. Grădina Castelului: Necropolis and/or Sacred Enclosure? Dealul Sânpetru: The Settlement*. Alba-Iulia: Editura ALTIP.

Sîrbu, V., Ștefan M.-M., Ștefan, D., Jugănaru, G. and Bochnak, T. 2013. The necropolis from Telița-Celic Dere (6th–3rd *c.* BC), Tulcea County, Romania. The study case of Tumulus T44, in Sîrbu, V. and Ștefănescu, R. (eds.), *The Thracians and their Neighbors in Bronze and Iron Ages. Necropolises, Cult places, Religion, Mythology. Proceedings of the 12th International Congress of Thracology*, II, 347–372. Brașov: Editura Istros.

ThesCRA VIII. 2012. *Thesaurus Cultus et Rituum Antiquorum. Private Space and Public Space. Polarities in Religious Life. Religious Interrelations Between the Classical World and Neighbouring Civilizations and Addedum to vol. VI. Death and Burial*. Los Angeles, CA: The J. Paul Getty Museum. Available at <https://www.scribd.com/doc/114936664/ThesCRA-VIII>

Tonkova, M. 1997. Un champ de fosses rituelles des Ve–IVe s. av. J.-C. près de Gledacévo, Bulgarie du Sud, in Roman, P., Diamandi, S. Alexianu, M. (eds.), *The Thracian World at the Crossroads of Civilisations, I*, Proceedings of the Seventh International Congress of Thracology, 592–611. Bucharest: Institutul Român de Tracologie.

Tonkova, M. 2010. On human sacrifice in Thrace (on archaeological evidence), in Cândea, I. (ed.), *Tracii și vecinii lor în Antichitate / The Thracians and Their Neighbours in Antiquity. Studia in Honorem Valerii Sîrbu*, 502–522. Brăila: Istros.

Trohani, G. 2005. *Locuirea getică din partea de nord a popinei Bordușani, comuna Bordușani, județul Ialomița*, vol. I. Târgoviște: Cetatea de Scaun.

Zanoci, A. 2004. Traco-geții din bazinul Răutului Inferior. Cetatea Mășcăuți 'Dealul cel Mare', in Niculiță, I., Zanoci, A., Băț, M. (eds.), *Thracians and Circumpontic World II*, Proceedings of the Ninth International Congress of Thracology, Chișinău-Vadul lui Vodă, II, 45–81. Chișinău: Free International University of Moldova.

Zirra, Vl. V., Babeș, M., Ganciu A. and Matei Gh. 2014. Stelnica, com. Stelnica, jud. Ialomița, Punct: Grădiștea Mare, in *Cronica Cercetărilor Arheologice din România. Campania 2013*, 131–132. București: Institutul Național al Patrimoniului.

Chapter 12

Out of the Cradle and into the Grave: The Children of Anglo-Saxon Great Chesterford, Essex, England

Christine Cave and Marc Oxenham[1]

Abstract: This paper seeks to shine a light on the lived experience of Anglo-Saxon children, especially the infants. Unlike most Anglo-Saxon sites, Great Chesterford appears to be the final resting place of the whole community, providing an opportunity to examine usually invisible children. As children are buried by adults, inferences can be made about their attitudes to the dead child, as well as community concepts of children and childhood. Where and how they are buried not only reflect adult points of view, but also provide a glimpse, albeit through the distorted lens of the grave, of the life of that child. We found that although some were buried with exceptional treatment, in general children were supplied with fewer, less expensive grave goods than adults; some were buried in ways that marked them as unusual. Nonetheless, most usual adult grave goods are represented in juvenile graves. While the funeral tableau provides an aid to remembering the dead, the burying of artefacts also functions as an aid to forgetfulness. Therefore, we conclude that the unwillingness or inability to commit scarce resources to a dead child's grave is not necessarily a sign that their deaths were without meaning or that the child was not missed.

Keywords: burial, infant, juvenile, childhood, mortuary, grave goods, Anglo-Saxon England

Introduction

I was once a child. You were once a child. All of the scholars cited in this paper were once children. Childhood, or socio-biological immaturity, is an experience shared by all adult humans, sometimes remembered sometimes forgotten, but always having shaped our lives. This stage of the life course may be conceptualised in different ways by different societies, but it is essentially that period where a child grows and learns how to be an adult, through socialisation, formal education or imitation.

The study of children is undoubtedly important in itself, but most aspects of childhood also involve adults, so understanding childhood is also essential to understanding the adult world (Orme 2009, 110). Every culture creates and defines the nature of childhood (Baxter 2006, 79), but adults are not the only influence on children; there exists a self-maintaining children's culture out of reach to outsiders (Sánchez Romero 2008, 20). Although subject to their parents' desires and agency, children also have their own agency; while they embrace cultural customs, they also struggle against them (Crawford and Lewis 2008, 11). Thus, children learn formally and informally from adults and peers, interpreting messages, ultimately transmitting refashioned culture to the next generation (Scott 1997; Baxter 2005).

While children are buried by adults, inferences can also be made about the life of the child; where they are buried and how they are buried reflects attitudes of adults to children, which affect the life of the child. Grave goods, orientation, depth and positioning all provide clues to infer cultural norms – like the age a child is considered a full member of society, or the age a child is expected to become adult.

This paper seeks to shine a light on the lived experience of Anglo-Saxon children, especially those less than two years of age. We start by introducing Anglo-Saxon funerary practices, and the cemetery at Great Chesterford, Essex, which was in use AD 450–600. Then we will examine the burials of infants and children and compare them with others. We will also present an illustrative case study, before a short discussion and conclusion.

Anglo-Saxon Archaeology

Anglo-Saxons are visible in their distinctive gendered funerary ritual – women buried with brooches and jewellery, men with weapons (Lucy 2000). Consequently, Anglo-Saxon burial archaeology has often used grave goods to determine the sex (or more accurately gender) of skeletons (Lucy 1997; Stoodley 1999); some even favour the determination of sex by grave goods over biological sexing if the two should disagree (see, for example, Hirst 1985; Evison 1987). Great Chesterford contains 20 children 'sexed', or gendered, through grave goods, 17 female (or feminine) and three male (or masculine) (Evison 1994).

Few infants are found in Anglo-Saxon cemeteries in England (Crawford 1999, 170) – the cemetery at Berinsfield, Wallingford, contains only one infant among 118 graves, while the youngest burial at Mill Hill, Deal, Kent, was two years old at death (Parfitt and Brugmann 1997). Stoodley (1999) developed a dataset of 46 Anglo-Saxon cemeteries containing 1095 aged burials, but only 28 (2.6%) individuals were aged less than one year.

The Missing Children

This absence of infants is not unusual and the differential burial of infants has a long history in time and space (Lucy 1994; Scott 1999; Kamp 2001; Lewis 2007; Murphy 2011).

This may be because a child requires particular treatment due to cultural reasons, or that they have not yet achieved personhood. Many cultures fear dead children – they may either become malevolent sprits or be too young to ward them off (Barretto-Tesoro 2008; Tsaliki 2008). Notwithstanding, quite contrary views have also been reported with dead, or even dying children, having been seen as transforming into angels (e.g. see Scheper-Hughes 1989).

Under-representation could also be the result of the fragility of infant bones, due not only to bone mineral density but also burial position (Manifold 2012); however, infants are found in larger numbers in comparable sites from Roman Britain and the Christian Anglo-Saxon period (Crawford 1993; 1999; 2007; Stoodley 1999; 2000). Infants may have been buried in shallow graves since disturbed by animals or the plough (Crawford 1999; 2000; Stoodley 1999; 2000) or, as Stoodley (1999) suggests, (although unlikely) child mortality may have been as low as the figures suggest. Alternatively, 'other methods may have been used to dispose of the quantity of dead infants that must have been cluttering up the Anglo-Saxon world' (Crawford 1993, 3). These 'other methods' may have been less visible archaeologically (Härke 1997; Crawford 2000; Stoodley 1999; 2000). Infant bones have been found in at least 11 Anglo-Saxon settlements, in pits, ditches, under floors, disregarded as rubbish and ignored in the site reports, perhaps missed by archaeologists focused on pot sherds or weaving tools (Hamerow 2006; Crawford 2008; Manifold 2011). Infants may also be interred as deliberate foundation or termination deposits associated with entrances and liminal spaces (Hamerow 2006).

One Anglo-Saxon cemetery provides an excellent opportunity to study the life of the Anglo-Saxon child. Great Chesterford, Essex, includes the graves of 88 subadults (including 63 infants less than two years of age), alongside the burials of 83 adults. Therefore, it appears to be the final resting place of the whole community; this study involves a re-examination of its mortuary archaeology.

Methods, Materials and Results

The Anglo-Saxon cemetery at Great Chesterford was excavated under rescue conditions during 1953–5, and 161 graves were discovered which included 171 inhumations, 33 cremations, two horse burials and two dog graves (Fig. 12.1). Evison (1994) dated the site through grave good typologies to AD 450–600. He concluded that it represents a community of normal Anglo-Saxon economic status, with only two swordsmen and a few rich women present, no gold and very little silver; the jewellery present is largely of bronze.

Analytical Procedure

Data from the cemetery were analysed using chi-squared statistics and independent t-tests with a 0.05 significance level. Three basic comparisons were made between subadults (0–15 years) and adults (16+ years); with subadults also assessed as the

182 Christine Cave and Marc Oxenham

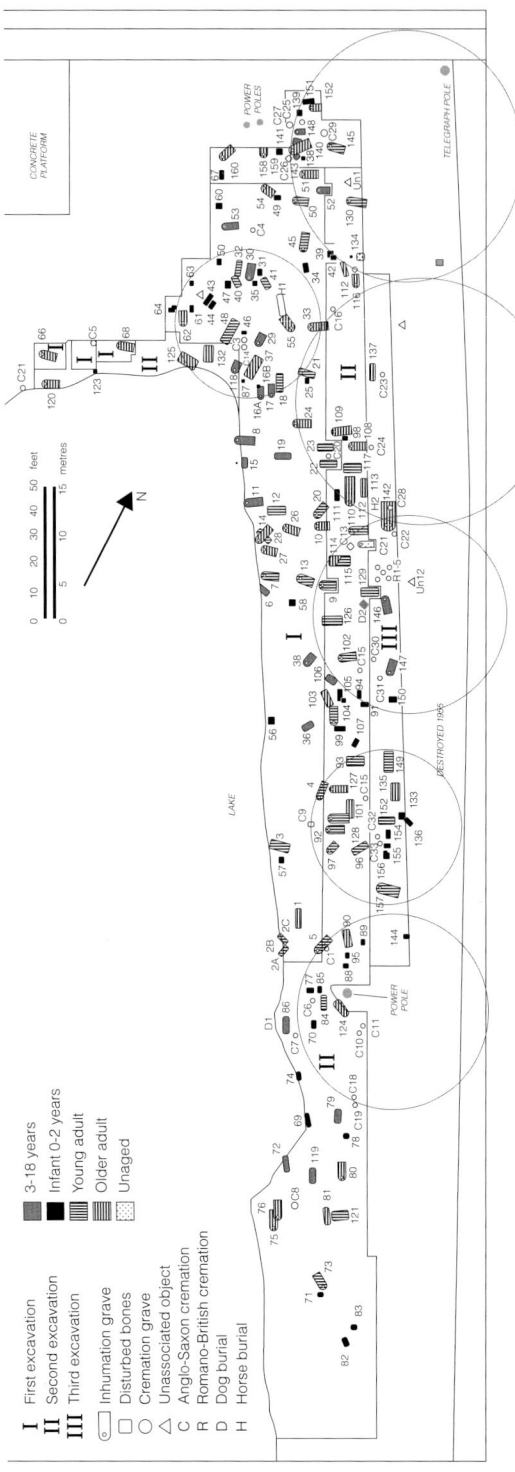

Figure 12.1. The Anglo-Saxon cemetery of Great Chesterford Essex: Age (after Evison 1994).

separate categories of infants (0–2 years) and children (3–15 years); with a further category of combined children and adults (3+ years) that excludes infants.

Demography

The excavated part of the Anglo-Saxon cemetery at Great Chesterford contained the remains of 173 individuals, including two perinates found within the pelvic region of their mothers (Table 12.1). An abridged life table of the burials (Table 12.2) shows that life expectancy at birth was 18.4 years, increasing to a maximum of 26.4 years in the 5–9.9-year age category.

Table 12.1. Great Chesterford Demography (ages in years) (data after Waldron 1994).

	Pre-term	0–2 years	3–5 years	6–9 years	10–15 years	'Juvenile'	15–25 years	26–35 years	36–44 years	45+ years	Adult	Unaged	Total
Female	–	–	–	–	–	–	9	10	18	6	1	1	44
Male	–	–	–	–	–	–	7	8	8	8	6	0	37
Unsexed	17	48	7	10	2	1	0	2	0	0	2	2	6
Total	17	48	7	10	2	1	16	20	26	14	9	3	173
Sex ratio							0.78	0.8	0.44	1.33			0.84

Total Subadults (0–15 years) 85; Total Adults (16+ years) 88

Table 12.2. Abridged Life Table for Great Chesterford (N=160) (data after Waldron 1994). Abbreviations – x: age category in years; nx: years in age category; Dx: number of deaths; dx: percentage of deaths; lx: number of survivors entering: indicates what percentage of a theoretical original population of 100 people remains alive at the end of each five-year period [calculated by subtracting dx during the preceding interval from lx in the same interval]; qx: probability of death: calculated by dividing the dx during an interval by the lx entering that interval; tx: total years lived between X and X+5: total number of years lived by all; individuals during each interval [formula: $Lx=nx(lx+lo)/2$; where lx is number of survivors entering interval x and lo is the number of survivors entering the following interval]; Tx: total years lived after lifetime: total number of years remaining in the lifetimes of all individuals entering each age interval [add values in Lx column for that interval and all succeeding intervals]; $e°x$: life expectancy: average number of years an individual entering age interval x can expect to continue to live [$e°x=Tx/lx$].

x	nx	Dx	dx	lx	qx	tx	Tx	e°x
0–4.9	5	72	45	100	0.45	387.5	1837.5	18.4
5–9.9	5	10	6.3	55	0.114	259.4	1450	26.4
10–14.9	5	2	1.3	48.8	0.026	240.6	1190.6	24.4
15–24.9	10	16	10	47.5	0.211	425	950	20
25–35.9	10	20	12.5	37.5	0.333	312.5	525	14
35–44.9	10	26	16.3	25	0.65	168.8	212.5	8.5
45–59.9	10	14	8.8	8.8	1	43.8	43.8	5

To assess fertility, a series of commonly employed measures were calculated, although the source report employed non-standard age categories – the juvenile/adult ratio (normally 5–15 years: 20+ years had to be modified to 5–15 years:25+ years) is 12:60 or 0.20; the D20+:D5+ ratio modified and reported as a D25+:D5+ ratio, is 60:88 or 0.68; the $15P_5$ ratio (normally 5–19 years:5+ years, modified to 5–25 years:5+ years) is 28:88 or 0.32. The adult sex ratio of 44 females to 37 males is 0.842 (not statistically significant from a 1:1 binomial distribution, two tailed p=0.505), while the sex ratio for the potential female fertile period (15–35 years, 15 males, 19 females) is 0.79 (again, not statistically significant from a 1:1 binomial distribution, two tailed p=0.608) (Bellwood and Oxenham 2008; Domett and Oxenham 2011; Willis and Oxenham 2013).

Waldron (1994) osteologically determined the sex of the adult skeletons, but Evison (1994) also 'sexed' or gendered individuals through their grave goods: children with brooches or beads (n=17) deemed to be female, while those with weapons (n=3), were recorded as male. The ratio of feminine to masculine subadults, based on grave good determinations was found to be statistically significantly different to a hypothesised 1:1 distribution of male and female subadults (binomial test, two tailed, p=0.003).

Burial Practices

The mean burial depth for adults was 1.07 m, and for subadults (0–15 years) it was 0.89 m. An independent samples t-test found this difference to be significant (t=3.271, p=0.001). With the exception of the age category 3–5 years, where 3/7 (42.9%) of individuals were buried at depths of at least 1.3 m, burial depth increases with advancing age until adulthood is reached.

Subadults (0–15 years) were interred in multiple burials significantly more often than adults (16+ years) (Tables 12.3 and 12.4). Subadults were buried on their sides (either right or left sides) significantly more than adults, while infants were buried on their sides significantly more often than both children (3–15 years) and a combined sample of children and adults (3+ years).

Most individuals at Great Chesterford were interred south–north (head to the south) or west–east (head to the west); reverse orientations are those between 116°–251° from True North (Evison 1994, 38–9). Significance testing suggests that infants are significantly more likely to be buried in reverse orientation compared to the rest of the population, but differences between infants and children (3–15 years), subadults and adults were not significant (see Tables 12.3 and 12.4).

Some individuals were buried in a way that marked them as different. These include prone burials, those without surviving grave inclusions, those buried with amuletic artefacts and burials containing large rocks. The largest category of non-normative burial is that without surviving grave goods (26.3%; 45/173); there are multiple examples of non-normative orientation and graves containing large rocks, as well as a mass grave (#83) containing six pre-term babies (see Table 12.3) (Connor 2009). Testing suggests that the difference between subadults and adults, infants (0–2 years)

Table 12.3. Contingency table data for burial practices and grave good types by age category (data after Evison 1994).

Age category	0–2 years			0–15 years			3–15 years			3+ years			16+ years		
	With	Without	%	With	Without	%	With	Without	%	With	Without	%	With	Without	%
Multiple burial	14	1		13	70	15.9					80		2	83	2.35
Side burial	7	29	22.2	23	11	27.7	9	10	45	21		20	12	70	14.1
Reverse Orientation	43	20	11.1	8	48	9.6	1	18	5	6	97	5.7	5	79	5.9
Non-normative burials	28	35	68.3	49	34	59	6	14	30	27	78	25.7	21	64	24.7
Any grave goods	13	12	44.4	44	39	53	16	4	80	91	14	86.7	75	10	88.2
Gendered goods	7	56	20.6	21	21	25.3	7	9	35	61	31	58.1	53	22	62.4
Weapons and jewellery	6	57	11.1	12	71	14.5	5	15	25	56	49	53.3	51	34	60
Jewellery	11	52	9.5	9	74	10.8	3	17	15	37	68	35.2	34	51	40
Beads	3	60	17.5	16	67	19.3	4	16	20	31	74	29.5	26	59	30.6
Polychrome glass beads	4	59	4.76	4	79	4.8	1	19	5	22	83	21	19	66	22.4
Amber beads	5	58	6.4	6	77	7.2	2	18	10	20	85	19.1	18	67	21.2
Multiple bead types	7	56	7.9	7	76	8.4	2	18	10	23	82	21.9	21	64	24.7
Containers			11.1	11	72	13.3	4	16	20	22	83	21	18	67	21.2
Total in age category	63				83			20			105			85	

Table 12.4. Results of significance testing of burial practices and grave good types by age category. χ2 Pearson's uncorrected. Definitions – subadult <15 years; adult >15 years; infant <2 years; non-infant >2 years; child 2–15 years. Comparisons: first value (e.g. subadult) greater than second (e.g. adult). Values in bold signify statistically significant differences – p<0.05.

	Comparison	χ2	p
Multiple burial	subadult>adult	7.585	**0.006**
	subadult>adult	32.060	**0.000**
Burial on side	infant>non-infant	29.268*	**0.000**
	infant>child	6.125*	**0.013**
	subadult>adult	2.770	0.096
Reverse orientation	infant>non-infant	5.836	**0.016**
	infant>child	0.533*	0.465
	subadult>adult	20.363	**0.000**
Non-normative burial	infant>non-infant	29.316	**0.000**
	infant>child	9.186*	**0.002**
	adult>subadult	25.219	**0.000**
Grave goods present	non-infant>infant	33.976	**0.000**
	child>infant	6.343	**0.012**
	adult>subadult	4.947	**0.026**
Gendered goods	non-infant>infant	1.730	0.188
	child>infant	0.266	0.606
	adult>subadult	37.162	**0.000**
Weapons or jewellery	non-infant>infant	29.268	**0.000**
	child>infant	6.128	**0.013**
	adult>subadult	18.746	**0.000**
Jewellery	non-infant>infant	13.671	**0.000**
	child>infant	0.008	0.784
	adult>subadult	2.865	0.090
Beads	non-infant>infant	3.056	0.080
	child>infant	0.000	1.000
	adult>subadult	9.492*	**0.002**
Polychrome glass beads	non-infant>infant	6.921*	**0.009**
	child>infant	0.000*	1.000
	adult>subadult	6.672	**0.010**
Amber beads	non-infant>infant	4.200*	**0.040**

(Continued)

Table 12.4 (Continued)

	Comparison	χ2	p
	child>infant	0.003*	0.957
	adult>subadult	8.006	**0.005**
Multiple bead types	non-infant>infant	5.531	**0.019**
	child>infant	0.000*	1.000
	adult>subadult	1.846	0.174
Containers	non-infant>infant	2.670	0.102
	child>infant	0.416*	0.520

*Yates corrected χ2 used when any value <5

and all aged 3+ years (combined children and adults), and infants and children (3–15 years) in receiving a non-normative burial is statistically significant (see Table 12.4) with the chance of receiving a non-normative burial decreasing with increasing age.

Grave Goods

Only 11.8% of adults were buried without any surviving grave inclusions, while 45.9% subadults had no goods (see Table 12.3). The differences between subadults and adults, between infants and all others (3+ years) and between infants and children (3–15 years) in relation to the presence of grave goods are significant (see Table 12.4).

The average number of beads in a necklace (21.07 overall) was also examined and found to peak in the 10–15 years age group. An independent samples t-test compared the mean number of beads worn by subadults compared to those worn by adults (t=3.130, p=0.004), as well as differences between infants and the rest of the population (3+ years) (t=3.500, p=0.002), and infants and children (t=2.157, p=0.049), finding significant differences between these categories.

To determine if any particular grave good was more likely to be placed with a child, χ^2 values were calculated for burials with or without weapons, jewellery, containers and beads. The popularity of polychrome glass beads and amber beads was also tested, as well as the likelihood of receiving more than one type of bead (see Tables 12.3 and 12.4). The results show that both subadults (0–15 years) and infants are treated significantly differently to older individuals in the placement of weapons, jewellery, polychrome glass beads, amber beads and of more than one type of bead. They are not treated significantly differently in the placement of containers or monochrome beads. Infants were treated significantly differently to children (3–15 years) in the placement of weapons and jewellery only.

Among those who received goods, the placement of apparent gender-indicating items (brooches, pins, girdle hangers, beads and weapons) was compared to those who received only gender neutral goods (likely to be deposited with both males and females, including knives, containers and nails) (see Table 12.3). Subadults received significantly fewer gendered grave goods relative to neutral goods than adults, but infants were not treated significantly differently to children (3+ years) or all older individuals (3+years) (see Table 12.4). The types and numbers of all grave goods deposited with infants (0–2 years) are presented in Table 12.5.

Case Study – Infant Grave #99

Three children were buried with apparent masculine grave good assemblages of weapons. Skeletons #16a (8–9 years) and #86 (7–8 years) were each buried with a spearhead, shield, knife and buckle; #86 also had a dog sitting approximately 10 cm above the feet. The third 'masculine' child was skeleton #99, aged at 1–2 years on the basis of dental eruption and long bone measurement, and buried with a spear, knife, buckle and a bronze ring; this child will be the focus of the case study.

Although weapons are associated with sexed, thus adult males, and the two older children could possibly have been apprentice warriors, or able to play warrior games, this does not provide proof that these children were male. These considerations cannot apply to #99 who, both physically and developmentally, would have been unable to wield a spear. At 19.7 cm the spearhead recovered from this burial is shorter than the average of the 19 Great Chesterford spears (25.1 cm), but it is still well within the range, being longer than five of them, three the property of adults. The spear was also placed in the grave in a way that it extended past the child's head, with the tip touching the end of the grave. This would tend to suggest that it was longer than the height of the child, a finding that may indicate it was not a specially

Table 12.5. All grave goods for infants (0–2 years) (data after Evison 1994).

	Pre-term (n=16)	0–2 years (n=49)
Brooch		2
Bracelet/anklet/finger ring		4
Pot/container	1	6
Amber beads		5
Monochrome glass beads		10
Polychrome glass beads		3
Roman coins		4
Spear		1
Knife		1
Buckle	1	1
Key		2
Ring	1	2
Pin		1
Miscellaneous bronze	1	2
Miscellaneous iron	1	4
Nail		5
Hobnail		2
Glass fragment		1

manufactured toy and that its placement is symbolic. It could be symbolic of the child's warrior role in the afterlife, or symbolic of the role the child did not live to assume, or placed to invoke the care of warrior ancestors (Waterman and Thomas 2011). It may represent the status of the child's family, or some other role or position. Archaeologists may ascribe an adult role to #99 due to the nature of the grave goods, although the individual had died in infancy, too young to participate in community activities (Evison 1994; Waterman and Thomas 2011). Yet, a spear may have a different meaning when carried by a child to that when it is carried by an adult.

The child was buried in a normal orientation of 60°, while the closest nearby graves were oriented at right angles to it, although still 'normal' (normal range is all excluding 116°–251° from true north). This grave was surrounded by burials largely lacking high status gendered grave goods – including three weaponless males and only one individual with a brooch. The infant #99 appears to be the highest status individual in the cluster (see Fig. 12.1). Was this child a 'foundation burial', the first burial in the group? Were great hopes held for the child that were shattered by early death? This child demanded notice at the time of burial, and still demands notice today. This glimpse of Anglo-Saxon life is short and shadowed by doubts and difficulties.

Discussion

Demography

The children of Great Chesterford were born into a society that buried everyone in the community cemetery. The demographic profile indicates high childhood mortality, with a lower life expectancy for women compared to men, a situation common for early populations (Evison 1994, 31; Waldron 1994, 59). The life table shows that 45% of Great Chesterford inhabitants died before their fifth birthday, but if an individual survived to reach that age, then they could expect to live another 19 years. This calculation is affected, however, by the difficulties of ageing mature skeletons and the entrenched under-ageing of older individuals (Lucy 2005; Cave and Oxenham 2016). It is reasonable to infer that, while there was high childhood mortality at Great Chesterford, any individual who survived the early dangerous years could hope for a reasonably long life.

Most fertility measures undertaken here suggest unremarkable levels of fertility, expected in an established agricultural community (Bocquet-Appel and Dubouloz 2004; Bellwood and Oxenham 2008). The modified $_{15}P_5$, conversely, can be interpreted as indicating a somewhat elevated fertility when compared to the data of Bocquet-Appel and Dubouloz (2004). As sub-five year olds make up 45% of the total assemblage, a relatively elevated level of fertility appears reasonable.

Burial Practices

While subadults aged 0–15 years were generally buried in shallower graves than adults, there are possible reasons for this other than lack of care or reduced energy expenditure on children. Few graves have dimensions other than depth recorded, but 'most graves were the minimum size necessary to receive the body' (Evison 1994, 28). Given that it is more awkward to dig a deep small grave than a deep large grave (Crawford 1993, 85), it follows that grave depth should increase with size and therefore with age at death. Except for the 3–5 years age category (three deep 2–5 years graves are found in a cluster of deep burials, probably due to a slightly easier to dig soil matrix), this was the case. Shallow subadult burials may be solely related to the mechanics of grave digging (Crawford 1993), rather than an indication of lesser care or energy expenditure, but could also be caused by ambivalence due to the high infant mortality rate, and/or a reduced willingness to spend time and energy on children.

Although double burials of Anglo-Saxon children are uncommon, Great Chesterford is an exception (Crawford 2007, 86), with more children than adults interred in multiple burials and mostly with other children. Three multiple burials, mass grave #83a–f and double graves #95a–b and #150a–b, were listed as 'disturbed bones', but the reasons for the disturbance are not discussed. Perhaps the act of reopening the grave to inter another individual disturbed the contents or maybe the disturbance happened during the excavation. Whatever the reason, the fact that large numbers of such graves belong to the very young is suggestive of differential mortuary treatment.

Lewis (2007, 90) notes that clusters of infants have been interpreted as evidence for infanticide, suggesting grave #83, which contained six preterm individuals of 36–40 weeks gestation, as evidence, but it is difficult to entertain a definitive verdict for infanticide from the evidence for this burial. Evison (1994) suggests that the grave was marked and reopened when needed for the burial of a stillborn baby. It is possible that this grave was used by a single family or household, or even by one woman unable to carry children to full term. Each subsequent burial may have then disturbed the previous inhumation. Given that these multiple burials were not recognised until post-excavation it is difficult to know whether these, and the other disturbed infant burials in the cemetery, were caused by Anglo-Saxon burial practices or were a consequence of a hurried rescue excavation.

Despite the large numbers of disturbed infant burials, burial position clearly changes with age with most infants buried on their sides. Rather than seeing this as a form of indifference, we suggest that it may indicate a level of care, almost tenderness, in the placement of these very young individuals in their graves.

Non-Normative Burials

Throughout time and space, individuals have on occasions been buried in a way that marks them as different; these include prone burials, those without surviving grave inclusions, those buried with amuletic artefacts and burials containing large rocks. Children are the group most likely to receive non-normative treatment (Aspock 2008, 20). Non-normative burials occur at Great Chesterford, and both children (32%) and infants (49%) were found more likely to have been given a non-normative burial than adults (20%).

The largest category of non-normative burial is interment without grave goods, and all except three of the non-normative infants were in this category (although some displayed other forms of difference as well). Fifteen children were buried with nails and/or hobnails, suggesting the presence of an object that has otherwise decomposed and although such burials are suggestive that graves may have contained other items that have not preserved, those buried without surviving grave goods still deviate from the Anglo-Saxon norm. More children than others were buried with amuletic items, perhaps used to protect the child, rather than mark them as dangerous. The observation that more non-normative burials are those of children simply suggests they were treated differently to adults – it does not suggest reasons for this difference.

Grave Goods
The dearth of grave goods interred with children and infants in comparison to older individuals aligns Great Chesterford with other Anglo-Saxon cemeteries (Crawford 1999; Stoodley 1999). Subadults received the same types of goods as their older contemporaries, albeit fewer, and usually less prestigious types of brooches, necklaces and weapons. Only three infants had brooches (n=2) or weapons (n=1 spearhead; #99 see Table 12.5) but, with the exception of swords, shield bosses and sleeve clasps, the usual adult grave goods are represented in the infant graves. Contrary to Gowland's (2007, 59) finding that goods buried with Anglo-Saxon children less than four years old are usually 'gender neutral', the Great Chesterford infants and children were buried with gendered items as often as with neutral items; although a large proportion of children received no surviving grave goods. Neither children nor infants were treated significantly differently to older individuals in the deposition of pottery, suggesting that it was a universal grave good.

The situation regarding children and grave goods is complex. The first step for newly grieving Anglo-Saxon parents may have been to decide whether the deceased child should be afforded a cemetery burial, a step overwhelmingly taken at Great Chesterford (Lucy 1998, 48). Then they needed to decide what, if any, grave goods should be deposited. Symbolism plays an important role in the masculine burial rite (Härke 1990), but it can be argued that all grave goods are symbolic, as they have no practical use (Lucy 1998), although this does not take into account beliefs held by the burying community regarding the use of these items in the afterlife. Artefacts were consciously selected for burial in the knowledge that this would mean their 'death' (Crawford 2004, 91). It is possible that items, such as swords, were symbolic items in the living world, however, as well as in the realm of the dead, meaning that the burial ritual did not remove useful items from the material world.

The burial ritual may also have functioned as an aid to forgetting (Williams 2005). The visible role of deposited grave goods may have been symbolic and/or for display, but the removal of items from the sphere of the living may have been used to eliminate reminders of the dead person from everyday life. Most grave goods appear to be personal items – jewellery, dress accessories, knives and weapons – all reminders of the deceased and the bereavement. The burial ritual not only provides a spectacle for congregated mourners, a final display to be remembered, but also removes symbols of loss from the everyday world while endowing the dead with their personal possessions. This means that low status is not necessarily the reason for the low numbers of grave goods deposited with the young. Young children, especially infants, have not created multiple relationships or established their individuality within the wider community. They also need little personal material culture – they have no tools of employment, no personal grooming equipment and few responsibilities. Infants are fed by their mothers, and have no need of showy adornments, although high status parents may endow their children with prestige items.

Crawford (1991, 18) suggests that in the burial ritual, a child's individuality may be subsumed by the status of the family and thereby become a 'text' by which the associated adult status can be read, but children are not given high status goods to the extent that adults are. This suggests that only some children are considered suitable to represent their high-status families, a representation downplayed or discounted, or that the highest status children are not displayed as such in the burial ritual. Alternatively, the child's individuality may be represented in the burial ritual.

Lucy (1998, 48) notes that burials without gender-designating items are rarely analysed as they 'cannot be accommodated in the traditional bipolar way of thinking'. One problem with such analyses is the non-specificity of such items. The most popular gender neutral items for infants are the container, or miscellaneous fragments of iron and nails. Containers may have contained food offerings, but it is difficult to infer meaning from fragments of rusty iron or to ascertain the use that nails were put to, other than to suggest that they are part of something almost completely decayed.

Gendered grave goods assemblages have reportedly been confirmed osteologically in most cases (Härke 1990; Stoodley 1999), although Lucy (1998, 34) suggests that skeletal analyses of burials have been rarely carried out or published in original reports. Although some individuals from Anglo-Saxon cemeteries are buried with the 'wrong' grave goods no such conflicts were identified at Great Chesterford. Despite this strong gendering, it is possible gender is not actually being displayed in the infant burial ritual, and the grave goods simply reflect the need to provide the child with their possessions while removing them from the realm of the living.

Sex determination of the Great Chesterford children through their grave goods has resulted in numbers of males and females which differ markedly from those expected and without expensive a DNA studies there is no way to confirm the patently skewed subadult sex ratio of 17 'females' to three 'males'. The cemetery population is also made up of individuals with gender-neutral assemblages or no surviving goods (Lucy 1997). Regarding adults, while it is claimed that their sex determination on the basis of material culture is osteologically supported in 99% of cases (Härke 1997), this approach is fraught with difficulties and gives rise to circular reasoning, while also surrendering to biases (Lucy 1997; Stoodley 1999). For subadults especially, sex determination through grave goods is highly problematic – we do not know how pre-pubescent Anglo-Saxon children or Anglo-Saxon infants were conceptualised, let alone gendered, and this is not the way to find out. The relationship between sex, gender and grave goods should be investigated rather than assumed.

Conclusions

We have sought to shine a light on the lived experience of Anglo-Saxon children, with an emphasis on infants less than two years of age. The children of the Anglo-Saxon cemetery of Great Chesterford are uniquely placed to assist us as they, unlike most Anglo-Saxon children, were buried in their community cemetery.

The study of children in archaeological contexts is important because it is through children that a society is reproduced physically, culturally and socially. Childhood is more than biological age; it is a series of cultural and social episodes leading towards adulthood. Children are people, they are not things, not extensions of the adult world; they have thoughts, feelings, experiences, emotions, activities, spaces, material culture, negotiations, worldviews and lives of their own; they have agency and even the youngest infant is able to communicate through eye contact, movement and crying (Halcrow and Tayles 2008).

The large numbers of infant graves at Great Chesterford is a demonstration of the fragility of young lives, a fragility which would have been known and felt by the adult community. Great Chesterford parents buried their children in the cemetery, where they were able to express, disguise, deny or transform their feelings of sorrow or otherwise for their loss. The frequency of infant death may have affected attitudes; some parents may have gone through the motions to comply with community expectations, while for others the burial ritual may have been the expression of acute anguish. Nevertheless, the ritual must have had meaning for the community. Notwithstanding the multiplicity of possible emotional responses, the fact that these infants were buried in the neighbourhood cemetery suggests that they were considered worthy of burial and fit for inclusion among the adult dead. For some, this message conveyed all they needed to say or were capable of expressing; for others, the addition of grave goods was necessary.

Acknowledgements

This paper was made possible, in part, by an Australian Research Council Grant: FT 120100299. Thanks to Geraldine Cave for redesigning the map of Great Chesterford Cemetery.

Note

1. Australian National University, Canberra, ACT 2601, Australia. Emails: Christine.Cave@anu.edu.au; Marc.Oxenham@anu.edu.au.

References

Aspock, E. 2008. What actually is a deviant burial? Comparing German-language and Anglophone research on deviant burials, in Murphy, E. M. (ed.), *Deviant Burial in the Archaeological Record*, 17–34. Oxford: Oxbow Books.

Barretto-Tesoro, G. 2008. Early death: an exploration of the treatment of foetuses, infants, and children in the Philippines, in Jago-on, S. C. B., De Leon, A. S. and Cuevas, N. T. (eds.), *Proceedings of the Society of Philippine Archaeologists* 6, 37–47. Manila: The Katipunan Arkeologist ng Pilipinas Inc.

Baxter, J. E. 2005. *The Archaeology of Childhood: Children, Gender, and Material Culture*. Walnut Creek, CA: Altamira Press.

Baxter, J. E. 2006. Making space for children in archaeological interpretations. *Archeological Papers of the American Anthropological Association* 15, 77–88.

Bellwood, P. and Oxenham, M. 2008. The expansions of farming societies and the role of the Neolithic demographic transition, in Bocquet-Appel, J.-P. and Bar-Yosef, O. (eds.), *The Neolithic Demographic Transition and its Consequences*, 13–34. London: Springer.

Bocquet-Appel, J.-P. and Dubouloz, J. 2004. Expected paleoanthropological and archaeological signal from a Neolithic demographic transition on a worldwide scale (Neolithic Studies 10), *Documenta Praehistorica* 31, 25–33.

Cave, C. M. and Oxenham, M. 2016. Identification of the archaeological 'invisible elderly': an approach illustrated with an Anglo-Saxon example. *International Journal of Osteoarchaeology* 26, 163–175.

Connor, M. A. 2009. Mass grave investigation, in Jamieson, A. and Moenssens, A. A. (eds.), *Wiley Encyclopedia of Forensic Science*. John Wiley & Sons, Ltd. Available at <http://onlinelibrary.wiley.com/doi/10.1002/9780470061589.fsa615/abstract>

Crawford, S. 1991. When do Anglo-Saxon children count? *Journal of Theoretical Archaeology* 2, 17–24.

Crawford, S. 1993. Children, death and the afterlife in Anglo-Saxon England. *Anglo-Saxon Studies in Archaeology and History* 6, 83–91.

Crawford, S. 1999. *Childhood in Anglo-Saxon England*. Stroud: Sutton Publishing.

Crawford, S. 2000. Children, grave goods and social status in Early Anglo-Saxon England, in Derevenski, J. S. (ed.), *Children and Material Culture*, 169–179. London: Routledge.

Crawford, S. 2004. Votive deposition, religion and the Anglo-Saxon furnished burial ritual. *World Archaeology* 36, 87–102.

Crawford, S. 2007. Companions, co-incidences or chattels? Children in the early Anglo-Saxon multiple burial ritual, in Crawford, S. and Shepherd, G. (eds.), *Children, Childhood and Society* (BAR International Series 1696), 83–92. Oxford: Archaeopress.

Crawford, S. 2008. Special burials, special buildings? An Anglo-Saxon perspective on the interpretation of infant burials in association with rural settlement structures, in Bacvarov, K. (ed.), *Babies Reborn: Infant/Child Burials in Pre- and Protohistory* (BAR International Series 1832), 197–204. Oxford: Archaeopress.

Crawford, S. and Lewis, C. 2008. Childhood studies and the Society for the Study of Childhood in the Past. *Childhood in the Past* 1, 5–16.

Domett, K, and Oxenham, M. F. 2011. The demographic profile of the Man Bac cemetery sample, in Oxenham, M. F., Matsumura, H. and Nguyen, K. D. (eds.), *Man Bac: The Excavation of a Neolithic Site in Northern Vietnam. The Biology* (Terra Australis 33), 9–20. Canberra: Australian National University Press.

Evison, V. I. 1987. *Dover: The Buckland Anglo-Saxon Cemetery*. London: Historic Buildings and Monuments Commission for England.

Evison, V. I. 1994. *An Anglo-Saxon Cemetery at Great Chesterford, Essex*. York: Council for British Archaeology.

Gowland, R. L. 2007. Beyond ethnicity: symbols of identity in fourth to sixth century AD England, in Semple, S. and Williams, H. (eds.), *Early Medieval Mortuary Practices*, 56–65. Oxford: Oxford University School of Archaeology.

Halcrow, S. E. and Tayles, N. 2008. The bioarchaeological investigation of childhood and social age: problems and prospects. *Journal of Archaeological Method and Theory* 15, 190–215.
Hamerow, H. 2006. 'Special deposits' in Anglo-Saxon settlements. *Medieval Archaeology* 50, 1–30.
Härke, H. 1990. 'Warrior graves'? The background of the Anglo-Saxon weapon burial rite. *Past & Present* 126, 22–43.
Härke, H. 1997. Early Anglo-Saxon social structure, in Hines, J. (ed.), *The Anglo-Saxons from the Migration Period to the Eighth Century: An Ethnographic Perspective*, 125–170. Woodbridge: The Boydell Press.
Hirst, S. M. 1985. *An Anglo-Saxon Inhumation Cemetery at Sewerby, East Yorkshire*. York: York University Archaeological Publications.
Kamp, K. A. 2001. Where have all the children gone? The archaeology of childhood. *Journal of Archaeological Method and Theory* 8, 1–34.
Lewis, M. E. 2007. *The Bioarchaeology of Children: Perspectives from Biological and Forensic Anthropology*. Cambridge: Cambridge University Press.
Lucy, S. 1994. Children in medieval cemeteries. *Archaeological Review from Cambridge* 13, 21–34.
Lucy, S. 1997. Housewives, warriors and slaves? Sex and gender in Anglo-Saxon burials, in Moore, J. and Scott, E. (eds.), *Invisible People and Processes: Writing Gender and Childhood into European Archaeology*, 150–168. London: Leicester University Press.
Lucy, S. 1998. *The Early Anglo-Saxon Cemeteries of East Yorkshire an Analysis and Reinterpretation* (BAR British Series 272). Oxford: Archaeopress.
Lucy, S. 2000. *The Anglo-Saxon Way of Death*. Stroud: Sutton Publishing.
Lucy, S. 2005. The archaeology of age, in Diaz-Andreu, M., Lucy, S., Babić, S. and Edwards, D. (eds.), *The Archaeology of Identity: Approaches to Gender, Age, Status, Ethnicity and Religion*, 43–66. London: Routledge.
Murphy, E. M. 2011. Children's burial grounds in Ireland (*Cillini*) and parental emotions towards infant death. *International Journal of Historical Archaeology* 15, 409–428.
Orme, N. 2009. Medieval childhood: challenge, change and achievement. *Childhood in the Past* 1, 106–119.
Parfitt, K. and Brugmann, B. 1997. *The Anglo-Saxon Cemetery on Mill Hill, Deal, Kent*. London: The Society for Medieval Archaeology.
Sánchez Romero, M. 2008. Childhood and the construction of gender identities through material culture. *Childhood in the Past* 1, 17–37.
Scheper-Hughes, N. 1989. The human strategy: death without weeping. *Natural History Magazine* (October), 8–16.
Scott, E. 1997. Introduction: on the incompleteness of archaeological narratives, in Moore, J. and Scott, E. (eds.), *Invisible People and Processes: Writing Gender and Childhood into European Archaeology*, 1–12. London: Leicester University Press.
Scott, E. 1999. *The Archaeology of Infancy and Infant Death* (BAR International Series 819). Oxford: British Archaeological Reports.
Stoodley, N. 1999. *The Spindle and the Spear: A Critical Enquiry into the Construction and Meaning of Gender in the Early Anglo-Saxon Burial Rite* (BAR British Series 288). Oxford: British Archaeological Reports.
Stoodley, N. 2000. From the cradle to the grave: age organization and the early Anglo-Saxon burial rite. *World Archaeology* 31, 456–472.
Tsaliki, A. 2008. Unusual burials and necrophobia: an insight into the burial archaeology of fear, in Murphy, E. M. (ed.), *Deviant Burial in the Archaeological Record*, 1–16. Oxford: Oxbow Books.
Waldron, T. 1994. The human remains, in Evison, V. I. (ed.), *An Anglo-Saxon Cemetery at Great Chesterford, Essex*, 52–66. York: Council for British Archaeology.
Williams, H. M. R. 2005. Keeping the dead at arm's length. *Journal of Social Archaeology* 5, 253–275.
Willis, A. and Oxenham, M. 2013. The Neolithic demographic transition and oral health: the Southeast Asian experience. *American Journal of Physical Anthropology* 152, 197–208.

Chapter 13

Emotional Act, Superstition or Ritual? – Evidence from Child Burials in the Medieval period. A Case Study from St Clemens Churchyard, Copenhagen, Denmark

Jane Jark Jensen[1]

Abstract: This is the first article to present the results of the excavation that took place in 2008 at the medieval churchyard of St Clemens in Copenhagen, Denmark. The aim of the archaeological investigation was to interpret the signs of grave markings and burial rituals within the grave. It soon became clear that the child burials were notably interesting and had the potential to provide insights relating to the beliefs of the Copenhageners in the medieval period. It was also discovered, by using a combination of historical sources and archaeological remains, that the churchyard represented the less prosperous levels of society. It was evident that much care and effort was invested in the burials of the children which might show us a little more of the mindset of the medieval Copenhagener when it came to issues of life and death.

Keywords: medieval churchyard, children's graves, burial rituals

Introduction

In 2008 the Museum of Copenhagen excavated the northern part of the churchyard belonging to the oldest known church in Copenhagen, that of St Clemens (Fig. 13.1). The churchyard was in use from *c.* AD 1000 until the Reformation (AD 1536) when the church was torn down (Johannsen 1987, 13–4). Research of this material is still in its infancy and all the results and interpretations presented in this article are based on primary observations made during the excavation and post-excavation work paid for by the contractor – which in Denmark by law, does not include research. Due to economic issues caused by changes in the project whilst undertaking the excavation,

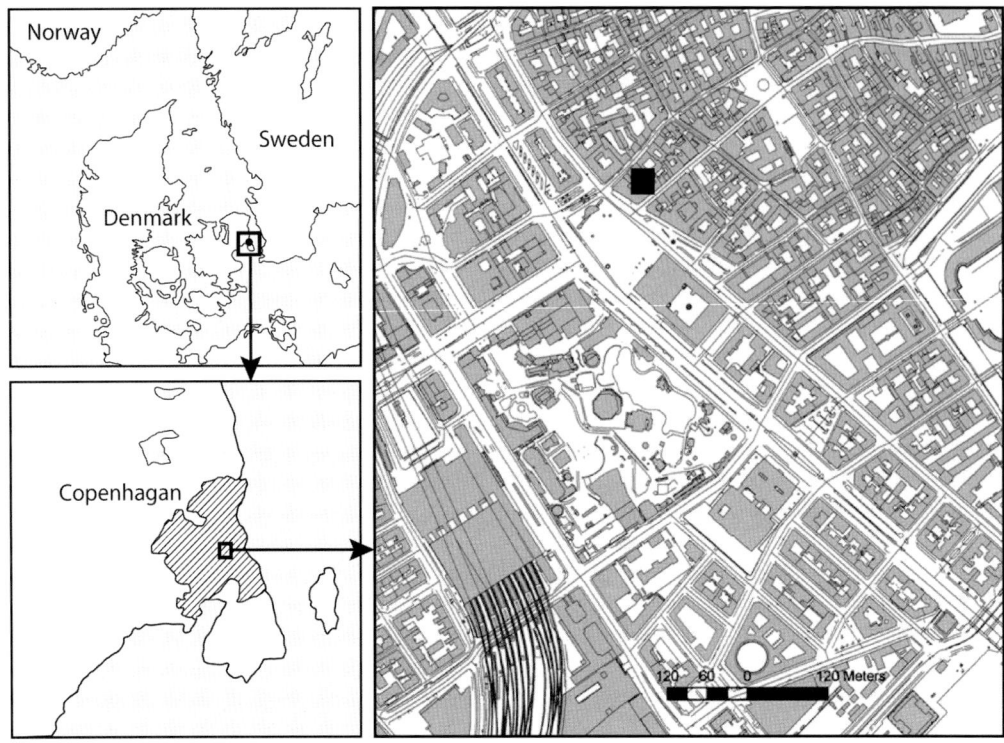

Figure 13.1. Location of the excavation (Museum of Copenhagen).

the level of recording of the graves decreased from its original aims in order to complete the project on time.

The Church and the Churchyard

Very few written sources have survived about the church and the archaeological records are very diverse in terms of both chronology and the methodology employed. The church is mentioned in a letter of Bishop Absalon, dated AD 1192–1201, and from this record it can be interpreted that it had been in use for some time at that stage (Johannsen 1987, 13). The remaining written records of the Middle Ages refer to the church as having been quite poor, struggling to survive and relying on contributions and goodwill (Ramsing 1940, 29). The church was torn down at the time of the Reformation in AD 1536 due to the poor condition of the church building (Jark Jensen and Dahlström 2009, 14) and the stones were then reused to restore the Church of Our Lady, the present day Cathedral of Copenhagen. In AD 1568 a written source on land tax refers to the existence of an empty plot in the place where the former church had been located (Johannsen 1987, 14).

The foundations of the church were rediscovered in the 19th century at a time when levels of interest in the early history of Copenhagen were awakening. The then presumed oldest part of the town was referred to as *Clemensstad* – 'The town of St Clemens'. This underlined the importance that was placed on this church as possibly having been the first church of the Copenhageners. Parts of the walls have been observed throughout the years – the last time in 1991 – and they are described as having comprised red bricks and limestone. These bricks were not found in Denmark until the 1150s (Liebgott 1989, 127) and the church was therefore dated to after this time. Some wooden structures were observed in 1906 (Rosenkjær 1909–10, 4) which led to the belief that the church had previously existed as a wooden building, but this remains to be proven.

Burials have been observed in the St Clemens area for many years as a result of construction work in the area having exposed many graves within the substantial former churchyard. This continuous construction work is the reason why the approximate extent of the churchyard has been known for at least 100 years. Not many written sources deal specifically with the churchyard but the surviving documents reveal precious information. A quote from a now disappeared account book from St Clemens Church describes a married couple, Ascaer Uddebroddesøn and his wife, who donated land/soil to the churchyard of St Clemens (Jark Jensen and Dahlström 2009, 16). The source is interesting as it implies that they not only gave physical land or a plot to the churchyard but also additional soil which was to be spread in the plot. Excavation undertaken in 2008 revealed that soil had to have been brought into the area – probably as a consequence of the intensive use of the churchyard – and the stratigraphy was found to contain up to seven layers of burials. Another source describes that in 1568 large parts of the former churchyard were sold as plots thereby suggesting that the churchyard had been decommissioned at the same time as the church was demolished (Johannsen 1987, 14).

Excavation Methodology and Results

The excavation in 2008 was undertaken in the northern area of the former church. No remains of the church were evident but plans of previous excavations suggest that the excavation went right up to and almost abutted the northern walls of the church building. A ditch was discovered to the north of the area and this was interpreted as the northern boundary ditch of the churchyard. No evidence for the original eastern or western boundaries or ditches was discovered so the extent of the burial ground in these directions remains unknown. No less than 1,048 burials were excavated and a small proportion of these contained more than one individual. The burials were quite disturbed/truncated due to the extensive use of the churchyard, and a large number of disarticulated bones were observed in the associated soil. The graves were found to date from the early medieval period (11th century) up until the late medieval period, ending with the Reformation in 1536. The phasing of the burials was based on a combination of arm positions (Positions A, B, C, D),[2] stratigraphic relations and a few datable finds (Jark Jensen and Dahlström 2009, 23-4).

The osteological material for many of the individuals was quite poorly preserved. This was mainly due to the extensive use of the churchyard and the constant disturbance of the burials since the natural soil comprised clay and the preservation conditions should have been very good. The aim of the excavation was to focus on burial practices with particular attention to the layout and shape of the graves, as well as on the occurrence of grave goods and evidence that material had been deliberately added to the base of the grave prior to interment preparation layers. An additional focus was to be on whether it was possible to detect any traces of grave markings and deliberate organisation of the burials within the churchyard. It was very clear from the beginning that some burials were more carefully arranged than others. This was not necessarily in terms of the inclusion of valuable grave goods, but rather in the nature of the arrangements of the burials, for example, how carefully a grave had been dug so that it was a perfect size to accommodate the remains of the deceased. It was also noted that stones – of both sandstone and limestone – amongst other items were used to make a burial 'special'. Furthermore it was evident that in multiple burials the arrangements of the deceased were undertaken in a way that is suggestive of care – this was particularly apparent in relation to the burials of children who were placed in a sleeping position or on the chest of the associated adult. In total around 25% of the burials showed signs of special care in their arrangement and it was only in very few cases that the treatment could be interpreted as definite evidence of higher social status (see below). The following section will summarise and then discuss the various burial types and funerary customs evident in the St Clemens churchyard. It will begin with a general description of all graves in order to put the children's burials into context.

The Burials

The different types of burials observed during the excavation are as follows:

- *Brick built graves with a separate room for the head.* Only six of these burials were excavated at the site, and they were all placed close to the church building. Both the placement close to the church and the fact they were brick built indicates that they are high status burials. This is a type that was common in the 13th century and partly continued in the 14th century. The graves had traces of having been reused several times (Jark Jensen and Dahlström 2009, 29).
- *Inhumation burials with a trapezium shape and a separate room for the head.* These were stratigraphically the oldest burial type in the churchyard. They were mainly dug quite deep into natural ground with the shape and the base levelled carefully. Most of these burials were of women, due to a likely gender-related organisation of the churchyard in the earliest period, and only two out of 79 of these graves comprised children. This variant is interpreted as an early version of the brick built

graves described above and was in use until approximately the mid-13th century (Kieffer-Olsen 1990, 99–100).
- *Shroud burials.* These were the most common type of burial on the site. Textiles or needles were only preserved in a very small number of these cases. The interpretation is therefore mostly based on the alignment of legs and feet and the fixed positioning of the arms.
- *Coffin burials.* These were very few in number and in some cases the wood used in the coffin appeared to have been reused from other objects/structures.
- *Abnormal layers within burials.* In some cases these were noted either beneath or over the deceased and they comprised crushed limestone, charcoal, sand or other materials, such as macrofossils. Note was also made of burials which contained fill that differed from that of the surrounding soil.
- *Objects or grave goods in the burials.* These comprise stones, which were found in graves either placed on different parts or adjacent to the deceased, or other objects such as items made from metal. The latter included the presumed remains of dress accessories, as well as coins, a comb and two pieces of jewellery.
- *Burials with grave markers.* Some of the graves were marked with large stones that were located on top of, or adjacent to, the deceased. In some instances the presence of an unusual fill in the burials was interpreted as an effort to mark the grave by filling it with soil that differed from the surrounding matrix. Lastly the deliberate use of older disarticulated bone material as markers was registered – larger bones or skulls were placed lying across the deceased in a manner that clearly left the impression they had been put there on purpose. In one case, for example, a skull was placed on top of a coffin lid facing the deceased (Jark Jensen and Dahlström 2009, 37–8).

This is a brief overview of the different burial types at the churchyard. A few other graves displayed individual differences but these will not be described here. The churchyard therefore contains a variety of common burial forms that have been previously documented throughout medieval Scandinavia (Kieffer-Olsen 1990, 107). The burials are suggestive of a less prosperous parish and churchyard with the interpretation made on the basis of the intense use of the burial area, the reuse of some of the graves and, to some extent, the fact that some burial plots comprised multiple graves. Lastly characteristics apparent in both the osteological material (Jark Jensen and Dahlström 2009, appendix 22 on the osteological material by Lise Harvig) and the written sources are also suggestive that St Clemens was located in a poor parish.

The Child Burials

The burials of children were spread evenly all over the excavated area of the churchyard, with only the area close to the church containing a slightly higher proportion of child burials (Fig. 13.2). In the medieval period, it was normal to use certain areas of the churchyard only for burying children (Kieffer-Olsen 1993, 92–3) but

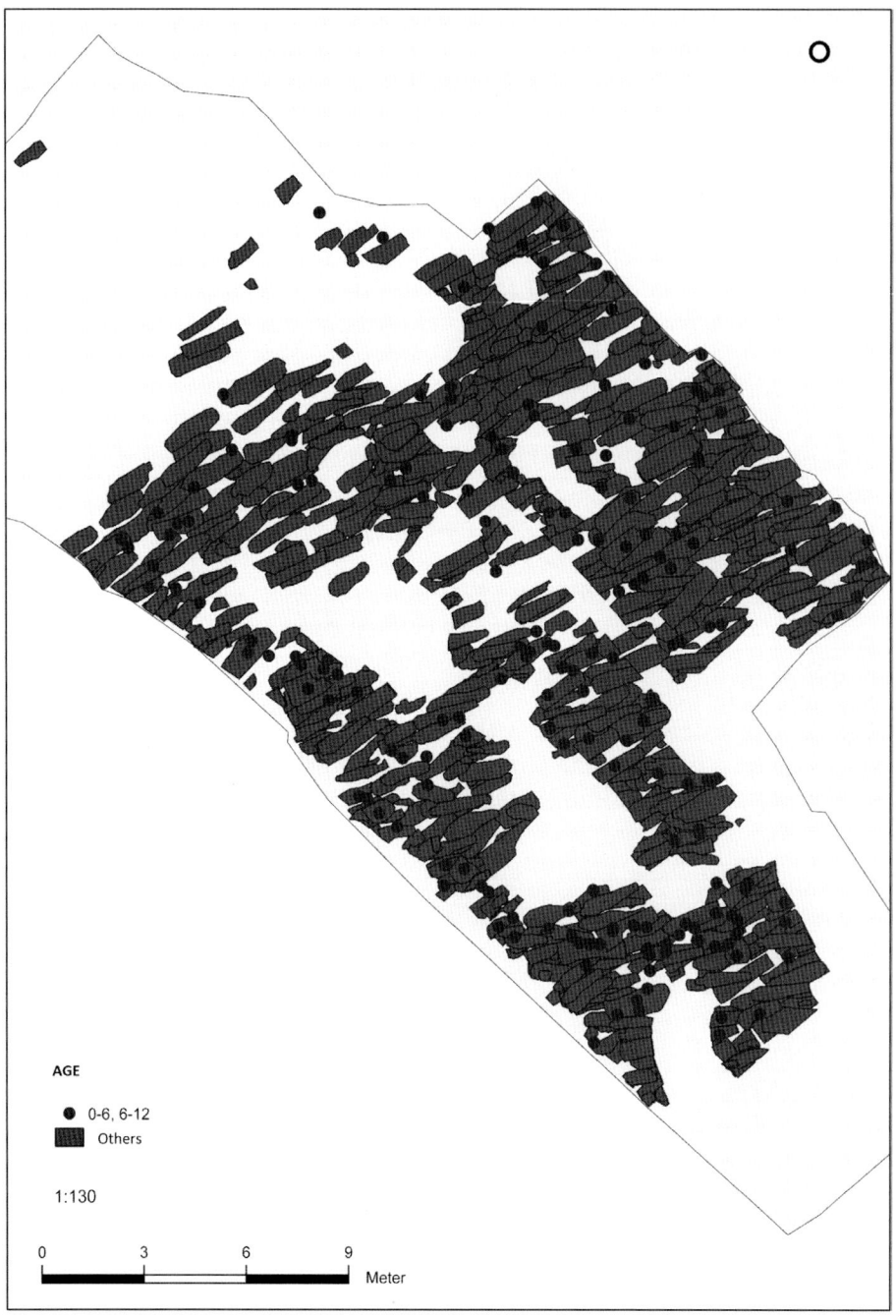

Figure 13.2. Plan showing all the graves. Child burials are marked with dots (Museum of Copenhagen).

as the excavation was restricted to the northern part of the churchyard, this cannot be confirmed at St Clemens. Approximately one-fifth of the total number of skeletons were categorised as those of children but the exact number is still uncertain since only a limited amount of the osteological material has been examined in detail to date.

At the beginning of the excavation it was observed that the upper layers contained a significant percentage of children – more than 50% of the burials excavated were children. This was quite a surprising situation but, as the excavation continued, the number of children lowered considerably. The conclusion was that more children were preserved in the upper and later layers of the churchyard and that these later burial activities had disturbed the earlier burials. It is also known that children are traditionally buried in shallower graves (Kieffer-Olsen 1990, 97). The intensive use of the churchyard also makes it probable that the remains of children would be less well preserved since they may have been more vulnerable to disturbance. The children were mainly dated on the basis of stratigraphy since the methodology of registering arm positions was difficult to undertake during excavation.

Of the 264 burials at the churchyard that were categorised as having been more carefully arranged, approximately 20% (59 graves) of these comprised child burials (Jark Jensen and Dahlström 2009, 48). These figures relate to the graves where special arrangements were visible during excavation – the exact quantity of these burial types may very well have been greater if preservation conditions had been better.

The Arrangement of the Body

As mentioned previously, children did not necessarily have the similar uniform placement in the grave as was the case for the adults. Besides the multiple burials, among others, the adult burials were only differentiated on the basis of their arm positions. In very many cases the arm positions of the children seemed to be more random, probably because the children's smaller and shorter limbs were difficult to arrange in a stable position. Furthermore, in some burials the child appeared to have been laid to rest in a manner that imitated a sleeping position.

In more than one case the bodies of children were placed with folded hands under one cheek, and a number of children were placed in a hocker burial type position.[3] This prehistoric type burial position is where the deceased is placed on the side with their legs pulled up. Both positions indicate a sleeping gesture, and this seems to have been the intention. The child was presented in a sleeping situation, which for obvious reasons provides a picture of innocence and vulnerability. Most of the children were presented lying on their backs with an east–west orientation as was the case for the adult burials. Some, but only a few, had been placed in wooden coffins and this was, as mentioned previously, very rare at the churchyard. In one of the coffins a child, age 0–6 years, had been buried with its head lying on a pillow, a situation that was only noted for a single adult burial.

Figure 13.3. Grave no. 128. Photo showing the burial of premature or new born twins. Note the stones positioned on the skeletons (Museum of Copenhagen).

Unbaptised Children, Stillborn and Infants

Unbaptised children – generally those that were stillborn or premature – were in accordance with the religious world of the medieval period considered to be in a period of limbo after birth and prior to baptism. If the child died during this time it was considered to be lost between heaven and hell (Lund 2010, 107). In some cases these children have been found buried outside the fences of churchyards or in other areas on non-hallowed ground, since they were not permitted burial in churchyards according to the rules of the Church (Kieffer-Olsen 2002, 28–9; Olsson 2002, 46). This was not the case at the St Clemens churchyard, as the remains of premature and stillborn children were identified during the osteological analysis and they appear to have been treated in the same manner as baptised individuals in the population (Fig. 13.3).

These infants displayed no evidence of having been buried in secrecy at night when nobody was watching. These very young deceased children were all buried in normal individual graves or they were buried with an adult in a multiple burial. They were also interred with care and placed in a careful manner within their graves. Only the northern boundary of the churchyard was investigated and no infants were found outside it – although it was only possible to investigate a limited area outside the burial ground due to the presence of upstanding buildings. It is not possible to say if infants had been buried outside the boundary ditch of the church to the south, east or west since the churchyard was not fully excavated but this is considered to be unlikely given the situation for the northern side of the church.

Figure 13.4. Grave no. 56. Photo of two children aged 6–12 buried together. A stone is placed over the right eye of the child buried to the right. This child also had a polished stone positioned at the left forearm. The child to the left is 'holding' a stone under its chin. Two larger stones are placed between the children (Museum of Copenhagen).

Multiple Burials including Children

A total of 25 simultaneous multiple burials were discovered, each of which contained up to three individuals. Twenty-three of these burials included the remains of an adult buried with one or two children or children buried together (Fig. 13.4). As was the case for the burials of very young infants discussed above, there were no indications that any of these children had been interred in secrecy in an open grave or on top of the lid of the coffin. All of the children in multiple burials with an adult were placed in a careful fashion either lying on the chest of the adult or close to it and a sleeping position was in some cases observed for the children. It is possible, but perhaps unlikely, that all of the individuals buried together may have derived from a single family. Due to the perception that the St Clemens churchyard was used by the poorer members of society it could be argued that some of the children were buried with an adult stranger to save on the expenses required for an individual burial. This would mean, however, that it was socially acceptable to bury strangers together and, moreover, that it was done with care. It is also suggestive that being buried with a child may have been an honour for an adult. The placement of children in multiple graves also gives the impression that there was a degree of concern to secure the child a 'safe journey' to the afterlife – the adult in the burial may have been viewed as capable of securing this.

Stones within Graves and Other Special Features

Stones play a major role at the St Clemens churchyard and they are found in numerous types of graves as a grave good, marker or decoration. St Clemens church is located in an area not known to have a high natural stone content in the ground. Therefore it soon became quite obvious to the archaeologists excavating the graveyard that a lot of the stones observed during the excavation were not accidental but, rather, purposeful associations. Again it needs to be noted that the economic constraints of the excavation meant it was not possible to continue the high standards of recording of burial stones undertaken at the start of the excavation for the duration of the project and it is quite likely that the number of stones deliberately associated with graves was a lot higher.

The graves had stones placed in their corners, alongside the body, on top of the body and also beside the head. The association of stones with the head is known from other burial sites in Denmark, England and France amongst others (Kieffer-Olsen 1993, 143–5) and is interpreted as a variation of another form of burial in which a special area is made for the head. It is thought to be related to an early medieval burial form that placed the head of the deceased looking upwards to meet their Saviour (Cinthio 2002, 73). It became more noticeable as the excavation progressed that a large number of stones were not placed by the head but instead functioned as grave goods. These types of stones ranged from 3–10 cm in diameter, were fashioned from all types of available stone and differed in shape, being flat, rounded, uneven or polished. Small polished stones were found on several occasions in children's graves, mainly placed on top of the body, particularly on the chest and arms. This suggests that the placement of the stone was deliberate and may have had a symbolic value (Fig. 13.4). A suggestion is that the polished stones may have been used, and placed in a special position, as a type of amulet but this may also have been the case for the less decorative stones.

Stones and their links with burials are known from different cultures to have a religious or ceremonial significance and stone is a material that everybody – rich or poor alike – has access to. The interpretation of the use of stones suggests several possible differences of function. Some of them were quite large and these were mostly interpreted as grave markers. Other minor stones were placed on the bodies – for example, in at least two cases on the eye. The placement of stones on the bodies seems to have been deliberate and it is natural to think that it was not a random act. Could it be that a stone was placed on a diseased or injured part of the body that had perhaps been related to the death of the individual? It would be interesting to test this theory in conjunction with the osteological findings in a future study. Figure 13.4 displays the remains of two children, both of whom were aged 6–12 years, and were buried together in Grave no. 56. The burial contains examples of both what has been interpreted as a large marking stone and a stone that was deliberately placed on the bodies. A polished stone appears to have been deliberately placed on the forearm of one of the children. The polished stones found in the children's burials possessed

Figure 13.5. Grave no. 189. Photo showing the placement of charcoal inside the cranial vault of a child age 6–12 years. Unfortunately, the skull was crushed before excavation and therefore it was not possible to observe how this had been added to the cranium (Museum of Copenhagen).

qualities that are suggestive of a more personal item, such as an amulet or grave good, but they could also have served the same function as the plainer stones.

In addition to the presence of stones in graves other special features were also observed. One of the more spectacular burials is that of a child, aged 6–12 years, which contained charcoal inside the cranial vault (Fig. 13.5). The skull had somehow been deliberately opened and filled with charcoal, a substance known for its symbolic cleansing properties (Kieffer-Olsen 1990, 107). This was quite a drastic act, and perhaps the child had suffered from a disease that affected the head – maybe a brain tumor or a mental illness – that the bereaved wanted to remove before the journey to the afterlife.

Emotional Act, Superstition or Ritual?

The differentiating features presented in this article represent more than ritual. The archaeologists who excavated the site were in no doubt that the carefulness of the child burials was an emotional act. The way the children were placed in multiple burials, arguably showed a placement of the dead child near an adult chest or neck as a special sign of affection. The fact that premature or stillborn infants were buried within the churchyard in carefully arranged graves also supports this interpretation. It is hard to argue against this view, unless one considers signs of affection to have substantially changed across time.

The act of placing stones over specific parts of the body or within graves may be indicative of a more symbolic character since stones have always played such a role. They have also been viewed as the 'foresaker of evil' (keeping evil away) or, in a practical capacity, as having the ability to hold something down (Stefánsson 2009, 466–9). This could point to the fact that evilness had to be dealt with in certain children's graves. The evilness could be illness – mental or physiological. It can also be interpreted as a reverse emotional act – an act to protect the living. The polished stones seem special in this respect since they were deliberately worked and it is possible they served as amulets or good luck charms for the journey to the afterlife.

Were the burials of the children a result of emotional acts, superstition or rituals is a difficult question to answer. The selection of child burials presented here shows individuality in the burials that could represent all three possibilities. The medieval church demanded certain uniformity in both ritual and burial practices but most of all it demanded a humble approach to life and death. This is also displayed at the St Clemens cemetery where in the overall picture, the vast majority of burials were simple shroud burials in pits and orientated west–east. Although little variety was evident in the medieval burial ritual at St Clemens, those variations that were present provide a glimpse of diversity in the mindset of the medieval people of Copenhagen and the emotional nature and actions of their society. The evidence left for us from 500 to 1,000 years ago represents only a small fraction of the burials but these traces reveal that individuality did indeed thrive in the medieval period.

Notes

1. Museum of Copenhagen, Vesterbrogade 59, 1620 København, Denmark. Email: janejj@kff.kk.dk.
2. The method of dating burials through arm positions has been used in Scandinavia since the 1950s but was fully developed in a thesis by Lars Redin in 1976 (Kieffer-Olsen 1993, 21). The date ranges are still under discussion but in general the connection between arm position and dating is as follows – Position A (arms against the sides) – AD 1000–1250; Position B (hands over pelvis) – dominates in AD 1250–1350; Position C (lower arms parallel over stomach) – AD 1350–1450; Position D (hands over chest) – AD 1450 to the present day.
3. It is not possible to provide final numbers at this stage because the analysis is still ongoing.

References

Cinthio, M. 2002. *De första stadsborna. Medeltida gravar och människor i Lund*. Stockholm: Brutus Östlings bokf Symposion.

Harvig, L. 2009. Appendix 22 in Jark Jensen, J. and Dahlström, H. 2009 KBM 3621. Beretning for Skt. Clemens I og III. Udgravning af den nordlige del af kirkegårdetilhørende den middelalderlige Skt. Clemens kirke, København – februar til juli 2008. Unpublished report prepared for the Museum of Copenhagen.

Jark Jensen, J. and Dahlström, H. 2009 KBM 3621. Beretning for Skt. Clemens I og III. Udgravning af den nordlige del af kirkegårdetilhørende den middelalderlige Skt. Clemens kirke, København – februar til juli 2008. Unpublished report prepared for the Museum of Copenhagen.

Johannsen, B. B. 1987. S. Clemens Kirke. *Danmarks Kirker. København* 6, 13–19.

Kieffer-Olsen, J. 1990. Middelalderens gravskik i Danmark – en arkæologisk forskningsstatus. *Hikuin* 17, 85–112.

Kieffer-Olsen, J. 1993. *Grav og gravskik i det middelalderlige Danmark*. Ph.D. thesis, University of Aarhus. Aarhus: Afd. for Middelalder-arkæologi og Middelalder-arkæologisk nyhedsbrev.

Kieffer-Olsen, J. 2002. Gravskikken som kilde til Danmarks kristning, in Arneborg, J., Kieffer-Olsen, J. and Nyberg, T. (eds.), *Kristningen af Norden – et 1000 års jubilæum*, 25–34. Odense: Center for Middelalderstudier, Syddansk Universitet.

Liebgott, N-K. 1989. *Dansk middelalder arkæologi*. Copenhagen: G.E.C GAD.

Lund, A. A. 2010. *Magi og hekseri*. Gylling: Gyldendal.

Olsson, M. 2002. De döpta och de udöpta barnen, in Tegnér, M. and Ödman, C. (eds.), *S:t Jörgen mitt i medeltiden*, 43–47. Malmo: Elanders Skogs Grafiska.

Ramsing, H. U. 1940. *Københavns Historie og Topografi i Middelalderen* (vol. 2). Copenhagen: Munksgaard.

Rosenkjær, H. N. 1909–10. *Historiske Meddelser om København*. Copenhagen: G.E.C. GAD.

Stefánsson, F. 2009. *Symbolleksikon*. Slovenia: Korotan Ljubljana.

Chapter 14

Interpreting Cultural and Biological Markers of Stress and Status in Medieval Subadults from England

Heidi Dawson-Hobbis[1]

Abstract: The aim of this paper is to explore the presence of stress to the growing child, by studying the prevalence of stress indicators (enamel hypoplasia, cribra orbitalia and periostitis), in association with the archaeological evidence for funerary status, within three medieval (AD 1086–540) cemeteries from England. Data is presented from 97 adult and 93 subadult skeletons of individuals buried at the late medieval priory of St Peter and Paul, Taunton. The prevalence rates of cribra orbitalia and enamel hypoplasia are both seen to be higher among the subadult population, while evidence for periostitis is seen to be fairly equal. Data collected from subadult skeletons from two other late medieval sites, the priory of St Oswald, Gloucester (65 individuals), and the priory of St Gregory, Canterbury (104 individuals), were also compared to that for the subadults from Taunton. Prevalence rates for stress indicators were higher at all sites for individuals of high status – those buried within church buildings. Prevalence rates were also higher for the Taunton children in comparison to the Canterbury children, with the funerary evidence indicating the burials at Taunton may have been of higher status.

Keywords: subadult, status, burial, cribra orbitalia, enamel hypoplasia, periostitis

Introduction

This paper aims to explore the presence of stress to the growing child and to question some of the assumptions that are regularly made in the interpretation of the manifestation of stress indicators seen in both subadult and adult skeletons. Stress is defined by Selye (1956, 55) as 'the non-specific response of the body to any demand', and markers, commonly termed *stress indicators*, can be recognised in the skeleton

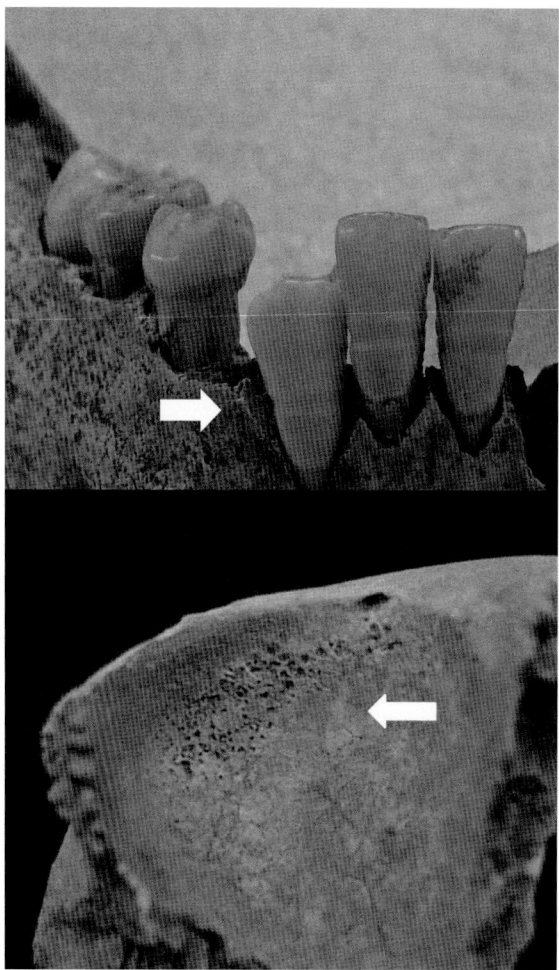

Figure 14.1. Top: Linear enamel defects on the permanent incisor teeth of a 7–8-year-old (SK 23) from Canterbury. Bottom: Cribra orbitalia in the right orbit of a 7–9-year-old (SK B106) from Gloucester (photographs by Heidi Dawson-Hobbis).

due to the ability of an individual to recover from the causal insult (Bush 1991, 11). Stress affecting the growth of a child can be caused by many factors, including nutritional deficiencies (Brickley 2000), infectious disease (Mensforth et al. 1978), parasites (Stuart-Macadam 1991) and psychological disturbance (Bush 1991, 16). The term *stress indicator*, therefore, incorporates various different lesions that can be seen in the skeleton caused by stress; some of these will remain present throughout adult life. The prevalence of three of these indicators will be explored in this study – enamel hypoplasia, cribra orbitalia and periostitis.

Enamel hypoplasia is the term used to describe defects on the teeth caused by a disruption to the growth of the enamel (Fig. 14.1). These defects include single or multiple pits, cusp deformation, and lines, which can range from slight marks to deep furrows within the enamel. The tooth crowns start to develop in-utero and continue developing until about seven years of age, with the exception of the later forming third molar (Hillson 1996, 125). Hypoplasia will therefore reflect disruption to growth of the enamel during the period of formation of the tooth. Enamel hypoplastic lines tend to only occur on the permanent teeth and therefore are related to postnatal stress to the child. Some defects, usually in the form of pits rather than lines, do occur on the deciduous dentition and these tend to most commonly involve the deciduous canine tooth (Skinner 1986). Enamel defects can occur on any tooth, however, and for this study all teeth present were analysed.

Disruption to the growth of the tooth enamel can be caused by nutritional deficiency, childhood illness (Hillson 2003, 7), and even emotional stress (Roberts and Manchester 1995, 164). These markers will remain on the teeth throughout life, and episodes of stress as a child can be inferred from the dentition of adults as well as children. Clinical studies have demonstrated that there are strong links between childhood infection and malnutrition (Rice *et al.* 2000), and that enamel hypoplasia is a good indicator for studies on the links between socioeconomic conditions and health in past populations (Hillson 2003).

Cribra orbitalia manifests as porosity within the orbits of the skull (see Fig. 14.1) and is the result of marrow expansion for increased red blood cell production (Lewis 2007, 111). The porosity consists of small holes in the outer table, which may increase in size and eventually link together as the outer table thins, exposing the trabecular bone underneath; the orbits are usually affected bilaterally (Stuart-Macadam 1991).

Traditionally interpreted as being associated with iron deficiency anaemia (Mensforth *et al.* 1978; Stuart-Macadam 1985), this assumption has recently been questioned by Walker *et al.* (2009). They state that recent haematological research shows that iron deficiency alone cannot account for the red blood cell production that causes the expansion of the marrow responsible for these lesions, although this claim is refuted by Oxenham and Cavill (2010). Walker *et al.* (2009) propose that it is megaloblastic anaemia caused by a vitamin B12 deficiency that is the likely cause of these pathological changes, in association with poor sanitary conditions leading to parasitism and diarrhoea. The presence of bacterial infection may play a part in the appearance of anaemia (Lewis 2007, 113). Supplies of iron in the blood are needed for bacteria to thrive, and therefore the body may try to withhold iron if under attack by certain bacterial infections. This suggests that a mild iron deficiency (hypoferremia) is not necessarily a negative condition but may be one of the body's defensive mechanisms against disease (Stuart-Macadam 1992). More recently, Rothschild (2012) has suggested we need to move away from the idea that iron deficiency causes the lesions associated with marrow hyperplasia, but that iron deficiency is the result not the cause. He suggests parasitic infection may be the important focus, and not dietary deficiency; the recent study by Steyn *et al.* (2016) appears to support this view.

In periostitis, periosteal bone formation occurs as a new layer of bone under the periosteum caused by the inflammation of this area in response to a variety of causes. The main cause has been traditionally interpreted as non-specific infection or trauma to the area (Ortner 2003, 207). More recently, Weston (2008) has discussed many categories of pathology that can lead to the manifestation of periostitis including metabolic diseases such as scurvy. Scurvy is caused by a lack of vitamin C, which in turn causes defects in the formation of connective tissues, which can lead to susceptibility to haemorrhage (Brickley and Ives 2008, 47). Periostitis can appear either as active woven bone, which is disorganised and porous, or healing/healed lamellar bone which is smoother and more organised in structure. A mixture of the two types of bone can be indicative of a chronic, active infection (Lewis 2007, 135). Non-specific infections can be caused by bacteria such as staphylococci or streptococci, whilst specific infections such as tuberculosis can also incorporate such lesions. There is a synergy between infection and malnutrition, where malnourished individuals will be more susceptible to disease, and infection can worsen nutritional status (Rice *et al.* 2000; Lewis 2007, 100). A caveat is that pathological lesions will only show on the skeleton if the immune system has reacted and the individual survived for some time. If a child has succumbed quickly to a disease, no evidence will be present on their skeletal remains.

One of the problems when dealing with periosteal reaction in subadult skeletons is differentiating between the pathological and normal growth that is occurring on the skeletal elements. Non-adult periostitis is most often characterised as a unilateral isolated patch of bone raised above the original cortex (Lewis 2007, 135). However, the lesions recorded by Mensforth *et al.* (1978) frequently exhibited bilateral expression and were most common on the tibia, humerus and femur. Unilateral lesions are more likely caused by injury, whereas bilateral lesions may be more likely to reflect infectious causes (DeWitte 2014). If the deposits are due to widespread inflammation they may be indistinguishable from rapid appositional growth; this has meant that certain conditions seen in infants, such as birth trauma and cortical hyperostosis, are rarely recognised archaeologically (Lewis and Roberts 1997).

One of the initial studies that attempted to look at the biological evidence for episodes of stress to the growing skeleton, and relate this to changes in population status over time focused on prehistoric American populations (Goodman *et al.* 1984). An increase in the prevalence rates of the three stress indicators described above was observed over time, relating to the shift from a hunter-gatherer subsistence strategy to an agricultural one. Many previous studies have tended to involve only adult skeletal remains, such as that by Robb *et al.* (2001) who found no link between location of burial and stress indicators or stature, and Pechenkina and Delgado (2006) who determined two distinct social groups. Two studies that focused on medieval subadult remains are those by Bennike *et al.* (2005), who compared Danish subadult skeletons from the medieval period and found a higher prevalence of stress indicators in a lower status group, and Craig and Buckberry (2010) who concluded that the southeastern area of the medieval cemetery at Raunds Furnells, England, was of lower status, both in terms of grave furnishings and because of a higher presence of stress indicators.

The data presented here derives from the author's analysis of skeletal remains from three priory church and cemetery sites from the south of England, dating from AD 1086–540 (Fig. 14.2). The first skeletal collection used for this analysis came from the excavation of the cemetery of the priory of St Peter and St Paul, Taunton, Somerset (Dawson 2014). The skeletons of both subadults and adults have been analysed and number 93 and 97 individuals respectively. Comparative data for the subadult population was collected from two other cemetery sites – the remains of 65 children excavated from the cemetery of the priory of St Oswald, Gloucester (Heighway and Bryant 1999), and 104 subadults excavated from the priory of St Gregory, Canterbury (Hicks and Hicks 2001). Throughout this paper the skeletons will be referred to as having derived from Taunton, Gloucester or Canterbury.

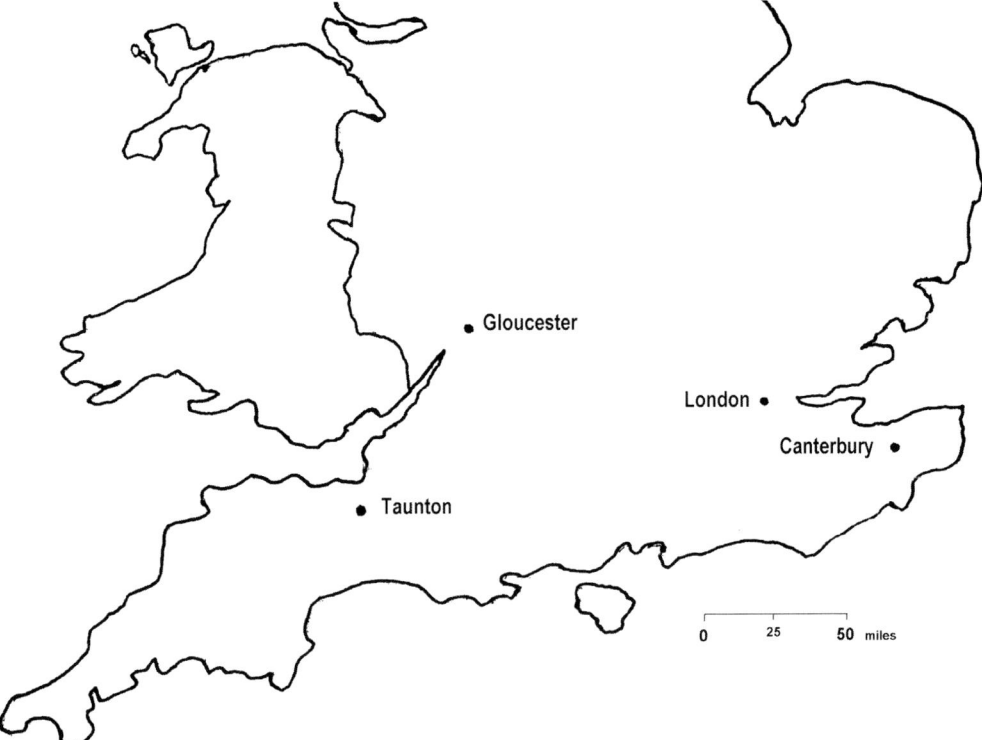

Figure 14.2. Map of southern England indicating the location of the three cemetery sites included in the study (prepared by Heidi Dawson-Hobbis).

Funerary Context

Both adults and subadults were present in all areas used for burial at the three priory cemetery sites, and these were labelled, church, west of the church, north of the church and south of the church. None of the sites had been excavated to the east and therefore no burials fell within this category. The limited and incomplete nature of all three excavations makes it difficult to comment on the preferred burial location for subadults as a whole, however, it was noted at Taunton that burial appeared to be more popular for subadults, than adults, close to the west side of the front of the church. Gloucester was the only site where the north side was excavated and it was shown to be a popular burial location for subadults, in particular infants less than two years of age. The funerary context was explored in more detail for the cemetery at Taunton. The adults and subadults varied in terms of burial position, and the adults were most often supine, with their arms across the pelvis or stomach area, whilst the subadults were laid with their arms by their sides. Only 13 adult and two child burials appear to have been interred within coffins, the majority of which were located within the church buildings. Evidence for coffined burial came from the remains of coffin nails surrounding the body or, in one case, the preservation of a charred wooden plank. The plank burial had no coffin nails present and suggests that care needs to be taken in assuming coffins are only present if nails are found. The majority of excavated burials in the church were of adult males, whilst in the area to the west of the entrance they mostly comprised children and adult females. Dawson (2014) gives a more detailed discussion on the funerary context of all three sites.

Stress and Burial Status

Enamel Defects

Enamel hypoplasia was common on both the subadults and adults from Taunton. For the adult population, 40% (22/55) of individuals with dentitions had enamel defects present on one or more permanent teeth. For the subadults, 46% (29/63) of individuals had enamel defects present on either, or both, the deciduous or permanent teeth; with 49% (22/45) on the permanent teeth, and 22% (12/55) prevalence on the deciduous teeth. When the data from the subadults of the three sites was collated together for the permanent teeth the most common enamel defect was found to be that of hypoplastic lines, whilst on the deciduous teeth lines were rare and defects tended to consist of either single pits or a series of pits in the enamel. The overall prevalence of defects scored per individual, 46%, was the same at Taunton (29/63) and Gloucester (16/35), whilst it is much lower for the Canterbury population at 26% (24/92). At Gloucester the defects often appeared to be more severe and to involve more teeth per individual. When enamel defects were tested by site using the chi-square test for independence, significance was seen for the permanent dentition (χ^2 9.522, df 2, p 0.009). Table 14.1 illustrates the difference in prevalence by burial location with a high prevalence of church burials scored for enamel defects. There was also a statistically significance difference between those buried to the west, compared to those buried to the south which is mainly reflecting the difference seen between the sites at Taunton, where the majority of burials were located to the west, and Canterbury, where the majority of burials were to the south (χ^2 9.439, df 3, p 0.024).

Table 14.1. Prevalence of enamel defects by burial location. N = number of skeletons analysed; Dec = deciduous teeth; Perm = permanent teeth.

	N Dec	Dec defects	N Perm	Perm defects	N all teeth	All teeth defects
Church	10	3 (30%)	9	4 (44%)	12	7 (58%)
West	92	16 (17%)	73	33 (45%)	103	44 (43%)
North	9	2 (22%)	5	1 (20%)	10	3 (30%)
South	59	7 (12%)	47	8 (17%)	65	15 (23%)
Total	170	28 (16%)	134	46 (34%)	190	69 (36%)

Cribra Orbitalia

In the Taunton population the subadults had a higher prevalence of cribra orbitalia than the adults, with 52% (26/50) presence, as opposed to 38% (20/52). The frequency of cribra orbitalia is very similar for the subadults from the three cemetery sites, with 43% (18/42) prevalence at Gloucester, and 44% (28/63) prevalence at Canterbury. No significant difference was evident when the three sites were tested against each other using the chi-square test for independence on the presence or absence of cribra orbitalia (χ^2 0.939, df 2, p 0.625). The highest prevalence of cribra orbitalia at Taunton for both the adult (57%) and subadult population (63%) was obtained from those buried close to the west side of the front of the church, and was lowest for the adult intramural burials (10%).

Table 14.2. Prevalence of cribra orbitalia by burial location. N = number of skeletons analysed; CO = cribra orbitalia presence.

	N	CO
Church	9	8 (89%)
West	82	44 (54%)
North	21	5 (24%)
South	43	15 (35%)
Total	155	72 (46%)

Table 14.2 illustrates the prevalence rates of subadult individuals with cribra orbitalia by burial location for all three sites. The prevalence of cribra orbitalia is high for those subadult burials from within the church buildings at 89% (8/9), compared to 44% (64/146) for the cemetery burials. The Fischer's exact test was performed, due to low numbers of data for the church burials, and gave a significant result (χ^2 14.793, p 0.002). A higher prevalence was also seen for those individuals buried to the west of the church, in comparison to those buried to the north and south.

Periostitis

Both woven and lamellar bone lesions (Fig. 14.3) were recorded for the adult and subadult population from Taunton. Of the 97 adults analysed from Taunton, 20 had some form of periostitis present – four (4%) as woven bone formation and 16 (16%) as lamellar bone. Of the 93 subadults, 18 had some form of periostitis visible; woven bone was seen in 16 individuals (17%) and lamellar bone in six individuals (6%). Therefore, a difference was evident in the type of bone formation present, with the subadults tending to display woven bone formation, while lamellar bone was apparent in the adults. This could be interpreted as an indication that children are more susceptible to dying quickly from stress-related disease, but it may also reflect the natural process of the growing bone in subadults (Lewis 2007). The higher prevalence of lamellar (healed) bone formation, as opposed to that of woven (active) bone formation, on the adult population at Taunton also supports DeWitte's (2014) suggestion that the presence of healed periostitis may be associated with increased longevity. Grauer (1993) found that evidence for periostitis on medieval subadults, from the cemetery of St Helen-on-the-Walls, Yorkshire, was low, with remodelled lesions being associated with adulthood.

For the subadult population from all sites periostitis was recorded on 19% (18/93) of the individuals from Taunton, 12% (8/65) of the individuals from Gloucester, and

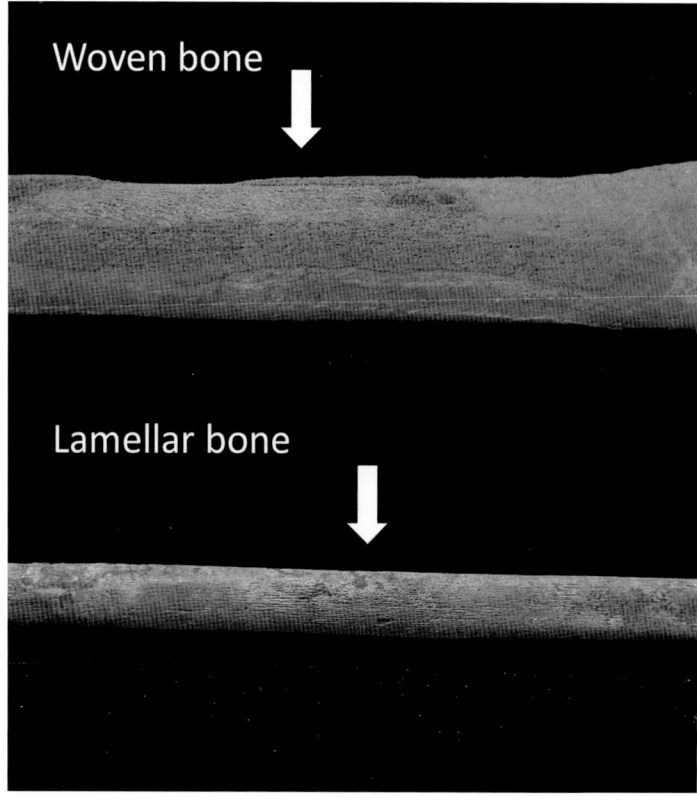

Figure 14.3. Top: Example of woven bone formation seen on both femora of a 4–5-year-old (SK 136) from Gloucester, Bottom: Example of lamellar bone formation on the right fibula of a 7–8-year-old (SK 800) from Canterbury (photographs by Heidi Dawson-Hobbis).

Table 14.3. Number of individuals with periostitis by burial location.

	N	Woven or lamellar bone	Porosity and/or striations
Church	15	3 (20%)	1 (7%)
West	143	24 (17%)	15 (10%)
North	34	3 (9%)	7 (20%)
South	70	5 (7%)	18 (26%)
Total	262	35 (13%)	41 (16%)

7% (7/104) of the individuals from Canterbury. Of the 71 bones from 33 individuals with periostitis present, 37% affected the tibia, 23% the femur, 11% the fibula, 6% the humerus, 4% the ribcage, 3% the mandible, and 3% the temporal bone. Other bones affected with less than 2% prevalence each comprised the parietal, occipital, sphenoid, frontal, zygomatic, radius, ulna and ilium. Porosity and/or striations were also recorded on the subadults as it has been suggested they may indicate non-specific

Table 14.4. Prevalence of multiple stress indicators by site and age groupings. N = number of skeletons analysed; I = number of skeletons with one stress indicator present; CO = cribra orbitalia; EH = enamel hypoplasia; P = periostitis; All = all three stress indicators present.

	N	I	CO+EH	CO+P	EH+P	All
Taunton 0–2 years	14	6	–	1	–	–
Taunton 2–7 years	38	19	4	2	2	3
Taunton 7–12 years	23	17	6	–	1	–
	75	42	10	3	3	3
Gloucester 0–2 years	13	1	–	–	–	–
Gloucester 2–7 years	15	10	2	2	–	2
Gloucester 7–12 years	12	12	5	–	–	1
	40	23	7	2	–	3
Canterbury 0–2 years	15	4	–	–	–	–
Canterbury 2–7 years	57	25	4	2	–	–
Canterbury 7–12 years	25	17	4	1	–	1
	97	46	8	3	–	1
Total all sites	212	111	25	8	3	7

infection (Ribot and Roberts 1996, 70). At Taunton and Gloucester porosity and/or striations were seen on 10% (9/93) of these populations, whilst at Canterbury they were apparent on 24% (25/104) of the population.

For this sample of subadults it is the bones of the leg that are most commonly involved and the tibia in particular. Porosity and striations were the most common lesions seen, with most of these being recorded on the tibia. Woven bone growth is more common than lamellar bone, as would be expected both in growing children, and if the individual did not recover from an infectious episode. The high prevalence of periostitis on the medial side of the tibiae in particular may indicate that this is a sign of normal bone growth. However, previous studies indicate that the tibia is often most frequently involved for adults as well as children (DeWitte 2014).

When the presence of woven and/or lamellar bone was tested for individuals by site using the chi-square test for independence a significant result was obtained (χ^2 7.086, df 2, p 0.029). Taunton has a high frequency of woven or lamellar bone recorded, whilst Canterbury has a high prevalence of porosity or striations recorded. Table 14.3 illustrates the prevalence of woven and/or lamellar bone by burial location where a high prevalence is evident for the church burials as well as those buried to the west.

Presence of multiple stress indicators

The relationship between the three stress indicators in individuals within three age groups (0–2 years, 2–7 years and 7–12 years) was explored and the results are

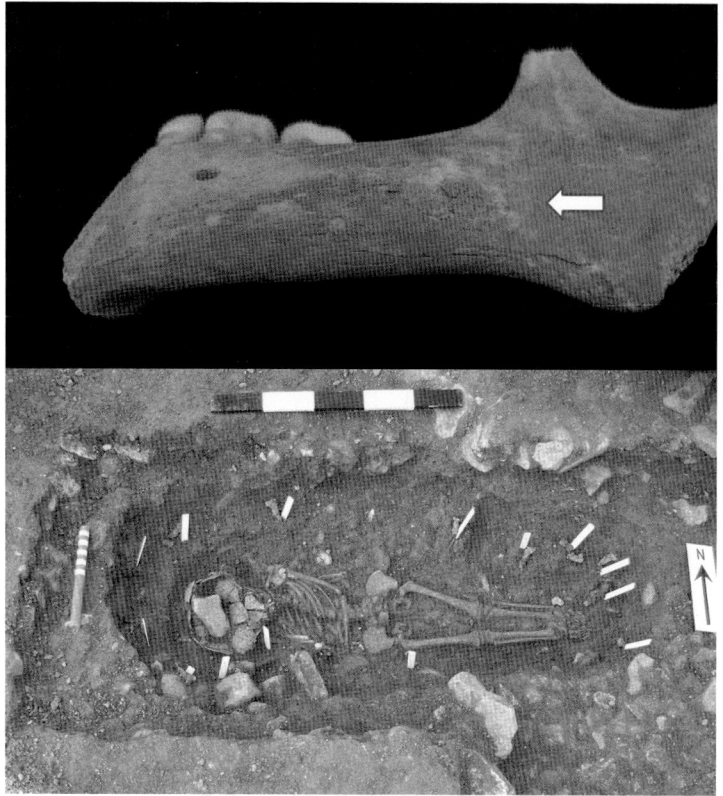

Figure 14.4. Top: Periostitis on the left side of a mandible from SK 914, a 3–4-year-old child from Taunton (photograph by Heidi Dawson-Hobbis); Bottom: SK 914 a coffined church burial from Taunton (the nails are marked by the white tags) (photograph by Context One Archaeological Services).

presented in Table 14.4. The individuals were grouped into these age ranges to reflect the definitions of different stages of childhood seen in the late medieval period (Demaitre 1977; Dawson 2014, 15). Cribra orbitalia and enamel hypoplasia were the indicators most commonly seen together, with 12% of the subadults presenting both. The chi-square test for independence showed that this was a statistically significant relationship (χ^2 5.387, df 1, p 0.024). The presence of multiple stress indicators was also seen to be high for the intramural burials. Of the nine subadults, with at least one stress indicator present, four had at least two stress indicators, and two presented all three indicators. Therefore, 67% (6/9) of those subadults buried within church buildings, with a manifestation of stress present in the skeleton, showed this in more than one way. This is in comparison to a figure of 39% (43/111) for the population analysed in total. Skeleton 914, aged around three to four years of age, who was buried within a coffin in the nave at Taunton priory, showed manifestation of all three stress indicators including periostitis on the mandible (Fig. 14.4). It is possible that the periostitis in this case may reflect haemorrhagic lesions due to impaired

Table 14.5. Prevalence rates of cribra orbitalia and enamel hypoplasia by age grouping. TPR = True prevalence rate.

	Cribra orbitalia		Enamel hypoplasia		Periostitis	
	N	TPR	N	TPR	N	TPR
0–2 years	7/22	32%	1/34	3%	3/42	7%
2–7 years	37/69	54%	27/98	28%	18/110	17%
7–12 years	25/37	68%	34/53	64%	8/60	14%

collagen development associated with scurvy, as it is manifest as an area of woven bone formation adjacent to the region of the masseter muscle.

Age-at-death appears to be a contributing factor to the presence of stress indicators, especially for cribra orbitalia and enamel hypoplasia (Table 14.5). The effect of age on the presence of enamel hypoplasia on the permanent teeth is an expected result as young children will not have the permanent tooth crowns fully formed until as early as two years for the first molar and as late as 13 years for the third molar (Smith 1991). The higher ages apparent for those with cribra orbitalia is more interesting; in the first months of life infants would not be expected to have an iron deficiency due to stores from the mother but this supply only lasts for about six months (Lewis 2007, 113). As such, it appears that infant diet may have been adequate to stave off deficiency but, in later childhood, nutrition may have become poorer. Rothschild (2012) and Steyn et al. (2016) have associated the presence of cribra orbitalia with parasitic infection, and it is easier for children to acquire parasites once the weaning process begins. Grauer (1993) found a higher prevalence of cribra orbitalia in the older age children (10–15 years) from St Helen-on-the-Walls, and stated that earlier mortality was not associated with this lesion.

Discussion

Data on the stress indicators from the three sites analysed for this study suggest, contrary to several other studies (Bennike et al. 2005; Pechenkina and Delgado 2006; Craig and Buckberry 2010), that it may be the higher status subadults that tend to have a higher presence of stress indicators, such as cribra orbitalia, enamel hypoplasia and periostitis.

Prevalence rates for cribra orbitalia, enamel hypoplasia and periostitis were all high for subadult burials from within the church buildings, a finding which is in contrast to the adult burials from Taunton, and they were statistically higher for those buried to the west of the church. Most of the burials to the west come from the Taunton population but as they appear to come from a well-used, and therefore popular, part of the cemetery, where burials were layered up to eight individuals deep in places, it may be postulated that they are of a higher status proportion of the population, at least when compared to the burials from the main south cemetery at

Canterbury. The statistical differences calculated often appear to fall between those to the west and to the south; the fact that the Canterbury cemetery was attached to a local hospital, to which the priory was obliged to allow free burial for the sick (Duncombe and Battely 1785), may suggest that some of the subadults fall within this category. However, as this cemetery was used for both the hospital inmates and the local lay community this can only be a supposition. Utilising the stages of interpretation suggested by Tilley (2012) representing the 'bioarchaeology of care' in further research into the adult skeletons buried in this cemetery, alongside the data on the subadults, may aid in our further understanding of this cemetery site. Evidence for the 'bioarchaeology of care' in the case of a child with tuberculosis has been observed at the Taunton cemetery (Dawson and Robson Brown 2012; Dawson 2016).

Wood et al. (1992) have suggested that it may be the advantaged groups in society that are more likely to survive a stress assault and therefore live long enough for a skeletal lesion to become visible. It is possible that the most stressed individuals actually died when a stressful assault occurred, thereby not surviving long enough for the skeletal manifestations to be seen. This could explain why a higher prevalence of stress was evident in the assumed higher status burial areas. To determine the status of burial areas, however, without the presence of certain high status elements such as coffins is difficult. The presence of one ash burial associated with a silver coin of Edward III, and one burial on a charred wooden plank (Dawson 2014) may indicate that the burial area to the west of the church at Taunton is a high status one. The prevalence of stress indicators from the area excavated to the south of the church at Taunton in 1977, which is further from the church buildings, is much lower, although only crude prevalence data is available (Rogers 1984). The skeletons analysed for this study come from a wide time period, however, and the cemeteries were in use during the time of both the Great Famine (AD 1315–17) and during the several outbreaks of the Black Death in England (AD 1347–75). Historical sources suggest that surviving to adulthood was not easy in the late medieval period. Razi (1980, 104) found a high number of married adults dying 'without issue' in the town of Halesowen during the 14th century, whilst Hanawalt (1993, 48) notes the average family sizes, from the London Letters Book, were 1.5 children in the middle of the 14th century, rising to 2.2 by the middle of the 15th century. This indicates that life conditions in the period were likely to have been difficult and the majority of the subadult population would have been subjected to various malnutrition and disease stresses.

There was a difference in the manifestation of periostitis between Canterbury and Taunton, with a higher prevalence of bone formation present on the Taunton individuals. Whether the appearance of porosity and striations occurs prior to the new bone formation of periostitis, or whether it represents a milder form of inflammation, or even the normal growth process in subadults needs further research. This difference might relate to the possibility that lower status individuals, were potentially buried at Canterbury and were dying before the bone formation associated with periostitis had time to manifest.

The subadults from Taunton had a slightly higher presence of cribra orbitalia, and enamel defects than the adults. The results for cribra orbitalia agree with evidence derived from other medieval cemetery sites that suggests subadults tend to have a slightly higher prevalence of cribra orbitalia than adults (Stroud and Kemp 1993; Mays 2007; Magilton *et al.* 2008) and may confirm Stuart-Macadam's (1985) view that the lesions of cribra orbitalia are related to episodes of stress during childhood, even when recorded on adult individuals. In contrast, enamel hypoplasia and periostitis prevalence rates tend to be higher on adults than subadults (Cowal *et al.* 2008; Magilton *et al.* 2008).

In agreement with other authors (Grauer 1993; Bennike *et al.* 2005) the higher prevalence of some stress indicators (cribra orbitalia and enamel hypoplasia) was associated with older children. Whilst the enamel hypoplasia result is related to the fact that the younger children in this sample will not yet have their permanent dentition erupted, there is no such explanation for the presence of cribra orbitalia. It may be that infants and young children were buffered against the causes of cribra orbitalia, whereas older children were not. It may also be the case that older children are more likely to have had previous stressful assaults in the past before they died.

It is apparent that further work needs to be carried out to enable a clearer picture of the relationship between status, in terms of the funerary context, and status as defined by the presence of stress indicators in individual skeletons. A review of the literature on status and stress indicates an underlying assumption that lower status individuals in societies will show greater evidence for stress in their skeletons. While this assumption may appear true for the adult population it does not appear to be the case for the subadult remains analysed here. In fact the high prevalence of stress indicators seen on the highest status burials, those from within church buildings, suggests the opposite. From the available evidence it appears that the excavated areas at Taunton are, in general, of a higher status than those excavated at Canterbury, and the results show significantly higher stress indicator prevalence for the Taunton children. It may be that we need to interpret the prevalence of stress indicators seen in adults and children separately. The adults with enamel hypoplasia and cribra orbitalia seem to be the survivors of childhood stress, whilst those adults without the lesions may have avoided stresses in childhood, possibly due to their higher status. Conversely, the children with stress indicators present were survivors of the initial stress insult even though they still went on to suffer an early death, whilst those children with no stress indicators present may have been non-survivors of any initial stress insult. Therefore, this study suggests that it may be the advantaged children living in these medieval populations that are more likely to show a higher prevalence of stress indicators.

Note

1. Department of Archaeology and Anthropology, University of Bristol, 43 Woodland Road, Bristol, BS8 1UU. Email: Heidi.Dawson@bristol.ac.uk.

References

Bennike, P., Lewis, M. E., Schutkowski, H. and Valentin, F. 2005. Comparison of child morbidity in two contrasting medieval cemeteries from Denmark. *American Journal of Physical Anthropology* 128, 734–746.

Brickley, M. and Ives, R. 2008. *The Bioarchaeology of Metabolic Bone Disease*. London: Academic Press.

Brickley, M. 2000. The diagnosis of metabolic disease in archaeological bone, in Cox, M and Mays, S. (eds.), *Human Osteology in Archaeology and Forensic Science*, 183–198. London: Greenwich Medical Media Ltd.

Bush, H. 1991. Concepts of health and stress, in Bush, H. and Zvelebil, M. (eds.), *Health in Past Societies: Biocultural Interpretations of Human Skeletal Remains in Archaeological Contexts* (BAR International Series 567), 11–21. Oxford: Tempvs Reparatvm.

Cowal, L., Mikulski, R. and White, W. 2008. The human bone, in Grainger, I. Hawkins, D., Cowal, L. and Mikulski, R (eds.), *The Black Death Cemetery, East Smithfield, London* (MOLAS Monograph 43), 42–55. London: Museum of London Archaeology Service.

Craig, E. and Buckberry, J. 2010. Investigating social status using evidence of biological status: a case study from Raunds Furnells, in Buckberry, J. and Cherryson, A. (eds.), *Burial in Later Anglo-Saxon England c. 650-1100 AD*, 128–142. Oxford: Oxbow Books.

Dawson, H. 2014. *Unearthing Late Medieval Children: Health, Status and Burial Practice in Southern England* (BAR British Series 593). Oxford: Archaeopress.

Dawson, H. 2016. Precious things: Examining the status and care of children in late medieval England through the analysis of cultural and biological markers, in Powell, L., Southwell-Wright, W. and Gowland, R. (eds.), *Care in the Past: Archaeological and Interdisciplinary Perspectives*, 53–69. Oxford: Oxbow Books.

Dawson, H. and Robson Brown, K. 2012. Childhood tuberculosis: a probable case from late medieval Somerset, England. *International Journal of Paleopathology* 2, 31–35.

Demaitre, L. 1977. The idea of childhood and child care in medical writings of the Middle Ages, *The Journal of Psychohistory* 4, 461–490.

DeWitte, S. N. 2014. Differential survival among individuals with active and healed periosteal new bone formation. *International Journal of Paleopathology* 7, 38–44.

Duncombe, J. and Battely, N. 1785. *Bibliotheca Topographica Britannica No 30: The History and Antiquities of the Three Archiepiscopal Hospitals At and Near Canterbury*. London: J. Nichols.

Goodman, A. H., Lallo, J., Armelagos, G. J. and Rose, J. C. 1984. Health changes at Dickson Mounds, Illinois (AD 950-1300), in Cohen, M. N. and Armelagos, G. J. (eds.), *Paleopathology at the Origins of Agriculture*, 271–305. London: Academic Press.

Grauer, A. L. 1993. Patterns of anaemia and infection from Medieval York, England. *American Journal of Physical Anthropology* 91, 203–213.

Hanawalt, B. A. 1993. *Growing up in Medieval London: The Experience of Childhood in History*. Oxford: Oxford University Press.

Heighway, C. and Bryant, R. 1999. *The Golden Minster: The Anglo-Saxon Minster and Later Medieval Priory of St Oswald at Gloucester* (CBA Research Report 117). York: Council for British Archaeology.

Hicks, M. and Hicks, A. 2001. *St Gregory's Priory, Northgate, Canterbury Excavations, 1988-91*. Canterbury: Canterbury Archaeological Trust.

Hillson, S. 1996. *Dental Anthropology*. Cambridge: Cambridge University Press.

Hillson, S. W. 2003. Wealth, health, diet and dental pathology, in Metz, W. H. (ed.), *Wealth, Health and Human Remains in Archaeology*, 7–38. Amsterdam: Stichting Nederlands Museum voor Anthropologie en Praehistorie.

Lewis, M. 2007. *The Bioarchaeology of Children: Perspectives from Biological and Forensic Anthropology*. Cambridge: Cambridge University Press.

Lewis, M. and Roberts, C. 1997. Growing pains: the interpretation of stress indicators. *International Journal of Osteoarchaeology* 7, 581–586.

Magilton, J., Lee F. and Boylston, A. 2008. *Lepers Outside the Gate: Excavations at the Cemetery of the Hospital of St James and St Mary Magdalene, Chichester, 1986-7 and 1993* (CBA Research Report 158). York: Council for British Archaeology.

Mays, S. 2007. The human remains, in Mays, S., Harding, C. and Heighway, C. (eds.), *Wharram A Study of Settlement on the Yorkshire Wolds, XI: The Churchyard* (York University Archaeological Publications 13), 77–192. York: English Heritage.

Mensforth, R. P., Lovejoy, C. O., Lallo, J. W. and Armelagos, G. J. 1978. The role of constitutional factors, diet and infectious disease in the etiology of porotic hyperostosis and periosteal reactions in prehistoric infants and children. *Medical Anthropology* 2, 1–59.

Ortner, D. J. 2003. *Identification of Pathological Conditions in Human Skeletal Remains*. Oxford: Academic Press.

Oxenham, M. F. and Cavill, I. 2010. Porotic hyperostosis and cribra orbitalia: the erythropoietic response to iron-deficiency anaemia. *Anthropological Science* 118, 199–200.

Pechenkina, E. A. and Delgado, M. 2006. Dimensions of health and social structure in the early intermediate period cemetery at Villa El Salvador, Peru. *American Journal of Physical Anthropology* 131, 218–235.

Razi, Z. 1980. *Life, Marriage and Death in a Medieval Parish: Economy, Society and Demography in Halesowen 1270-400*. Cambridge: Cambridge University Press.

Ribot, I. and Roberts, C. 1996. A study of non-specific stress indicators and skeletal growth in two medieval subadult populations. *Journal of Archaeological Sciences* 23, 67–79.

Rice, A. L., Sacco, L., Hyder, A. and Black, R. E. 2000. Malnutrition as an underlying cause of childhood deaths associated with infectious diseases in developing countries. *Bulletin of the World Health Organization* 78, 1207–1221.

Robb, J., Bigazzi, R., Lanzzarini, L., Scaraini, C. and Sonego, F. 2001. 'Social' status and 'biological' status: a comparison of grave goods and skeletal indicators from Pontecagnano. *American Journal of Physical Anthropology* 115, 213–222.

Roberts, C. and Manchester, K. 1995. *The Archaeology of Disease*. Stroud: Sutton Publishing.

Rogers, J. 1984. Skeletons from the lay cemetery at Taunton Priory, in Leach, P. (ed.), *The Archaeology of Taunton: Part 2 The Finds* (Western Archaeological Trust Excavation Monograph 8), 194–199. Gloucester: Sutton Publishing.

Rothschild, B. 2012. Extirpolation of the mythology that porotic hyperostosis is caused by iron-deficiency secondary to dietary shift to maize. *Advances in Anthropology* 2, 157–160.

Selye, H. 1956. *The Stress of Life*. New York: McGraw-Hill Book Company.

Skinner, M. F. 1986. An enigmatic hypoplastic defect of the deciduous canine. *American Journal of Physical Anthropology* 69, 59–69.

Smith, B. H. 1991. Standards of human tooth formation and dental age assessment, in Kelley, M. A. and Larson, C. S. (eds.), *Advances in Dental Anthropology*, 143–168. New York: Wiley-Liss Inc.

Steyn, M., Voeller, S., Botha, D. and Ross, A. H. 2016. Cribra orbitalia: prevalence in contemporary populations. *Clinical Anatomy* 29, 823–830.

Stroud, G. and Kemp, R. L. 1993. *Cemeteries of the Church and Priory of St Andrew Fishergate, The Archaeology of York: The Medieval Cemeteries*. York: York Archaeological Trust/CBA.

Stuart-Macadam, P. 1985. Porotic hyperostosis: representative of a childhood condition. *American Journal of Physical Anthropology* 66, 391–398.

Stuart-Macadam, P. 1991. Anaemia in Roman Britain: Poundbury Camp, in Bush, H. and Zvelebil, M. (eds.), *Health in Past Societies: Biocultural Interpretations of Human Skeletal Remains in Archaeological Contexts* (BAR International Series 567), 101–113. Oxford: Tempvs Reparatvm.

Stuart-Macadam, P. 1992. Porotic hyperostosis: a new perspective. *American Journal of Physical Anthropology* 87, 39–47.

Tilley, L. 2012. The bioarchaeology of care. *The SAA Archaeological Record* 12, 39–41.

Walker, P. L., Bathurst, R. R., Richman, R., Gjerdrum, T. and Andrushko, V. A. 2009. The causes of porotic hyperostosis and cribra orbitalia: a reappraisal of the iron-deficiency-anemia hypothesis. *American Journal of Physical Anthropology* 139, 109–125.

Weston, D. A. 2008. Investigating the specificity of periosteal reactions in pathology museum specimens. *American Journal of Physical Anthropology* 137, 48–59.

Wood, J. W., Milner, G. R., Harpending, H. C. and Weiss, K. M. 1992. The osteological paradox: problems of inferring prehistoric health from skeletal samples. *Current Anthropology* 33, 343–370.

Chapter 15

Atypical Burial Practice and Juvenile Age-at-death in Later Medieval Gaelic Ireland: The Evidence from Ballyhanna, Co. Donegal

Eileen M. Murphy[1]

Abstract: This paper will examine the characteristics of the juvenile burials excavated at the later medieval Gaelic burial ground of Ballyhanna, Co. Donegal. It will seek to analyse subtle variations to typical Christian burial practice in relation to orientation, position of the body, and the inclusion of grave goods, in a systematic manner to see if reasons to account for such deviations can be determined. Osteological information is incorporated into the analysis for the purposes of ascertaining whether age may help explain why there was a departure from the Christian norm for certain juveniles in death. The implications of the findings in relation to the agency of the families of the dead children, as well as to the nature of management of the burial ground by local authorities will also be considered.

Keywords: non-normative burial; children; body position; orientation; grave goods; emotion; poverty

Introduction

Discussions of children are rare within the archaeological record for Ireland although it goes without saying that every adult would once have been a child and the island would have been full of youngsters throughout its past. It has recently been noted that 'the funerary treatment of children by medieval Christian communities has scarcely begun to be addressed' (Hadley and Hemer 2014, 5). This paper will focus on the funerary practices afforded to juveniles buried in the later medieval rural Christian graveyard at Ballyhanna. It is perhaps obvious to state that the living are responsible for burying the dead – the funerary rituals observed in the archaeological record provide an insight into the religious beliefs and agency of the living and the nature of the relationship they may have had with those whom they buried. This paper will explore the nature of deviations from typical Christian burial practice (an extended supine body, positioned with the head to the west and with no associated grave goods) at Ballyhanna in relation to different juvenile age categories. An attempt will be made to interpret the reasons which may have prompted these atypical burial practices.

The townland of Ballyhanna is located on the southern bank of the River Erne on the eastern outskirts of Ballyshannon in County Donegal (Fig. 15.1). The remains of some 1,296 individuals, including 427 children, were recovered during excavations in 2003–4, and an extensive programme of radiocarbon dating revealed that the majority of interments had occurred between AD 1200 and 1650 (Macdonald and Carver 2015, illus. 3.5). When taken in conjunction with the location of the site in south Donegal, the dating evidence is indicative that this was a later medieval Gaelic population, and the only major corpus of its type excavated in Ireland to date. The burial ground at Ballyhanna would have lain within church lands under the control of the Bishop of Clogher throughout the medieval period. He would have appointed an erenagh (steward) to manage the lands on his behalf and historical research has suggested the McGockquin lineage would have been responsible for the administration of this small estate (Donnelly 2015, 38); members of the clergy would have been derived from the erenagh lineage. The annals indicate that the local Gaelic elite were buried at the nearby Cistercian abbey of Assaroe and it is therefore probable that the graveyard at Ballyhanna was the final resting place for ordinary members of this society (Donnelly 2015, 39).

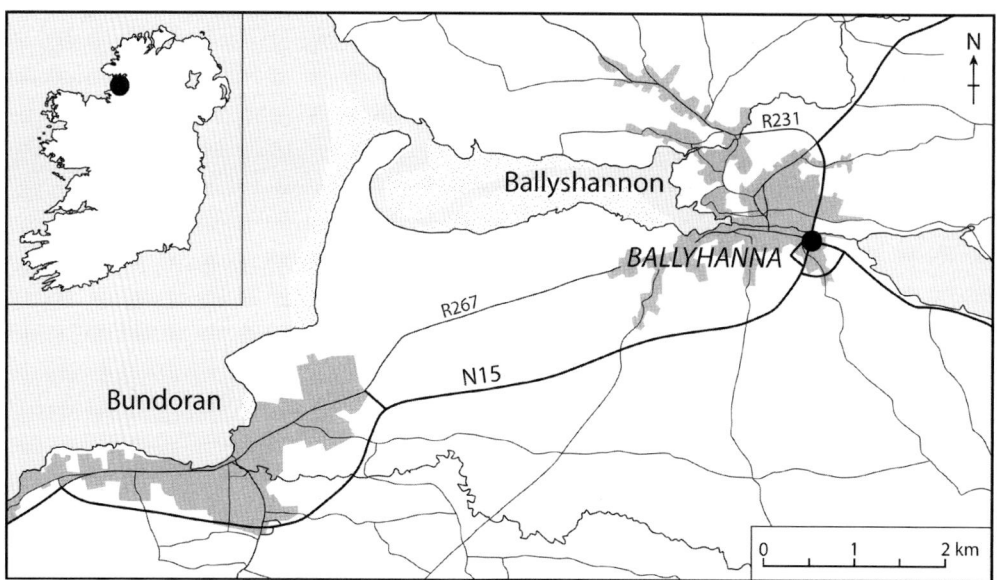

Figure 15.1. Map showing the location of the Ballyhanna burial ground (prepared by Libby Mulqueeny).

Macdonald and Carver (2015) undertook a general review of burial practices at Ballyhanna and this indicated that the majority of individuals, both adults and juveniles, had been given typical Christian burials. The bodies were orientated with the head to the west, so the dead would be ready to see Christ as he rises from the east at the Day of Judgement, and were mostly buried in simple, shallow, earth-lined individual graves. The majority of bodies were interred lying on the back with legs straight and the hands placed either by the sides or positioned over the pelvis. They concluded that the size of surviving grave cuts was incompatible with the use of stone- or timber-lining and that the majority of people would have been buried in a shroud (Macdonald and Carver 2015, 68, 73). They also recorded, however, the presence of smaller numbers of burials which deviated from this burial form in terms of body position and/or orientation (Murphy *et al.* 2014, 141). The following paper is an in-depth study of the characteristics of the juvenile burials at Ballyhanna in which deviations from normal Christian burial practices are analysed in relation to the age-at-death of the juveniles.

Methodology and data set

Previous osteological analysis of the juveniles from Ballyhanna indicated that the burial ground contained the remains of children of all ages. The biological age categories presented in Table 15.1, and based on Scheuer and Black (2000, 468–9), were used in the current study. While these sub-divisions may bear no relationship to how the young were viewed by their societies, they are recognised biological categories, and some may correspond to life course stages. This structured approach has the potential to enable insights to be gained in relation to age-related differences in mortuary practice, while also facilitating future comparative analysis. Only a small number of preterm babies (1.6%) were represented in the juvenile population, and it is evident that almost half of the juveniles had died between the ages of one and six years of age (46.4%) (Murphy 2015, 106). It was not possible to undertake an analysis of the funerary treatment afforded to the preterm infants in a comparable manner to that for the other juveniles due to their small numbers (n=7) and, as a consequence, this group has been omitted from the study.

The 1,296 skeletons from Ballyhanna were recovered from an area that only measured 30m east–west by 18.5 m north–south and one of the main limitations of the skeletal and funerary analysis has been the truncation of earlier burials by later interments. Skeletons less than 25% complete, or that lacked associated burial data, were excluded since it was impossible to reliably assess the nature of their original burial configuration. This approach resulted in the inclusion of 69.8% (298/427) of juveniles in the study (see Table 15.1), although it should be noted that not all of these skeletons were suitable for inclusion in each aspect of the analysis.

The original analysis of the excavation findings resulted in the production of an excel spreadsheet that contains information about the position, orientation

Table 15.1. Age-at-death terminology and categories used in the study (after Scheuer and Black 2000, 468–9), with details of the numbers of individuals of each age group represented in the Ballyhanna juvenile population (Murphy 2015, 106). Details of the proportions of suitably preserved skeletons (>25% completeness), with associated burial data, are also included (McKenzie and Murphy forthcoming).

Terminology	Age range	Proportions of juveniles by age	<25% or lack of burial data	>25% with burial data
Preterm	< 37 weeks from conception	1.6% (7/427)	28.6% (2/7)	71.4% (5/7)
Full term/neonate	37–44 weeks from conception	9.4% (40/427)	20.0% (8/40)	80.0% (32/40)
Infant	1 month to 1 year	5.2% (22/427)	27.3% (6/22)	72.7% (16/22)
Younger child	1–6 years	46.4% (198/427)	35.9% (71/198)	64.1% (127/198)
Older child	6–12 years	21.3% (91/427)	29.7% (27/91)	70.3% (64/91)
Adolescent	12–18 years	16.1% (69/427)	21.7% (15/69)	78.3% (54/69)
Total			30.2% (129/427)	69.8% (298/427)

and associated grave goods of each individual buried at Ballyhanna (Macdonald and Carver 2015, supplementary data on CD-ROM). The data contained in this spreadsheet for the juveniles formed the starting point for the current analysis and was augmented by recourse to excavation skeleton recording sheets, as well as *in situ* plans and photographs. The study involved an examination of five characteristics – orientation, position of the torso, position of the legs, position of the arms, the occurrence of grave goods or furnishings, and combinations thereof – for each juvenile age category.

Following the principles of archaeothanatology, only aspects of body position considered to have remained stable, or at least to have been obvious if disturbed, formed the focus of the study. The direction of the face was excluded because it was frequently not possible to determine from the plans and photographs if a head remained in its original burial position or had rotated as a consequence of taphonomic processes (see Duday 2009, 17–19). An examination of the precise positioning of the bones of the hands and feet was also omitted since they were frequently poorly preserved and differences in skeletal development would have made it impossible to directly compare the results across the different age groups (see Scheuer and Black 2000, 327–9, 446, 449–51).

Orientation

Macdonald and Carver (2015, 68) noted that 86.0% (697/810) of individuals at Ballyhanna were aligned on an approximate west–east orientation, in which the head is orientated to the west. When the data for juveniles was examined it was found that the expected west–east orientation predominated across all juvenile age groups, with the proportions ranging from 57.2% for full term/neonatal infants to

Table 15.2. Orientation of the head by age category in the Ballyhanna juvenile population.

Age	North	South	East	West	North-east	South-east	South-west	North-west	Overall deviation from West
Full term/ neonate (n=14)	0	7.1% (1/14)	7.1% (1/14)	57.2% (8/14)	0	7.1% (1/14)	21.5% (3/14)	0	21.4% (3/14)
Infant (n=11)	0	9.1% (1/11)	9.1% (1/11)	63.6% (7/11)	0	0	18.2% (2/11)	0	18.2% (2/11)
Younger child (n=106)	1.9% (2/106)	1.9% (2/106)	6.5% (7/106)	74.5% (79/106)	1.9% (2/106)	1.9% (2/106)	7.6% (8/106)	3.8% (4/106)	14.2% (15/106)
Older child (n=55)	1.8% (1/55)	0	1.8% (1/55)	83.8% (46/55)	1.8% (1/55)	3.6% (2/55)	3.6% (2/55)	3.6% (2/55)	9.1% (5/55)
Adolescent (n=47)	0	2.1% (1/47)	6.4% (3/47)	76.6% (36/47)	2.1% (1/47)	0	4.3% (2/47)	8.5% (4/47)	10.6% (5/47)
Total (n=233)	1.3% (3/233)	2.1% (5/233)	5.6% (13/233)	75.5% (176/233)	1.8% (4/233)	2.1% (5/233)	7.3% (17/233)	4.3% (10/233)	12.9% (30/233)

83.8% for older children and an overall proportion of 75.5% for all juvenile burials (Table 15.2).

When the deviations from the west–east orientation were scrutinised further it was sometimes possible to exclude such alignments as accidental. Due to the slight southerly orientation of the church when compared to the majority of burials it was suggested the grave diggers would have used the position of the sunrise to help them align the graves (Macdonald and Carver 2015, 68). As such, it is not entirely unexpected that southwest (7.3%) or northwest (4.3%) deviations occurred with next greatest frequency. A small number of cases were present, however, where it would seem that the head was deliberately positioned in a non-westerly direction. Full term/neonates (21.4%) and infants (18.2%) were most likely to have been afforded such an atypical orientation, followed in frequency by young children (14.2%), while older children (9.1%) and adolescents (10.6%) were affected to a lesser extent. A comparison of the frequencies of atypical orientations for full term/neonates and infants compared to older children and adolescents, however, revealed that the differences were not statistically significant (Fisher 2.02, p 0.31).

Body Position

Torso and Legs

Macdonald and Carver (2015, 56, 79) observed that the majority of individuals buried at Ballyhanna were in a supine position. It was possible to examine the positions of both the body and legs for 158 juveniles and it is clear that burial in a supine position with extended legs (81.6%) notably predominated across all juvenile age categories,

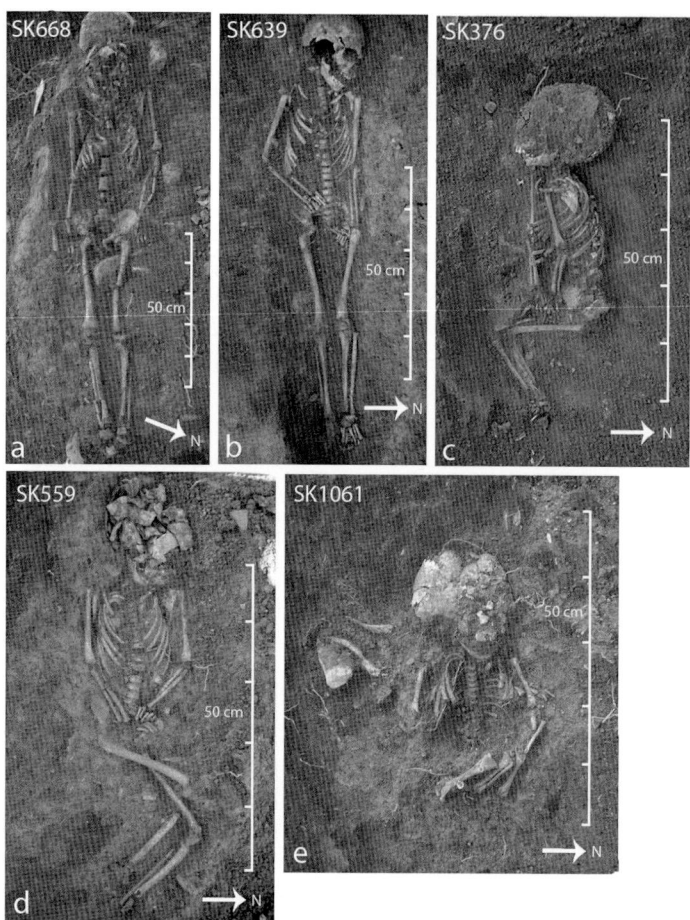

Figure 15.2. Example of the different torso, leg and arm positions identified among the Ballyhanna juveniles. a. SK 668, an older child, lying in a supine position with extended legs and both arms alongside the body; b. SK 639, an older child, lying in a supine position with extended legs, the left arm alongside the body and the right hand on the pelvis; c. SK 376, a younger child, lying on the right side with the legs to the right, the upper legs semi-flexed at the hips by 80° and the lower legs flexed at the knees by 110°; d. SK 559, a younger child, lying in a supine position with the legs to the left, the upper legs semi-flexed at the hips (R 80°; L 40°), the lower right leg flexed at the knee by 110°, the lower left leg semi-flexed at the knee by 90° and both hands on the pelvis; and e. SK 1061, an infant, lying in a supine position with the legs to the left, the upper legs flexed at the hips by 120° and the lower legs tightly flexed at the knees (R 160°; L 180°) (photographs by Irish Archaeological Consultancy Ltd, annotation by Libby Mulqueeny).

with burial in a supine position with flexion of the legs (12.0%) occurring with next greatest frequency, followed by burial on the side with flexion of the legs (4.5%; Table 15.3; Fig. 15.2). In practically all cases no obvious topographical reason that might have accounted for burial on the side, rather than the back, was evident.

Table 15.3. Position of the torso and legs by age category in the Ballyhanna juvenile population. Note that all of the individuals with flexion of the upper legs also displayed flexion of one or both lower legs so the generic term 'legs' is used.

Age	Supine; legs extended	Supine; flexion of legs	Side; flexion of legs	Side; legs extended	Prone; legs extended
Full term/neonate (n=7)	71.4% (5/7)	0	28.6% (2/7)	0	0
Infant (n=10)	50.0% (5/10)	20.0% (2/10)	20.0% (2/10)	10.0% (1/10)	0
Young child (n=66)	77.3% (51/66)	19.7% (13/66)	3.0% (2/66)	0	0
Older child (n=44)	97.7% (43/44)	2.3% (1/44)	0	0	0
Adolescent (n=31)	80.7% (25/31)	9.7% (3/31)	3.2% (1/31)	3.2% (1/31)	3.2% (1/31)
Total (n=158)	81.6% (129/158)	12.0% (19/158)	4.5% (7/158)	1.3% (2/158)	0.6% (1/158)

Full term/neonates (71.4%) and infants (50%) displayed notably lower proportions of supine extended leg positions compared to older children (97.7%) and adolescents (80.7%), with younger children lying in between (77.3%). Both infants (20.0%) and younger children (19.7%) were notably more likely to be buried in a supine position with flexion of the legs than either older children (2.3%) or adolescents (9.7%). Similarly, full term/neonates (28.6%) and infants (20.0%) were more likely to have been interred lying on their sides with flexed legs than younger (3.0%) or older children (0%), or adolescents (3.2%). A Fisher's Exact test revealed that full-term/neonates and infants were significantly more likely to be buried in a non-supine position compared to older juveniles (Fisher 8.11, p 0.004).

Position of the Legs
The remains of 161 juveniles had both legs well preserved which facilitated an analysis of leg position (Table 15.4; see Fig. 15.2).[2] In the majority of cases the legs of juveniles of all ages (81.4%) were extended straight. Full term/neonates (60.0%), infants (50.0%) and younger children (77.6%) were less likely to display extended legs, however, relative to older juveniles (97.7%) and adolescents (83.4%); but the differences were not statistically significant (χ^2 0.79, df 1, p 0.37).

Semi-flexion of the hips to the right (8.7%) or left (6.2%) was the second most common position for the juveniles. Flexed upper legs to the right (1.9%) or left (0.6%) occurred relatively infrequently, and only among children less than six years old. Semi-flexion of the hips to the right was the second most common position for full term/neonatal infants (20.0%), infants (30.0%) and young children (11.9%), while semi-flexion to the left was the second most common position for older children (2.3%)

Table 15.4. Extension of the legs and flexion of the upper legs by age category in the Ballyhanna juvenile population.

Age	Extended	Extended and lower legs crossed	Upper leg flexed to R	Upper leg flexed to L	Upper leg semi-flexed to R	Upper leg semi-flexed to L
Full term/neonate (n=10)	60.0% (6/10)	0	10.0% (1/10)	0	20.0% (2/10)	10.0% (1/10)
Infant (n=10)	50.0% (5/10)	10.0% (1/10)	0	10.0% (1/10)	30.0% (3/10)	0
Younger child (n=67)	77.6% (52/67)	0	3.0% (2/67)	0	11.9% (8/67)	7.5% (5/67)
Older child (n=44)	97.7% (43/44)	0	0	0	0	2.3% (1/44)
Adolescent (n=30)	83.4% (25/30)	3.3% (1/30)	0	0	3.3% (1/30)	10.0% (3/30)
Total (n=161)	81.4% (131/161)	1.2% (2/161)	1.9% (3/161)	0.6% (1/161)	8.7% (14/161)	6.2% (10/161)

and adolescents (10.0%). Although the numbers are small, it is tempting to suggest that age may have been a determining factor when deciding the side to which the legs were to be directed.

A practical reason for flexion at the hips could only be identified for one of the four affected individuals. In SK 1061, an infant, the femora were flexed from the axis with the torso by 120°, while the lower legs were tightly flexed at the knees by 160° (R) and 180° (L). The infant's grave pit abutted the remains of an adult (SK 1032) and it is considered possible the baby had been buried in a tightly flexed manner to avoid disturbing this grave.

All of the individuals with flexion of the upper legs also displayed a degree of flexion of one or both lower legs. When the degree of flexion of the lower legs was considered for individuals with flexed legs it was evident that semi-flexion (64.0%; 16/25) occurred more frequently across most of the age categories when compared to flexion – 100.0% (3/3) for full term/neonatal infants, 66.7% (2/3) for infants, 100.0% (1/1) for older children and 75.0% (3/4) for adolescents, the exception being for young children where equal proportions of flexed and semi-flexed lower legs were evident (50.0%; 7/14).

It is perhaps possible that some of the cases of semi-flexion at the hip and knees in the full term/neonates and infants may have been related to the natural configuration of infant lower limbs. At the later medieval burial ground of the Priory of St Peter and St Paul, Taunton, Somerset, for example, flexion of the legs of three neonates and one older infant was considered to have represented 'a natural position' (Dawson 2014, 83–4). Natural positioning of the body, however, cannot satisfactorily explain the cases of semi-flexion of the legs apparent in older juveniles or any of the examples of more complete flexion. In addition, laying a body on its side was clearly a deliberate choice. It may simply be the case that such body positions were considered to be more restful poses, viewed as more suitable for the interment, in particular, of the very young as opposed to the rigid extended, supine position. This has previously been suggested by Gilchrist and Sloane (2005, 155–6) in relation to burials on the side in their review of 8,000 British monastic burials where they noted that such positions were rarely used for adults. Studies of sleeping positions for modern people have demonstrated that, with the exception of neonates, lying on the side is the most common sleep position across all age groups, with prone and supine positions less favoured (De Koninck *et al.* 1992, 143, 147).

Prone Burial

SK 341, a probable male adolescent, was buried in an extended prone position, with his head to the east, although previous studies of funerary practice at Ballyhanna omitted this individual from the counts of prone burials (Murphy *et al.* 2014, 133; Macdonald and Carver 2015, 78–9). Previously, six adults – two females and four males – were identified as prone burials. Traditionally, the prone position is viewed as a disrespectful way in which to inter the dead (e.g. Tsaliki 2008, 2–3). When dealing with the context of a Christian burial ground, however, the situation is less clear – prone individuals were being marked out as different in the manner of their burial but they were still deemed suitable for inclusion within the burial ground of a Christian community. On the basis of contemporary accounts, Gilchrist and Sloane (2005, 154) suggested that some prone burials may have been penitential in nature to atone for the sins of either the deceased or their family. It has previously been suggested that certain adults from Ballyhanna were singled out for atypical burial because they were known to have suffered during life, or been particularly holy individuals (Murphy *et al.* 2014, 141). The same may also have been true for this teenage boy whose skeleton displayed lesions characteristic of the chronic debilitating condition of Scheuerman's disease (McKenzie and Murphy forthcoming).

Position of the Arms

It was possible to examine the position of both arms in 115 individuals. Those who were buried in a non-supine position were excluded from the analysis since the arms of these individuals were either poorly preserved or it was difficult to reliably interpret their position. The terminology of Sprague (2005, 30) was followed – arms along the sides, hands on the pelvis and hands to the shoulder – but an additional position was also observed in which the elbow was semi-flexed at 90° and the lower arm rested on the waist, while the hand touched the elbow of the other arm (Table 15.5; Fig. 15.3).

The positioning of the hands on the pelvis (40.0%), or the arms alongside the body (37.4%), occurred most frequently across all age categories and these were the only positions observed for full term/neonates and infants. A notable proportion of older

Table 15.5. Position of the arms by age category in the Ballyhanna juvenile population.

Age	Along the sides	Hands on the pelvis	One hand on pelvis; one at side	One hand on waist; one at side	One hand on waist; one on pelvis	One hand on shoulder; one on pelvis	One hand on shoulder; one on waist
Full term/ neonate (n=7)	71.4% (5/7)	28.6% (2/7)	0	0	0	0	0
Infant (n=2)	100.0% (2/2)	0	0	0	0	0	0
Younger child (n=52)	44.2% (23/52)	30.8% (16/52)	25.0% (13/52)	0	0	0	0
Older child (n=30)	36.7% (11/30)	50.0% (15/30)	10.0% (3/30)	0	0	0	3.3% (1/30)
Adolescent (n=24)	8.3% (2/24)	54.2% (13/24)	16.7% (4/24)	12.5% (3/24)	4.2% (1/24)	4.2% (1/24)	0
Total (n=115)	37.4% (43/115)	40.0% (46/115)	17.3% (20/115)	2.6% (3/115)	0.9% (1/115)	0.9% (1/115)	0.9% (1/115)

juveniles also displayed one arm alongside the body and one hand on the pelvis (17.3%). It is thought possible that some of these instances may represent cases whereby an arm positioned alongside the body had slipped from its original position on the pelvis as a result of decomposition processes. This association with decomposition is by no means certain, however, since a number of older children and adolescents were clearly buried with deliberate mixed arm positions (see below). Furthermore, researchers in England have also noted the occurrence of relatively high proportions of this form of mixed armed configuration, for example, it was evident in 12.0% of burials at Site 26 at Wharram Percy, Yorkshire, and 30.0% of burials at the Medieval Jewish burial ground at Jewbury, York (Heighway 2007, 235).

An additional mixed arm position observed amongst the adolescents occurred in four individuals, each of whom had one of their arms semi-flexed by 90° at the elbow so the lower arm crossed the waist and the hand touched the elbow of the opposite arm, which either lay alongside the body or had its hand on the pelvis. It is possible that both arms had originally crossed the waist but that one had slipped down towards the pelvis during decomposition although it is equally feasible that the mixed arm position was deliberate.

It is possible that movement of the lower arms during decomposition may account for some of the previously described pelvic/waist positions but a further variant, which involved the positioning of one hand at a shoulder near the chin, appears highly deliberate. In SK 732B, an older child, the left arm was semi-flexed at 90° across the abdomen so that it touched the right elbow which was tightly flexed by almost

180° with the hand positioned at its shoulder, beneath the right side of the chin. A similar position was evident in SK 772, an adolescent, and the right arm was tightly flexed so that its hand touched its shoulder, while the left hand was on the pelvis (see Fig. 15.3).[3]

It is interesting to speculate that the greater variety of arm positions amongst older children and, particularly, adolescents may have been related to their increasing age – in the majority of cases perhaps simply due to the increasing length of their arms. Statistical tests revealed that the differences in adolescent arm positions were not statistically significant when compared to data for the younger age groups (χ^2 0.23, df 1, p 0.62). The shoulder variation was considered appropriate for the occasional pre-teen and teenage individual, however, and perhaps

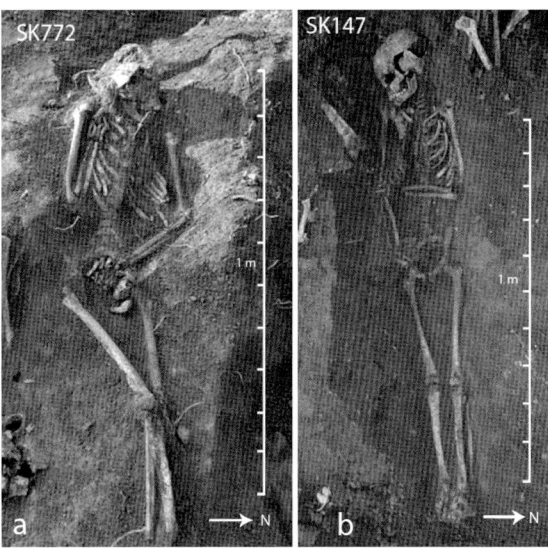

Figure 15.3. Unusual arm positions among the Ballyhanna juvenile. a. SK 772, an adolescent, with the right arm tightly flexed at the elbow so that its hand touched its shoulder and the left hand on the pelvis and b. SK 147, an adolescent, whose left arm was semi-flexed by 90° at the elbow so the lower arm crossed the abdomen and the hand touched the elbow of the opposite arm, which lay alongside the body (photographs by Irish Archaeological Consultancy Ltd, annotation by Libby Mulqueeny).

was an attempt to replicate a more mature position of repose in which the hand supports the head. A preliminary examination of adult arm positions at Ballyhanna is indicative that this position occurred in at least six cases, for both males and females. Gilchrist and Sloane (2005, 156) reported two individuals – an adult of indeterminable sex in a stone cist in the graveyard of St Saviour, Bermondsey (AD 1090–150), and another from St Mary Spital, London (no further details provided) – in which the arms were tightly flexed at the elbows so they could touch their respective shoulders. They suggested the arms could have moved outwards during decomposition from an original position of prayer, but this explanation does not seem compatible with the cases from Ballyhanna which genuinely do not seem to have been bilateral.

Position of the Arms and Legs
To enable an overview of arm and leg positions for the supine burials the data for side-related variations and degree of flexion of the legs was simplified and some eight variants were identified amongst those juveniles (n=86) in which all four limbs were well preserved (Table 15.6). The most common position was to have both arms and legs extended (38.3%) or the hands resting on the pelvis with the legs extended

Table 15.6. Position of the arms and legs by age category in the Ballyhanna juvenile population. Note that all of the individuals with flexion of the upper legs also displayed flexion of one or both lower legs so the generic term 'legs' is used.

Age	Arms along sides; legs extended	Arms along sides; flexion of legs	Hands on pelvis; legs extended	Hands on pelvis; flexion of legs	Arm along side and hand on pelvis; legs extended	Hand on waist and arm alongside; legs extended	Hand on waist and hand on pelvis; legs extended	Hand at shoulder and hand on pelvis; flexion of legs
Full term/neonate (n=6)	50.0% (3/6)	0	33.3% (2/6)	16.7% (1/6)	0	0	0	0
Infant (n=2)	100.0% (2/2)	0	0	0	0	0	0	0
Younger child (n=37)	35.1% (13/37)	13.5% (5/37)	13.5% (5/37)	10.8% (4/37)	27.1% (10/37)	0	0	0
Older child (n=21)	52.4% (11/21)	0	38.1% (8/21)	0	9.5% (2/21)	0	0	0
Adolescent (n=17)	23.5% (4/17)	5.9% (1/17)	29.3% (5/17)	5.9% (1/17)	11.8% (2/17)	5.9% (1/17)	11.8% (2/17)	5.9% (1/17)
Total (n=86)	38.3% (33/86)	10.5% (9/86)	23.2% (20/86)	7.0% (6/86)	16.3% (14/86)	1.2% (1/86)	2.3% (2/86)	1.2% (1/86)

(23.2%). The extended leg position with one arm alongside and one hand on the pelvis occurred with next greatest frequency (16.3%) but only amongst those older than infants. Individuals with their legs in various degrees of flexion, but with the two most common arm positions – along the sides (10.5%) and on the pelvis (7.0%) – were also well represented. Three further variants restricted to adolescents only occurred in small proportions and involved typical leg positions with anomalous arms – hand on waist with other hand on pelvis and legs extended (2.3%), hand on waist with arm alongside and legs extended (1.2%) and hand at shoulder with other hand on pelvis and flexion of the legs (1.2%). The level of variation in arm/leg position among the juveniles at Ballyhanna is notable compared to the situation at the burial ground of the Priory of St Peter and St Paul, Taunton, Somerset, where only three such variants were observed. The most common position involved the extension of both arms and legs (72.7%; 40/55), although instances in which the hands rested on the pelvis with the legs extended (20.0%; 11/55), and extended arms with flexion of the legs were also identified (7.3%; 4/55; Dawson 2014, 83).

Grave Goods and Furnishings

The final component of the analysis involved an examination of items that may have been deliberately placed within juvenile burials (n=293). Gräslund (1994, 19) has observed that in religions such as Christianity, where the soul is believed to leave the body at the moment of death, the inclusion of grave goods is deemed unnecessary. As would be expected, therefore, only a small proportion of Ballyhanna juveniles (6.1%) were associated with one or more possible grave goods (Tables 15.7 and 15.8). The presence of grave goods seemed to increase with age; 3.1% of full term/neonates were associated with objects and the proportion gradually increased to 9.3% for adolescents. A Fischer's Exact test, however, indicated no significant difference in the quantity of grave goods by age (Fischer 0.24, p 0.29). The most common grave items were quartz pebbles which occurred with a frequency of 2.7%, followed by nails or other metal objects (1.7%), pottery sherds (1.0%), beads (0.7%) and metal slag (0.7%).

An early medieval reliquary shrine which contained around 80 pieces of quartz was discovered during the excavations at Ballyhanna. A funerary practice that may have been related to the earlier shrine was the inclusion of white stones in the graves of some individuals. The inclusion of white stones in funerary contexts is a tradition of notable longevity and it has been associated with Christian burial grounds throughout Ireland and Britain (see Gilchrist and Sloane 2005, 145; Murphy 2011, 68–9). It is considered probable that, prior to their deposition in the burials as votive items, the stones had originally been placed in the reliquary shrine so they could become charged with spiritual power from the (unknown) saint associated with the graveyard and church at Ballyhanna. Macdonald and Carver (2015, 83) were of the view that the practice was largely restricted to adult females based on the association of quartz with the hand. This study has demonstrated, however, the likely association of white stone with some eight juveniles and it would seem the hand was not the only position selected for the deposition; in five of the eight juvenile cases the stones were associated with the head (62.5%), especially its left side (Fig. 15.4; 4/5).

A glass bead was recovered near the left side of the skull of SK 994, a younger child, while SK 1074, an older child, had one bead at the head and another at the right hip. Beads were also associated with two adults from Ballyhanna; SK 495, a middle-aged female, was found with 15 beads at her neck that are through to have been the remains of a paternoster or set of rosary beads, while a single bead was recovered near the left side of the pelvis of SK 634, a young adult female (McKenzie and Murphy forthcoming). Gilchrist (2008, 149) suggested that single beads were sometimes used in the graves of children in later medieval England to protect against the evil eye. Given the inclusion of quartz in certain burials at Ballyhanna it seems feasible the glass beads were being used in a similar votive manner. Perhaps they were considered holy and suitable burial inclusions since they had originally formed part of a set of religious beads. Two juveniles were associated with grave goods of particular interest. Three pins manufactured from a copper alloy wire were recovered with SK 100, a neonate. Two of the pins were retrieved from either side of the cranium, while the third was positioned under its posterior aspect. It seems possible the pins are the remnants of a head-dress and perhaps were related to the attachment of a chrisom band. During baptism, the head was anointed with chrisom oil and, to prevent it from rubbing off, a white chrisom cloth was wrapped around the head of the newly baptised infant.

Table 15.7. Details of the possible grave goods and their locations discovered with the Ballyhanna juveniles.

Context	Age	Artefacts	Location
SK 100	Full term/neonate	3 bronze pins	beneath back of head & evenly spaced
SK 1196	Full term/neonate	quartz	at pelvis, adjacent to R hand
SK 156	Young child	curved iron bar (15mm long)	beneath head
		pottery sherd	at pelvis, adjacent to L hand
SK 613	Young child	quartz	head
SK 770	Young child	2 pieces quartz	L side of head
SK 994	Young child	bead	L side of head
SK 1029*	Young child	nodule hammer slag (2.5mm long)	feet
		quartz	L side of feet
SK 1076*	Young child	3 pieces quartz	–
SK 1115	Young child	minute globular-shaped blob of iron	in pelvis
SK 1240	Young child	quartz	L side of head
SK 77	Older child	pottery sherd	–
SK 732B	Older child	Flat piece of iron with 3 circular depressions (22mm long)	–
SK 1074	Older child	bead	head
		bead	R side of pelvis
SK 1117	Older child	carpentry nail (41mm long)	L side upper femur
		quartz	–
SK 1243	Older child	quartz	L side mid femur
SK 32	Adolescent	pottery sherd	R side of feet
SK 245	Adolescent	iron nail	above R side of pelvis
SK 652	Adolescent	incomplete whittle-tanged knife (141mm long)	R side lumbar vertebrae
SK 796	Adolescent	quartz	L shoulder
SK 1052	Adolescent	2 pieces quartz	L side skull

*These two individuals were <25% complete and have not been included in the summary analysis.

If a young baby died it is thought he/she may have been buried wearing their chrisom cloth as evidence of baptism, with the blessed nature of the cloth also affording a degree of protection to help ensure they entered Heaven (Oosterwijk 2007, 339). SK 652, an adolescent, seems to have been deliberately buried with a relatively substantial (141mm long) incomplete iron whittle-tanged knife. The knife was located at the right side of the waist and may have been worn on a belt (see Fig. 15.4; Johnson 2015, 10).

15. Atypical Burial Practice and Juvenile Age-at-death in Later Medieval Gaelic Ireland   241

Table 15.8. A summary of the possible grave goods recovered in juvenile burials at Ballyhanna.

Age	Quartz	Pottery	Metal slag	Nails/Metal	Beads	Total N individuals affected
Full term/neonate (n=32)	3.1% (1/32)	0	0	0	0	3.1% (1/32)
Infant (n=16)	0	0	0	0	0	0
Younger child (n=127)	2.4% (3/127)	0.8% (1/127)	1.6% (2/127)	0.8% (1/127)	0.8% (1/127)	5.5% (7/127)*
Older child (n=64)	3.1% (2/64)	1.6% (1/64)	0	3.1% (2/64)	1.6% (1/64)	7.8% (5/64)*
Adolescent (n=54)	3.7% (2/54)	1.8% (1/54)	0	3.7% (2/54)	0	9.3% (5/54)
Total (n=293)	2.7% (8/293)	1.0% (3/293)	0.7% (2/293)	1.7% (5/293)	0.7% (2/293)	6.1% (18/293)

*Individuals with more than one grave good have only been counted once.

Apparently mundane, everyday objects – pottery sherds, metal slag, nails and fragments of metal objects – were included with seven juveniles (see Table 15.7). Such objects are often dismissed as accidental inclusions within a grave but some spatial patterning, which mirrored that of the quartz and other more definitive grave goods, was evident and is suggestive they may have been deliberate funerary inclusions (see below). Since a definitive association of quartz is clear for some of the individuals from Ballyhanna it seems unwise to completely dismiss the possibility that everyday objects had been deliberately included in the graves of some juveniles. Indeed, even today many children collect bits and bobs to play with and undoubtedly such items would have held even more value for children

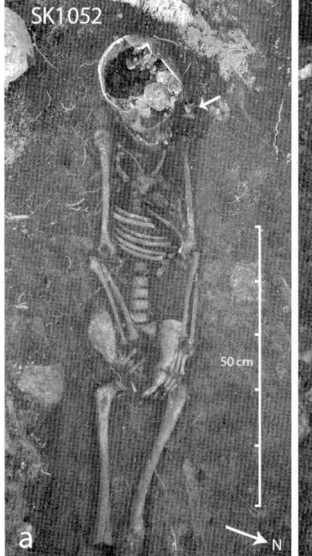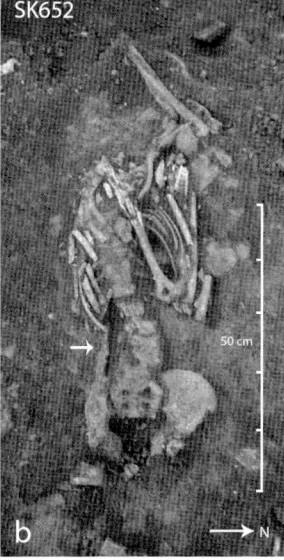

Figure 15.4. SK 1052, an adolescent, with two pieces of quartz positioned to the left of the skull and b. SK 652, an adolescent, with an incomplete iron whittle-tanged knife located at the right side of the waist (photographs by Irish Archaeological Consultancy Ltd, annotation by Libby Mulqueeny).

in the past before formal toys became abundant playthings. We might even imagine that many interesting organic objects, such as conkers or rush figurines, might also have been present in some of these burials, the remnants of which have long since disappeared (see Gilchrist and Sloane 2005, 228). It is possible that a shiny nail or a blob of slag may have been a treasured possession of either the dead child or of their friends or siblings who wanted to give them one last gift for the afterlife (see Murphy 2011, 68–9). An alternative explanation that may account for some of the items is that they are representative of folk magic and were sympathetic charms. Gilchrist (2008, 152–3) has noted a long folk tradition of women using magic and charms to protect their families and has suggested this protection extended into the burial sphere. Grandmothers and mothers, amongst others, may have included charms, particularly those related to domestic life, to protect those considered to be the most vulnerable dead – often the children. She proposed that: 'Medieval people placed charms and materials with occult powers with the dead to heal or transform the corpse, to ensure its reanimation on judgement day, and to protect the dead on their perilous journey through purgatory'.

While the numbers of objects buried with the juveniles at Ballyhanna is admittedly small a spatial patterning of deposition does seem to be apparent. Excluding the pins associated with SK 100 and the knife associated with SK 652, it was possible to map the location of 18 of the 22 objects.[4] Some 44.4% (8/18) were found at the head, with a further 33.3% (6/18) at the pelvis and 16.7% (3/18) at the feet, while a single item (5.6%) was positioned adjacent to the midpoint of the left femur. The head, pelvis and perhaps feet seem to have been deliberately selected locations for the deposition of grave items.

Two juveniles were associated with possible 'pillow stones'. The skull of SK 1117, an older child whose body was orientated west–east in an extended, supine position, lay on top of a substantial, rectangular-shaped pillow stone. SK 120, an adolescent whose body was orientated east–west in a supine position with semi-flexed legs, also seems to have been buried with a pillow stone but the skull had rolled off by the time of excavation. It has been suggested that pillow stones may have been used a form of punishment (since the individual's head may have felt uncomfortable and cold), or as a sign of penitence (Daniell 1997, 160). If we return to the suggestion that many of the juveniles in non-extended supine positions had been laid out in ways that mimicked natural sleeping positions, however, then perhaps the inclusion of a pillow stone is a variation of this concept. The remains of soft, organic pillows have been recovered from a number of later medieval burials in England (Gilchrist and Sloane 2005, 147), but these require exceptional preservation conditions.

Associations between Variables

It was possible to examine all five features of the analysis in 29 individuals and this was undertaken to determine if any association existed between the different variables

(Table 15.9). The overall trends indicated that flexion of the legs (20.7%) was the most common variation, closely followed by atypical orientation (17.2%), the presence of grave goods or furnishings (17.2%) and a combination of atypical orientation and flexion of the legs (17.2%). Less common deviations included atypical arm positions (10.3%) and the occurrence of more than two atypical features in a burial (6.9%). It was only possible to include small numbers of full term/neonates, infants and older children in this aspect of the analysis which made it impossible to ascertain reliable patterns. Among the young children a clear focus on flexion of the legs remained evident (see above), however, which occurred either in isolation, or alongside another variant, and accounted for 76.9% (10/13) of the atypical burials for this age group. Turning to the adolescents the position of the arms appears to have become more significant and 50.0% (4/8) of individuals had either isolated atypical arm positions or a combination of atypical arms with flexion of the legs. It should be noted that flexion of the legs was still an important deviation for this age group, however, since it occurred in isolation in one instance, while both adolescents, with more than two variations, displayed flexion of the legs (37.5%; 3/8).

The two adolescents who were associated with more than two burial variations warrant further mention. The head of SK 32 was orientated to the north-east, while the body lay on the left side with the upper legs semi-flexed at the hips (90°) and lower legs flexed at the knees (100°). A pottery sherd was recovered at the right side of the feet. The head of SK 120 was orientated to the east, the upper and lower legs were semi-flexed and the individual's head had been positioned on a pillow stone. Two of the lesser preserved younger children also displayed more than two variations[5] – SK 156 lay on the right side with semi-flexed upper and lower legs and a small curved iron bar was retrieved from beneath the head, while a pottery sherd was recovered at the pelvis adjacent to the left hand. The head of SK 994 was orientated to the north, the torso was supine with the upper legs semi-flexed at the hips (70°), while the lower legs were flexed at the knees (130°). A glass bead was retrieved near the left side of the child's head. The burial of these four individuals was clearly undertaken in a highly personal manner and there is much evidence of intentional thought in relation to the orientation, position of the body and inclusion of grave goods/furnishings by those who buried them.

Conclusions

Some 24.8% (74/298) of the Ballyhanna juveniles, with greater than 25% completeness, displayed one or more atypical funerary characteristics and there are hints of differences in the nature of the mortuary ritual applied depending on the age of the individual although statistical tests revealed that the only significant difference was in relation to body position. Full-term/neonates and infants were significantly more likely to be buried in a non-supine position compared to older juveniles and it is suggested that burial on the side and/or with flexed legs replicated a nature sleeping position that was considered appropriate for the younger children.

Table 15.9. Analysis of the five features included in the analysis in 29 relatively complete individuals to determine the existence of any potential associations.

Age	Atypical orientation	Atypical torso	Flexion of legs	Atypical arms	Grave goods/furniture	Orientation + flexion of legs	Orientation + atypical torso	Atypical torso + flexion of legs	Flexion of legs + atypical arms	>2 deviations
Full term/neonate (n=3)	0	0	0	0	66.7% (2/3)	33.3% (1/3)	0	0	0	0
Infant (n=1)	0	0	0	0	0	0	100% (1/1)	0	0	0
Younger child (n=13)	15.4% (2/13)	0	38.5% (5/13)	0	7.7% (1/13)	30.7% (4/13)	0	7.7% (1/13)	0	0
Older child (n=4)	50% (2/2)	0	0	0	50% (2/2)	0	0	0	0	0
Adolescent (n=8)	12.5% (1/8)	0	12.5% (1/8)	37.5% (3/8)	0	0	0	0	12.5% (1/8)	25% (2/8)
Total (n=29)	17.2% (5/29)	0	20.7% (6/29)	10.3% (3/29)	17.2% (5/29)	17.2% (5/29)	3.5% (1/29)	3.5% (1/29)	3.5% (1/29)	6.9% (2/29)

Macdonald and Carver (2015, 65) demonstrated that burial had not occurred randomly at Ballyhanna for the most part, but rather in alternating eastward and westward running rows. Presumably the erenagh and/or priest responsible for the graveyard would have regulated burial to some extent, monitoring the depth of graves as well as identifying suitable locations for interments. Could it be the case that some families deliberately circumvented the formal governance of the burial ground through necessity? Historical sources indicate that church fees were associated with death in later medieval Ireland. Within the area of the Diocese of Armagh (*inter Anglicos*) which operated under English common law, the 'canonical portion', for example, was a levy exacted on portable items in the deceased's estate. The accounts suggest this 'portion' could range from 10.0% to as much as 33.0% and it has been suggested that it was particularly burdensome on the poor (Jefferies 1997, 15, 31). There are indications in the historical sources that priests in Gaelic areas did not receive such 'mortuaries', although it is reported that the charge for performing the Last Rites was a sheep (Jefferies 1997, 70). Perhaps the very poorest members of Gaelic society could not afford to pay the fees associated with death and burial, preferring instead to invest their resources into the living members of their families. It is possible that atypical burials represent those that occurred outside the formal management of the graveyard; perhaps these were clandestine interments that took place early in the morning or under the cover of darkness. Poor families, who lost a child, may have been content in the knowledge that they had simply been interred within consecrated ground since this was essential for a successful transition to the afterlife (Daniell 1997, 103).

If this scenario is true what are the implications for the nature of the relationship between the local erenagh and/or priest and the Christian community? Did instances occur when a priest was unavailable to undertake a burial? The burials within the Priory of St Peter and St Paul at Taunton appear to conform more rigidly to typical Christian practices, with some 92.7% (51/55) of juveniles buried in a supine position with extended legs compared to a corresponding figure of 81.6% (129/158) for Ballyhanna. In addition, only three variations in body position were identified at Taunton compared to the eight noted for Ballyhanna (see Dawson 2014, 83). Historical sources suggest that the Augustinian Priory at Taunton had great wealth and it is reported to have had as many as 26 canons in AD 1339 (Dawson 2014, 45). One would imagine that burial in this high profile religious centre would have been much more controlled and regulated by religious authorities when compared to the small rural graveyard at Ballyhanna.

It is also possible that some members of society may have simply wanted to bury the dead children of their family themselves and, indeed, one might argue that many of the atypical characteristics evident in the juvenile burials are indicative of care and tenderness. The early Irish Church is generally viewed as having operated in a system of relative isolation, with the reformers of the 12th century attempting to bring it into line with Christianity elsewhere. They succeeded in matters including the establishment of diocesan and parish systems, the introduction of new religious orders, such as the Cistercians, and the payment of tithes. They were unsuccessful, however, in their efforts to enforce celibacy and remove the practices of the hereditary succession of clergy as well as the abolishment of married clergy. Historical sources have indicated that some priests did not practise celibacy in certain Gaelic areas as late as the early 17th century (Donnelly and Murphy 2008, 214–6). Given their willingness to flout central Church teaching it is perhaps possible that an attitude of tolerance was shown by the local erenagh or priest towards the emotive matter of burial. Maybe a blind eye was turned in those cases when a family was too poor to pay the priest burial costs and they were able to undertake the interment themselves. Alternatively, a priest may have officiated in these atypical burials but allowed family members to bend conventions and lay the dead out in their graves as they saw fit.

As discussed above, burial on the side and/or flexion of the legs are common sleep positions, while the atypical arrangements of the arms may also be an attempt to replicate positions of repose in older juveniles. Could it be the case that families wanted these dead children – particularly the youngest ones – to look as if they were sleeping peacefully in their graves? The inclusion of various objects and furnishings within burials is also compatible with this notion of tenderness; religiously charged items may have been included to afford the child with a degree of protection in the afterlife, while treasured collectables may have been regarded as a final gift to the dead child. Gilchrist (2008) observed that the material culture of later medieval burial grounds overlapped with the domestic sphere of the home and sometimes drew upon older folk practices. The presence of material remains within the juvenile burials at Ballyhanna seems to support this assertion and it is possible that the positioning of dead children in their graves as if asleep is a further connection to the domestic sphere and the comforts of home.

Atypical features may only have been apparent in the burials of a quarter of the Ballyhanna juveniles, with greater than 25% completeness, but they clearly demonstrate how this later medieval population should not be considered as a homogeneous, uniform mass in terms of their funerary practices. Evidence of agency is certainly at play that may relate to the identity of the dead child and/or their families. While it may be impossible for us to fully interpret these deviations from typical Christian burial practices it is important to try. Comparative studies of juvenile burials from contemporary Christian burial grounds which use a clear, systematic methodology, cognisant of age-at-death and archaeothanatological approaches, are necessary to test the validity of the findings of this study. Furthermore, the application of a similar approach to the burials of the Ballyhanna adults may provide further insights in relation to interment practices in operation at the site. The occurrence of the variations in mortuary ritual afforded to the Ballyhanna juveniles helps to remind us of the humanity of the people interred in this graveyard – those who buried the dead, and the dead themselves during life, would each have had a personality and an identity which we are privileged to catch glimpses of through the burial record.

Acknowledgements

The initial analysis of the Ballyhanna burials was undertaken as part of the Ballyhanna Research Project, funded by Transport Infrastructure Ireland (then NRA) through Donegal County Council, and I am grateful to these organisations for this support. Thanks are due to the staff of Irish Archaeological Consultancy Ltd who excavated the Ballyhanna burial ground and to the members of the Ballyhanna Research Project team whose original research enabled this further analysis to be possible. I am also grateful to Dr Colm Donnelly, Queen's University Belfast, for his guidance on Church fees and to Libby Mulqueeny, of the aforementioned institution, for her help with the illustrations. Finally, I am also appreciative of the help given by Dr Mélie Le Roy, University of Bordeaux, with the statistical tests and of the insightful comments made by the anonymous reviewer which have helped strengthen the paper.

Notes

1. Archaeology and Palaeoecology, School of Natural and Built Environment, Queen's University Belfast, Belfast BT7 1NN, Northern Ireland. Email: eileen.murphy@qub.ac.uk.
2. The descriptive terminology of Sprague (2005, 86–9) was followed for the analysis of leg position according to the following definitions: extended – the legs were straight and joined the torso at an angle of 0°; semi-flexed – the angle between the torso and the axis of the femur lay between 0° and 90°; flexed – the angle between the axis of the trunk and femur was greater than 90°; tightly flexed – the angle between the trunk and the axis of the femur approached 180°. This system was also applied to the lower leg to enable the degree of flexion at the knee to be ascertained.
3. A further adolescent, SK 652, displayed a similar arrangement but in this case the right arm was flexed at the elbow by 150° so that the hand was positioned at the left shoulder. Unfortunately, the left arm was not preserved so the individual was not included in the counts.
4. The total figure of 22 includes three sets of items that were recovered from SK 1029 and SK 1076, both of whom were <25% complete.
5. It was not possible to ascertain the position of both arms in these two individuals which is why they have been excluded from Table 15.9.

References

Daniell, C. 1997. *Death and Burial in Medieval England*. London: Routledge.
Dawson, H. 2014. *Unearthing Late Medieval Children: Health, Status and Burial Practice in Southern England* (BAR British Series 593). Oxford: Archaeopress.
De Koninck, J., Lorrain, D. and Gagnon, P. 1992. Sleep positions and position shifts in five age groups: An ontogenetic picture. *Sleep* 15, 143–149.
Donnelly, C. J. 2015. Ballyshannon and Ballyhanna during the medieval period, in McKenzie, C. J., Murphy, E. M. and Donnelly, C. J. (eds.), *The Science of a Lost Medieval Gaelic Graveyard - The Ballyhanna Research Project* (TII Heritage 2), 15–46. Dublin: Transport Infrastructure Ireland.
Donnelly, C. J. and Murphy, E. M. 2008. The origins of *cillíní* in Ireland, in Murphy, E. M. (ed.), *Deviant Burial in the Archaeological Record*, 191–223. Oxford: Oxbow Books.
Duday, H. 2009. *The Archaeology of the Dead: Lectures in Archaeothanatology*. Oxford: Oxbow Books.
Gilchrist, R. 2008. Magic for the dead? The archaeology of magic in Later Medieval burials. *Medieval Archaeology* 52, 119–159.

Gilchrist, R. and Sloane, B. 2005. *Requiem: The Medieval Monastic Cemetery in Britain*. London: Museum of London Archaeology Service.

Gräslund, B. 1994. Prehistoric soul beliefs in northern Europe. *Proceedings of the Prehistoric Society* 60, 15–26.

Hadley, D. M. and Hemer, K. A. 2014. Introduction: Archaeological approaches to medieval childhood c. 500–1500, in Hadley, D. M. and Hemer, K. A. (eds.), *Medieval Childhood: Archaeological Approaches*, 1–25. Oxford: Oxbow Books.

Heighway, C. 2007. The attributes of burials, in Mays, S., Harding, C. and Heighway, C. (eds.), *The Churchyard* (Wharram: A Study of Settlement on the Yorkshire Wolds, XI; York University Archaeological Publication 13), 216–242. York: University of York.

Jefferies, H. A. 1997. *Priests and Prelates of Armagh in the Age of the Reformations 1518-58*. Dublin: Four Courts Press.

Johnson, C. 2015. Small finds report. On accompanying CD in McKenzie, C. J., Murphy, E. M. and Donnelly, C. J. (eds.), *The Science of a Lost Medieval Gaelic Graveyard – The Ballyhanna Research Project* (TII Heritage 2). Dublin: Transport Infrastructure Ireland.

Macdonald, P. and Carver, N. 2015. Archaeological excavations at Ballyhanna graveyard – chronology, development and context, in McKenzie, C. J., Murphy, E. M. and Donnelly, C. J. (eds.), *The Science of a Lost Medieval Gaelic Graveyard – The Ballyhanna Research Project* (TII Heritage 2), 47–84. Dublin: Transport Infrastructure Ireland.

McKenzie, C. J. and Murphy, E. M. forthcoming. *Life and Death in Medieval Gaelic Ireland: The Ballyhanna Skeletons*. Dublin: Four Courts Press.

Murphy, E. M. 2011. Parenting, child loss and the *cillíní* of Post-Medieval Ireland, in Lally, M. and Moore, A. (eds.), *(Re)Thinking the Little Ancestor: New Perspectives on the Archaeology of Infancy and Childhood* (BAR International Series 2271), 63–74. Oxford: Archaeopress.

Murphy, E. M. 2015. Lives cut short – insights from the osteological and palaeopathological analysis of the Ballyhanna juveniles, in McKenzie, C. J., Murphy, E. M. and Donnelly, C. J. (eds.), *The Science of a Lost Medieval Gaelic Graveyard – The Ballyhanna Research Project* (TII Heritage 2), 103–120. Dublin: Transport Infrastructure Ireland.

Murphy, E., Macdonald, P., Donnelly, C., MacDonagh, M., McKenzie, C. and Carver, N. 2014. The 'lost' medieval Gaelic church and burial ground at Ballyhanna, County Donegal: An overview of the excavated evidence, in Corlett, C. and Potterton, M. (eds.), *The Early Church in Ireland*, 125–142. Dublin: Wordwell.

Oosterwijk, S. 2007. Swaddled or shrouded? The interpretation of 'Chrysom' effigies on late medieval tomb monuments, in Rudy, K. M. and Baert, B. (eds.), *Weaving, Veiling and Dressing: Textiles and their Metaphors in the Late Middle Ages*, 307–351. Turnhout: Brepols Publishers.

Scheuer, L. and Black, S. 2000. *Developmental Juvenile Osteology*. London: Elsevier Academic Press.

Sprague, R. 2005. *Burial Terminology: A Guide for Researchers*. Lanham, CA: AltaMira Press.

Tsaliki, A. 2008. Unusual burials and necrophobia: an insight into the burial archaeology of fear, in Murphy, E. M. (ed.), *Deviant Burial in the Archaeological Record*, 1–16. Oxford: Oxbow Books.

Chapter 16

Interring the 'Deserving' Child: The Archaeology of the Deaths and Burials of Children at the Kilkenny Workhouse during the Great Famine in Ireland, 1845–52

Jonny Geber[1]

Abstract: The Great Irish Famine (1845–52) had a particularly devastating impact on children. One of the social consequences of poverty, destitution and starvation was that hundreds of thousands of Irish children became institutionalised in the union workhouses. One of these institutions was the Kilkenny Union Workhouse which, during the height of the crisis, provided relief to over 4,000 individuals, of which almost 1,900 were children. Child mortality was high, and this became evident following an archaeological excavation of an intramural Famine-period burial ground at the workhouse. Of the 970 individuals interred on this site, 522 were aged less than 15 years at the time of their deaths.[2] The archaeological evidence suggests that burials were undertaken in an organised and structured manner and that children were treated equally to adults in death. Both the archival and archaeological records indicate that, despite a challenging crisis situation with severe economic difficulties, the dead were treated with respect and buried with care.

Keywords: bioarchaeology, 19th century, non-adult, poverty, Poor Law, Victorian

Introduction

In late 2005, human remains came to light during pre-development works on a site adjacent to the former union workhouse in Kilkenny City in Ireland (Ó Drisceoil 2005). An archaeological excavation, which took place the following year, revealed a mass burial ground comprising 63 sub-rectangular pits within an area of 39 m by 14.5 m. Only a small proportion of the burial ground (approximately 5%) expanded beyond the limit of the excavation. The burials were discovered underneath a 0.5 m thick layer of sterile soil (O'Meara 2006).

While human remains are not an unusual archaeological occurrence at the location of former 19th-century institutions in Ireland (e.g. McKee 2006; Rogers, *et al.* 2006; Lynch 2014), the vast scale of this burial ground, with nearly 1,000 interred individuals, was completely unanticipated (O'Meara 2006). Later archival research has confirmed that the burial ground was used by the Kilkenny Union Workhouse over a period of 44 months between August 1847 and March 1851 (Geber 2012; 2015). This period includes the height of the notorious Great Famine, which affected Ireland between 1845 and 1852.

The Famine was initiated by a potato blight *(Phytophtora infestans)*, which destroyed more or less the only source of food for a vast proportion of the populace at the time. Inadequate government responses and policies aggravated the situation, and an estimated one million people, who comprised primarily the labouring classes, the poor and the destitute, perished (Boyle and Ó Gráda 1986). Children undoubtedly suffered the most, not just in terms of relative and absolute mortality, but also in the way they were affected physically, socially and, also most likely, psychologically. For those children who survived, the trauma of what they had experienced and witnessed during these years may very well have stayed with them for the rest of their lives (see Amir and Lev-Wiesel 2003, for a discussion on the adult psychological health of Holocaust child survivors from the Second World War).

Poverty and Childhood in Mid-19th-Century Ireland

At the onset of the Famine, much of Ireland was suffering from extreme poverty and destitution. Typically described as a 'wretched hovel', and some as nothing more than 'mere gravel-pits or wide ditches' (de Beaumont 1839, 129; Anon. 1840, 94), the homes for nearly 517,000 Irish families in 1841 were mud cabins, comprising only a single scantily furnished room (Census of Ireland Commission 1843, xvi). Social and economic conditions were most harsh for the deprived populace in the west of Ireland (Kennedy *et al.* 1999). The county and city of Kilkenny, which are located in the south-east (Fig. 16.1), were relatively prosperous in comparison with other parts of Ireland, but the labouring and poorer classes in Kilkenny still endured considerable hardship. For them, as elsewhere across Ireland, potatoes and milk were more or less the only sources of subsistence (Tighe 1802). As long as the potato yield was good, however, people were able to sustain themselves. The potato diet, although very monotonous, was fairly nutritious, and it has been both argued and stated that the Irish poor were of relatively good health in the decades prior to the Great Famine (Mokyr and Ó Gráda 1994; Viator 1822). Other contemporaneous reports have noted, however, that the deprived nature of their living conditions caused many people to suffer from respiratory disease and other ailments aggravated by poverty and social conditions (Tighe 1802, 481).

For the Irish children who grew up in poverty during this period, life would undoubtedly have been very difficult (see Jordan 1998). While a diet of potatoes and dairy products may have supplied them with somewhat adequate levels of nutrients

16. Interring the 'Deserving' Child: The Archaeology of the Deaths and Burials   251

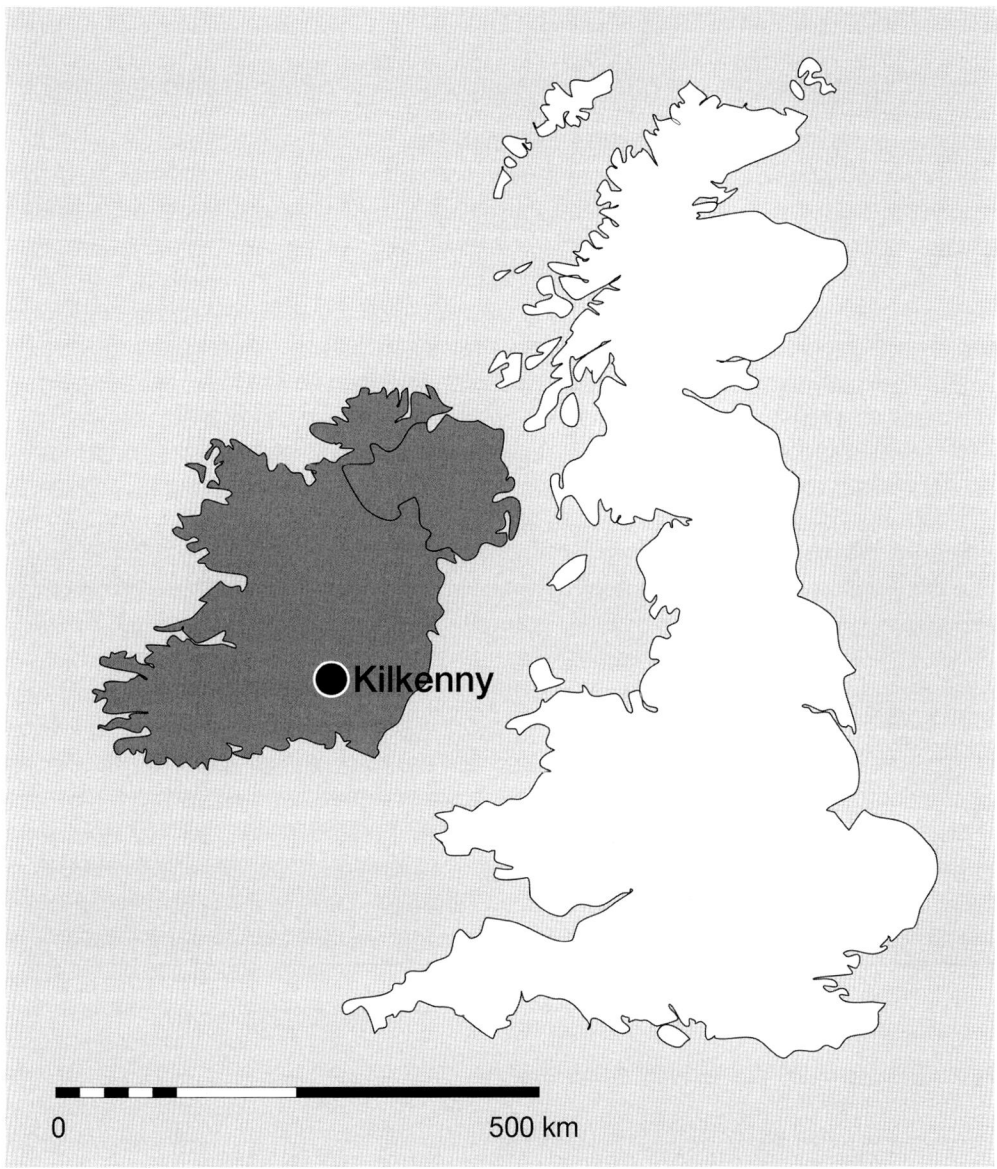

Figure 16.1. Map showing the geographical location of Kilkenny City, Ireland.

and calories, work and living conditions would have had a negative impact on their health. The skeletal evidence from the Kilkenny mass burials has indicated that 76% (174/230) of all children between the ages of six months and 12 years were retarded in growth (Geber 2014a). Poverty and deprived living conditions also caused many of them to have suffered from rickets and respiratory disease (Geber 2015; 2016a), which would have resulted in deformed limbs, dull pain and illnesses such as nasal congestions or discharge, constant coughing and most likely also recurrent headaches and fevers (Brickley and Ives 2008; Shrum *et al.* 2001).

Workhouse Children: Reformed and Disciplined
The Kilkenny Poor Law Union was formally declared in July 1839 as part of a greater national social and economic reform that sought to provide a 'more effectual relief of the destitute poor in Ireland' through legislation (Poor Relief (Ireland) Act 1838; O'Connor 1995). The reform involved institutionalisation of the 'deserving poor' in workhouses. A Board of Guardians, which comprised prominent local members of society, was elected to manage a regional or local workhouse. The Guardians adhered to the rules and regulations stipulated by the Law, and their work and the running of the workhouse institution were centrally overseen by Poor Law Commissioners who were based in Dublin. The New Poor Law did not simply seek to address the near chronic condition of poverty in Ireland, but also to initiate moral and social reform to make the poorer classes – as it was then stated – more industrious and beneficial members of society (see O'Malley 1837). The education and training of children were, therefore, an important function of the workhouse institution (see Thomas 2013). In the Kilkenny workhouse, the boys were taught agricultural labour and how to mend shoes, while the girls were occupied with learning needlework (Geber 2015). Classes in reading, writing, arithmetic and Christianity were mandatory for children for three hours every day (Irish Poor Law Commissioners 1849, 61–83), and the competence of the school teachers was constantly assessed and evaluated by both the Guardians and the Commissioners (KBGM/21/14K, 17 January 1856).

The nature of the discipline within the workhouse meant that punishments were resorted to when pauper inmates broke rules and regulations. For adult inmates, a reprimand usually involved additional hard labour or imprisonment in the gaol and reduced food rations (Anon. 1848). Children were also disciplined, although boys (between two and fifteen years of age) were the only class in the house that, according to the rules, could be subjected to corporal punishment. If a boy was over seven years of age, he would have been brought before the Board of Guardians and was then expected to explain the reason for his misconduct (Irish Poor Law Commissioners 1849, 72).

The Famine Workhouse: Overcrowding, Rampant Disease and Mass Death
From the very beginning, the workhouse system was considered by many to be ill-suited to solve the problems relating to poverty in Ireland (Crossman 2013), and it was evident that the system would struggle to cope with the situation further when famine took hold of the population. As people lost the ability to support themselves the workhouses of Ireland became critically overcrowded. In Kilkenny, the workhouse inmate population rose from a weekly average of 743 in 1845 to 3,348 in 1851. For the child population, the weekly average was 381 individuals in 1845 and 1,495 in 1851. By late June 1851, the

16. Interring the 'Deserving' Child: The Archaeology of the Deaths and Burials

Figure 16.2. Aerial photograph of the Kilkenny Union Workhouse complex, taken during the 1960s. Note the location of the burial ground which has been annotated on the top left of the image (photo courtesy of Karen Deegan, via the Kilkenny County Library Local Studies Section).

workhouse had exceeded its intended capacity (1,300 inmates) by over three times when 4,375 people were receiving indoor relief, and 1,887 of these were children (KBGM/17/10K, 26 June 1851). Additional accommodation was provided not only from sheds erected in the workhouse grounds but also in various buildings and premises in the city that were rented by the Union and set up as workhouse auxiliaries (Geber 2011).

Having been forced to provide relief to such a huge number of people put considerable economic strains on the institution and, in April 1847, the Union had overdrawn its bank balance by £312.8.1 (RLFC/3/2/14/49, 8 April 1847). The overcrowded nature of the workhouse not only affected the economy of the Union; the high concentration of poorly nourished and ill people had a devastating impact on the survival rates of those that were receiving indoor relief. People also had a general perception that the workhouses were 'death houses', as the risks of entering the institution were well-known (Geber 2015). The medical officers struggled to contain the spread of famine fevers (mainly epidemic typhus and typhoid fever), which were the main causes of death, and they were constantly wary of other threats, such as cholera. This disease eventually reached Kilkenny, in early 1849, and 172 people reportedly died as a consequence (Anon. 1849).

Mass death in the workhouse meant that an overflow burial ground became an unavoidable necessity. The city cemeteries in Kilkenny had become desperately overcrowded, even long before the Famine, and the Union had no other choice but to resort to burying its deceased inmates in a makeshift cemetery on the grounds of the workhouse itself (Fig. 16.2; Geber 2012; 2015). It is evident from the mortality statistics recorded in the workhouse minute books that children were particularly sensitive to famine-induced stress. Of an estimated 2,234 deaths between August 1847 and March 1851 of inmates belonging to the Kilkenny Union, 514 were infants aged less than two years, and 732 children aged between two and fifteen years (Geber 2016b). The archaeological evidence from the mass burials also supports this finding; more than half of all the skeletons (522/970) in the mass burial pits were those of children aged less than 15 years.

Burial of Children within the Mass Burial Ground

During the archaeological excavation, it became evident that the dead had all been interred in coffins. This evidence comprised nails, fragments of preserved wood and the presence of hexagonal wood stains surrounding the skeletons. The coffins were simple constructions made out of pine, and they conformed to three standardised sizes – about two and a half feet long used for infants, four feet for children and six feet for adults (Geber 2015). The archival records state that the dead were shrouded before they were placed in their coffins (KBGM/12/7K, 15 May 1847). The fact that the Kilkenny Union buried its dead in shrouds and coffins, despite the significant economic difficulties it struggled with during the Famine, is ultimately a sign of respect for the dead and a desire to uphold some form of decency even for the poorest members of society during the harshest of times. Nevertheless, the workhouse inmates were still interred in a regimental manner, and little evidence for individuality (which people would have lost the moment they entered the workhouse and were registered as inmates) can be discerned from an examination of how the dead were placed in their graves. Possible attempts at individuality are perhaps apparent in the burials of a few individuals who were interred with artefacts – a not too uncommon practice at the time (Mooney 1888, 295). The remains of four rosaries, four medallions and two finger rings were found in the mass burials. One medallion, a so-called 'Miraculous Medal', which at the time was believed by some to possess healing powers (Aladel 1880), had been placed in the coffin with a 15-year-old adolescent (Fig. 16.3). All of the other objects were found with adult females, with the exception of two medallions that could not be associated with a particular skeleton.

The surviving archival records that include details of the weekly mortality statistics for the Kilkenny Union are fragmented and incomplete, but they nevertheless indicate that substantial periodic peaks in mortality occurred during the Famine period, particularly in the winter months (Patterson 1997; Geber 2015, Fig. 2.7). Noticeable chronological differences in mortality are apparent within the age-classes of the infants and children, who suffered particularly high mortality rates in early 1847, 1848

16. Interring the 'Deserving' Child: The Archaeology of the Deaths and Burials

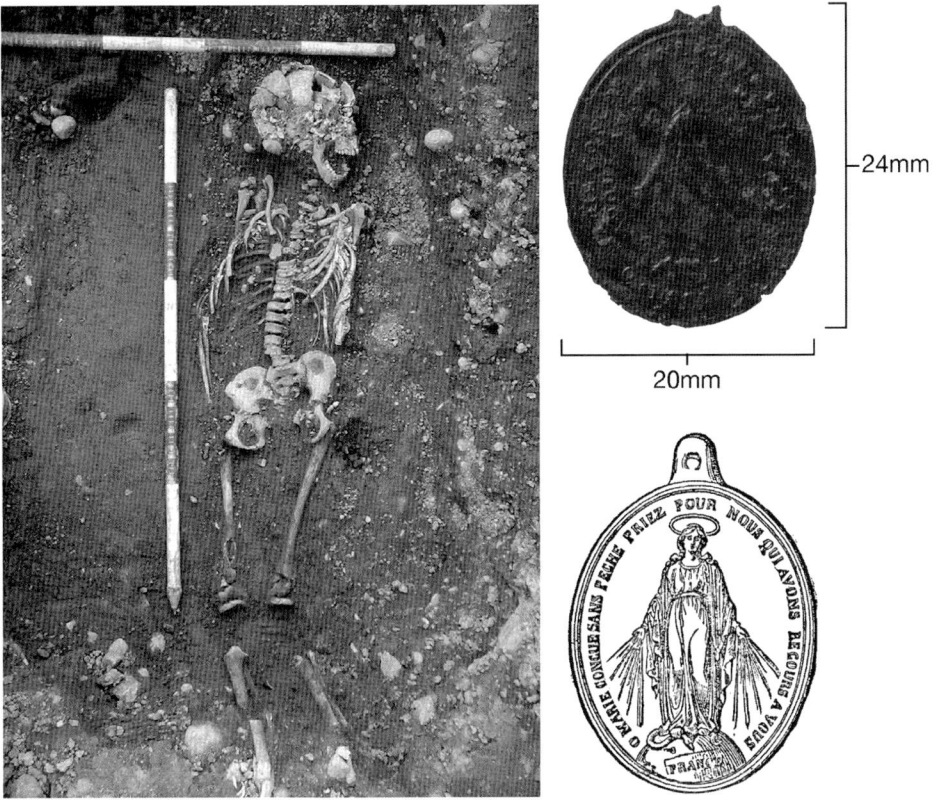

Figure 16.3. In situ *photograph (left) of the skeleton of a c. 15-year-old adolescent (Burial CCCLXXXIV), who was buried with a Miraculous Medal (upper right) (photo: Margaret Gowen & Co. Ltd. and Jonny Geber). An illustration of a contemporary Miraculous Medal (lower right) is included for comparative purposes (after Aladel 1880, 272).*

and late 1851 (Geber 2016b). The archaeological evidence, where each pit represents a separate chronological event, however, does not indicate such a clear pattern. There seems to be a slightly higher relative proportion of infants interred in the burial pits located in the middle and southern parts of the burial ground. A more noticeable cluster in terms of burial location is evident when assessing the two to five years and five to nine years child age categories; the highest relative proportion of these individuals were buried in the middle of the cemetery. The pit with the highest relative proportion of nine- to fifteen-year-old children was also located in this general area (Fig. 16.4). On a general level across the burial ground, it is evident that the higher the density of interments per pit, the higher proportion of children it is likely to contain. It seems possible that this is simply a reflection of practicalities rather than events associated with notably high child mortality rates – if two pits were opened simultaneously it would have been most practical to place the child coffins in the spaces between adult coffins to fill in the gap, so that as many as possible would fit.

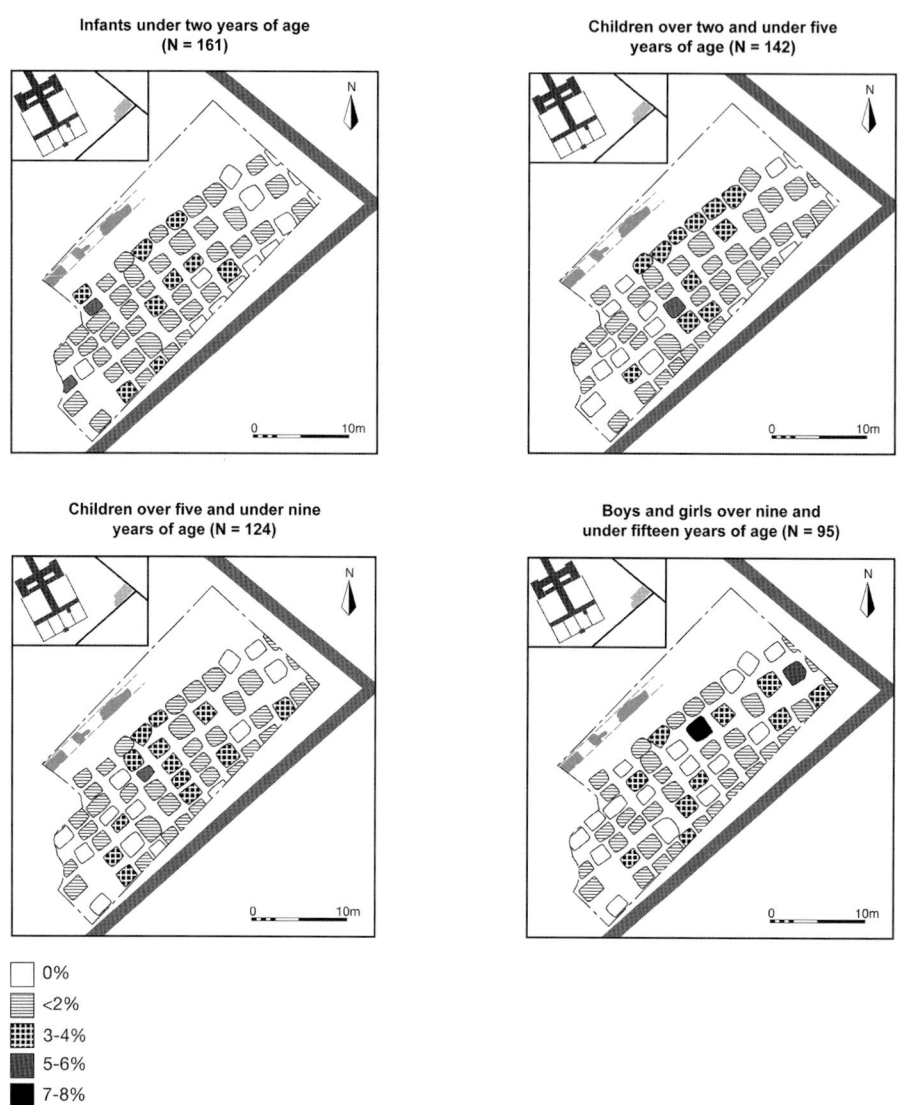

Figure 16.4. Spatial distribution showing the relative frequency of child interments on the basis of their respective age class categories in the pits within the Famine mass burial ground at the Kilkenny Union Workhouse.

The mass burials at the workhouse are a physical manifestation of the burial record, rather than the true mortality record of the institution. Discrepancies between the burial records and the mortality records are also evident in further direct comparisons between the data. The Census of Ireland 1851, published in 1856, provided details of the number of deaths by age that occurred in the public institutions of each county

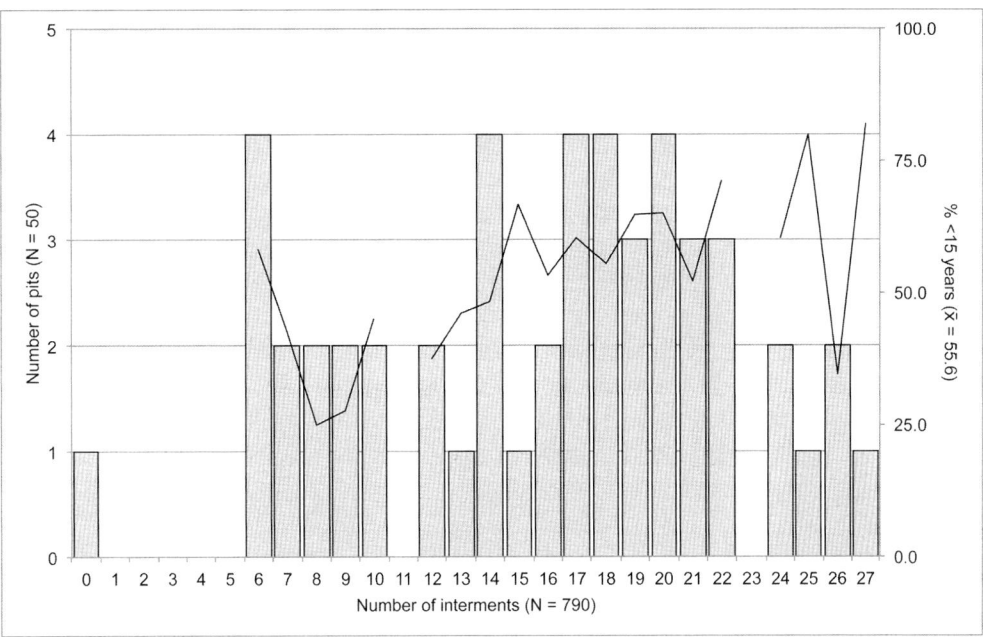

Figure 16.5. Number of interments (n=790) within non-truncated pits (n=50) at the Kilkenny Union Workhouse mass burial ground, and the percentage of non-adults per burial pit.

and city. In Kilkenny City, 419 boys and 394 girls aged less than 15 years are recorded as having died in public institutions (including the workhouse) between January 1848 and March 1851 (Census of Ireland Commission 1856). In the workhouse mass burial ground, however, 'only' 522 skeletons of children were discovered.

An Archaeology of Denial and Social Trauma?

The workhouse mass burial ground was never consecrated, and this is likely to have caused a lot of additional pain and grief for people who lived during a time when a 'decent' burial was extremely valued (Mooney 1888). The Guardians had no authority over how and when the burial ground would, or could, be consecrated; this was a decision to be made by the local clergy, who evidently decided against it. The Guardians could, however, at least provide the dead inmates with the decency of having been interred in coffins, which they did. Many other union workhouses across Ireland had depleted their funds to such an extent that coffined burials of their inmates were not possible (see Ó Gráda 1993, 129). One can imagine that there may have been a certain degree of relief in Kilkenny in the knowledge that those who died in the workhouse would at least be interred in a coffin.

It is clear from the historical evidence that some of the Kilkenny Union dead must have been buried elsewhere. Considering that many individuals were accommodated in various auxiliaries across the city, it seems likely that at least some of the children who died at other premises may have been interred in other burial grounds, such as the intramural mass burial ground that was simultaneously in use at the Kilkenny Fever Hospital, which during the Famine was operated by the Union as an annexe to the workhouse (Geber 2015). The relative proportion of children aged between two and nine years of age is higher in the skeletal record from the workhouse mass burials compared to the historical record and, conversely, infants and the older category of children were less represented in relative numbers when compared to the mortality statistics noted in the workhouse minute books (see Geber 2016b). While the under-representation of infant skeletons is likely to be related to poor bone preservation (see Guy *et al.* 1997), the overall pattern may suggest a selective bias in terms of who was buried at the intramural burial ground. It may be the case that the majority of younger children were cared for in the workhouse infirmary, while some of the older children who required medical care were treated and subsequently died at the Fever Hospital where they were then buried. If this interpretation is accepted, it may also be suggestive that the remains of some children were less likely to be claimed by friends or relatives outside the institution. It seems feasible that their parents (if still alive) would have been in the workhouse with them, and therefore had neither the means to leave the workhouse nor the ability to pay for the expense of their child's burial.

Choosing to Forget?

It is an interesting circumstance that the intramural burial ground at the Kilkenny Workhouse was eventually forgotten. Could this situation have occurred because a high proportion of the dead were children? Many of the children who died would have been orphans, and they would have already lost their families before they entered the workhouse. In addition, the trauma of having lost a child in the workhouse, and the knowledge that he/she was buried in an unmarked pauper's grave on its premises, may have been too difficult for parents or other relatives to come to terms with. How this situation may have been perceived is rather evocatively recalled from a story that was recorded as Famine folklore in County Leitrim, in the north-west of Ireland in the late 1930s:

> I heard your great grandfather telling this – it is perfectly true.
> Your great grandfather Owen Hunt owned the mill at Killukin. He knew a man, a big strong man whose child died in the workhouse during the Famine years. The child was buried without a coffin or else, in the 'Sand Bank' the burial place at Carrick workhouse.
> The poor man went in at night, with a big pardog[3] with errishes[4] on it, and he lifted the dead body out of the pauper's grave, and carried the body naked through Carrick, in the pardog on his back, and out to the Railway Bridge. He lowered the pardog, took off his torn coat and shirt, and wrapped the dead child's body in the shirt and coat and buried his child among his kith and kin in Killukin graveyard. He could not rest, he said, day or night, dead or alive, as long as his child lay among the paupers. I cannot think of his name now – I do forget names lately (Irish Folklore Commission, Schools' Folklore Collection 0210).

As discussed above, the pauper burial ground at the Kilkenny Workhouse remained unconsecrated, and this is perhaps the reason why it was not marked on any maps; it was never a cemetery in a formal sense, and would therefore not have been recorded as such in any official documents. When local awareness of the burials became completely lost is unknown. The Guardians had written to the Poor Law Commissioners in September 1866 and asked for permission to enclose the burial ground. The grave pits would still have been evident at that stage, either as elevated small mounds or depressions in the ground. The Commissioners gave a rather inconclusive reply to the query (Anon. 1866), and it seems likely the Guardians then decided that the burials were to be covered over with soil. This layer, which was discovered during the archaeological excavation is, in a sense, a representation of an intentional attempt to erase and ultimately forget about what had happened to Kilkenny's poor and destitute; individuals who had attempted to survive the Famine by resorting to the workhouse but who ultimately succumbed to malnutrition and disease.

Recognising the Child Victims of the Irish Famine

The last year of the blight was in 1850, but social and economic distress continued to be severe in parts of Ireland well into 1852 (Kinealy 2006). In Kilkenny, the end of the Famine can probably be set at around September 1851. At that time, the mortality rate in the workhouse was similar to the pre-Famine levels of 1845 (Geber 2016b). The number of children in the workhouse during that month averaged at 400 out of a population of *c.* 800 inmates. Four months earlier, during the last week of May 1851, the child population, of 1,909 out of a total of 4,312 inmates, appears to have reached its peak in the workhouse (KBGM/17/10K, 29 May 1851). Overall, it is clear, however, that despite substantial economic and logistic difficulties, the Kilkenny workhouse managed to function relatively well during the Famine crisis.

The Kilkenny workhouse mass burial ground, with its high proportion of children, illustrates the reality of how complete families were destroyed and social bonds severed during the Famine, and how the children were the most vulnerable on so many levels. Many of the youngest were foundlings and orphans who had to endure the Famine and enter the workhouse on their own. As nobody was there to care for them in life, they were to be buried anonymously with all the other workhouse inmates in death. Eventually, their burial place would be forgotten. This was possibly due to memories being too painful to retain. The decision by the Guardians to cover the burial ground with a thick layer of soil may be indicative of a deliberate intention to forget and not speak about the painful Famine years.

Although historical research into the Great Irish Famine was relatively rare until the mid-1990s (see Kinealy 2006, xv–xliii), there is now a substantial historiography relating to the subject (e.g. Daly 1986; Morash and Hayes 1996; Kinealy 1997; 2006; Kennedy, *et al.* 1999; Ó Gráda 1999; Donnelly 2001; Crowley, *et al.* 2012; Corporaal, *et al.* 2014). Archaeological studies of the Great Famine have, however, been much scarcer (Orser

1996; Geber 2014b; 2015). It is impossible to estimate how many more unknown Famine-period mass burial grounds, similar to the one discovered in Kilkenny, are located across Ireland. What is clear, however, is that archaeology has the potential to discover the existence of more of these poignant sites, thereby providing further insights into the true human experience of this tragic period in Irish history. This is particularly true for the children whose experience of the Famine has been scantily addressed by scholars to date, despite the fact that they represented more than half of its victims.

The true identity of the children buried in the mass burial ground at the Kilkenny Workhouse is not known. On the basis of their mortal remains, however, it has been possible to recreate a narrative of their lives, their experiences of the Famine, and how society dealt with their death and burial during one of the worst social disasters in world history.

Acknowledgements

The archaeological excavation of the Kilkenny Workhouse mass burials was directed by Brenda O'Meara of Margaret Gowen & Co. Ltd in 2006, under licence from the National Museum of Ireland (05E0435). This research was funded by the Irish Research Council (GOIPD/2013/36), and was undertaken at the Department of Archaeology, University College Cork, under the post-doctoral mentorship of Dr Barra O'Donnabhain. Thanks to Robert McPhee, Dr Charlotte King and Anne Marie Sohler-Snoddy of the Department of Anatomy, University of Otago, for help with the graphics and comments on the manuscript.

Notes

1. Department of Anatomy, University of Otago, P.O. Box 56, Dunedin 9054, New Zealand. Email: jonny.geber@otago.ac.nz.
2. In this article, individuals aged less than 15 years are considered as children as this was the age criteria employed by the workhouse system in accordance with the 1838 Poor Law. The infant age class comprised those under two years of age, and children were sub-divided into three additional categories – over two and under five years of age; over five and under nine years of age; and over nine and under 15 years of age.
3. A traditional wooden basket or creel that was generally used for carrying turf or manure, and hung over the sides of a donkey or small pony.
4. A strap, used to tie pardogs together.

References

Archive Sources
Famine Relief Commission: Incoming Letters, *National Archives of Ireland*, Dublin, RLFC/3/2/14/49.
Kilkenny Board of Guardians Minutes, *Kilkenny County Library, Kilkenny*, KBGM/12/7K.
Kilkenny Board of Guardians Minutes, *Kilkenny County Library, Kilkenny*, KBGM/17/10K.
Kilkenny Board of Guardians Minutes, *Kilkenny County Library, Kilkenny*, KBGM/21/14K.
Poor Relief (Ireland) Act, 'An Act for the More Effectual Relief of the Destitute Poor in Ireland', 2 & 3 Vic. c. 56, 31 July 1838.

Schools' Folklore Collection (1937–8), Volume 0210, Page 491: 'A Tale of the Famine', told by Mrs Hunt, Jamestown, to her granddaughter Mary Gorman, Jamestown, *Irish Folklore Commission/ Department of Education and the Irish National Teacher's Organisation*.

Publications

Aladel, M. 1880. *The Miraculous Medal: Its Origin, History, Circulation, Results*. Philadelphia, PA: H. L. Kilner & Co.

Amir, M. and Lev-Wiesel, R. 2003. Time does not heal all wounds: quality of life and psychological distress of people who survived the Holocaust as children 55 years later. *Journal of Traumatic Stress* 16, 295–299.

Anon. 1840. Cottages – thatching. *The Penny Magazine* 509, 93–95.

Anon. 1848. Board of Guardians. *Kilkenny Journal*, 17 June.

Anon. 1849. Board of Guardians. *Kilkenny Journal*, 14 March.

Anon. 1866. Kilkenny Union. *Kilkenny Journal*, 15 September.

Boyle, P. P. and Ó Gráda, C. 1986. Fertility trends, excess mortality, and the Great Irish Famine. *Demography* 23, 543–562.

Brickley, M. and Ives, R. 2008. *The Bioarchaeology of Metabolic Bone Disease*. Amsterdam: Academic Press.

Census of Ireland Commission. 1843. *Report of the Commissioners Appointed to Take the Census of Ireland for the Year 1841*. Dublin: A. Thom & Co.

Census of Ireland Commission. 1856. *Report of the Commissioners Appointed to Take the Census of Ireland for the Year 1851, Part V, Tables of Deaths, Vol. 2*. Dublin: A. Thom & Co.

Corporaal, M., Cusack, C., Janssen, L. and van den Beuken, R. (eds.) 2014. *Global Legacies of the Great Irish Famine: Transnational and Interdisciplinary Perspectives*. Oxford: Peter Lang.

Crossman, V. 2013. *Poverty and the Poor Law in Ireland, 1850-914*. Liverpool: Liverpool University Press.

Crowley, J., Smyth, W. J. and Murphy, M. (eds.) 2012. *Atlas of the Great Irish Famine, 1845-52*. Cork: Cork University Press.

Daly, M. E. 1986. *The Famine in Ireland*. Dublin: Historical Association of Ireland.

de Beaumont, G. 2006 [1839]. *Ireland: Social, Political, and Religious*. Cambridge: The Belknap Press.

Donnelly, J. S., Jr. 2001. *The Great Irish Potato Famine*. Stroud: Sutton Publishing.

Geber, J. 2011. Osteoarchaeological and archaeological insights into the deaths and intramural mass burials at the Kilkenny Union Workhouse between 1847–51 during the Great Famine. *Old Kilkenny Review* 63, 64–75.

Geber, J. 2012. Burying the famine dead: Kilkenny Union Workhouse, in Crowley, J., Smyth, W. J. and Murphy, M. (eds.), *Atlas of the Great Irish Famine, 1845-52*, 341–348. Cork: Cork University Press.

Geber, J. 2014a. Skeletal manifestations of stress in child victims of the Great Irish Famine (1845–52): prevalence of enamel hypoplasia, Harris Lines, and growth retardation. *American Journal of Physical Anthropology* 155, 149–161.

Geber, J. 2014b. Reconstructing realities: exploring the human experience of the Great Famine through archaeology, in Corporaal, M., Cusack, C., Janssen, L. and van den Beuken, R. (eds.), *Global Legacies of the Great Irish Famine: Transnational and Interdisciplinary Perspectives*, 139–156. Oxford: Peter Lang.

Geber, J. 2015. *Victims of Ireland's Great Famine: The Bioarchaeology of Mass Burials at Kilkenny Union Workhouse*. Gainesville, FL: University Press of Florida.

Geber, J. 2016a. 'Children in a Ragged State': Seeking a biocultural narrative of a workhouse childhood in Ireland during the Great Famine (1845–52). *Childhood in the Past* 9, 120–138.

Geber, J. 2016b. Mortality among institutionalised children during the Great Famine in Ireland: bioarchaeological contextualisation of non-adult mortality rates in the Kilkenny Union Workhouse, 1846–51. *Continuity and Change* 31, 101–126.

Guy, H., Masset, C. and Baud, C.-A. 1997. Infant taphonomy. *International Journal of Osteoarchaeology* 7, 221–229.

Irish Poor Law Commissioners. 1849. *Second Annual Report of the Commissioners for Administrating the Laws of Relief of the Poor in Ireland*. Dublin: Alexander Thom.

Jordan, T. E. 1998. *Ireland's Children: Quality of Life, Stress, and Child Development in the Famine Era*. London: Greenwood Press.

Kennedy, L., Ell, P. S., Crawford, E. M. and Clarkson, L. A. 1999. *Mapping the Great Irish Famine: A Survey of the Famine Decades*. Dublin: Four Courts Press.

Kinealy, C. 1997. *A Death-dealing Famine: The Great Hunger in Ireland*. London: Pluto Press.

Kinealy, C. 2006. *This Great Calamity: The Irish Famine 1845-52*. Dublin: Gill & Macmillan.

Lynch, L. G. 2014. Death and burial in the Poor Law union workhouses in Ireland. *The Journal of Irish Archaeology* 23, 189–203.

McKee, J. 2006. Archaeological Excavation Report: For the Development at the Old Donard School Site, Edenderry Road, Banbridge, Co. Down. Unpublished report, Archaeological Development Services.

Mokyr, J. and Ó Gráda, C. 1994. The heights of the British and Irish *c*. 1800–15: evidence from recruits to the East India Company's army, in Komlos, J. (ed.), *Stature, Living Standards, and Economic Development: Essays in Anthropometric History*, 39–59. Chicago, IL: University of Chicago Press.

Mooney, J. 1888. The funeral customs of Ireland. *Proceedings of the American Philosophical Society* 25, 243–296.

Morash, C. and Hayes, R. 1996. *'Fearful Realities': New Perspectives on the Famine*. Dublin: Irish Academic Press.

Ó Drisceoil, C. 2005. Archaeological Assessment of Workhouse Burial Site, MacDonagh Junction Development, Kilkenny. Unpublished report, Kilkenny Archaeology.

Ó Gráda, C. 1993. *Ireland Before and After the Famine: Explorations in Economic History, 1800-925*. Manchester: Manchester University Press.

Ó Gráda, C. 1999. *Black '47 and Beyond: The Great Irish Famine in History, Economy, and Memory*. Princeton, CA: Princeton University Press.

O'Connor, J. 1995. *The Workhouses of Ireland: The Fate of Ireland's Poor*. Dublin: Anvil Books.

O'Malley, T. 1837. *An Idea of a Poor Law for Ireland*. London: Henry Hooper.

O'Meara, B. 2006. A preliminary account of recent excavations adjacent to Kilkenny Union Workhouse. *Old Kilkenny Review* 58, 154–162.

Orser, C. E., Jr. 1996. Can there be an archaeology of the Great Famine?, in Morash, C. and Hayes, R. (eds.), *'Fearful Realities': New Perspectives on the Famine*, 77–89. Dublin: Irish Academic Press.

Patterson, T. 1997. Famine fever in Kilkenny. *Old Kilkenny Review* 49, 74–88.

Rogers, T., Fibiger, L., Lynch, L. G. and Moore, D. 2006. Two glimpses of nineteenth-century institutional burial practice in Ireland: a report on the excavation of burials from Manorhamilton Workhouse, Co. Leitrim, and St Brigid's Hospital, Ballinasloe, Co. Galway. *Journal of Irish Archaeology* 15, 93–104.

Shrum, K. M., Grogg, S. E., Barton, P., Shaw, H. H. and Dyer, R. R. 2001. Sinusitis in children: the importance of diagnosis and treatment. *Journal of the American Osteopathic Association* 101, S8–13.

Thomas, L. 2013. The evolving moral and physical geometry of childhood in Ulster workhouses, 1838–55. *Childhood in the Past* 6, 22–51.

Tighe, W. 1802. *Statistical Observations Relative to the County of Kilkenny made in the Years 1800 & 1801*. Dublin: Graisberry and Campbell.

Viator. 1822. On the character of the lower Irish; and their modern popular songs. *The Country Constitutional Guardian; and Literary Magazine* 1, 384–385.

Index

Because of inconsistency in definition, childhood age categories are not indexed separately. All burials are inhumations unless stated otherwise. Numbers in italics denote pages with images, 't' = table.

additional bones 91, 98, 102, 103, 104, *see also* disarticulated bone
aDNA 2, 30, 59–60, 104, 158
adult
    attitude to children/child death 2, 8, 11, 14, 15, 31, 35, 36, 37–41, 43–4, 57, 73, 140, 148, 150, 158, 175, 179, 180, 191, 193, 197, 200, 205, 207–8, 227–8, 229, 245–6, 254, 258–9
    body position 51, *51*, 85, 86, 87, 129–30, 141, 143, 170, 175, 184, 203, 205, 215, 229
    burial 8, 12–13, 15, 24–6, 29, 30, 36, 37, 38, 49, 50–3, *52*, 62, 63, 64, 71–2, 77, 78t, 80, 81t, 82, 85t, 85, 86, 87, 92, 98, 99, 102, 109, 113, 115, 118, 124t, 127, 127t, *128*, *130*, 133, 136, *137*, 141, 142, 143, 144 n5, 147, 151, *152*, 153, 154, 156–7, 159 n8, 167–8, 170, 173–4, 176, 181, 183–4, 183t, 186t, 187, 189–90, 200, 201, 211, 214, 215, 234–5, 237, 239–46, 249, 254–7, *257*,
    buried with children 26, 30, 62, 68, 69, 80, 82, 115, 129–30, *132*, 136, 151, *152*, 153, 156, 168, 170, 173, 175, 200, 201, 204, 205, 207
    compared with child burial 7, 8, 9, 12, 13–14, 15, 21, 28, 29, 38, 39, 43, 50–6, *52*, 64–6, 71–3, 77, 85–7, 102, 107, 109, 115, *128*, 129–30, 129t, *130*, 132–3, *134*, 136–8, *137*, 142, 143, 147, 150–2, 155–7, 163, 179, 183–4, 187–9, 203, 211, 216–9, 229, 235–46, 254–7, *257*
    grave goods 8–9, 50, 53, 68, 71, 115, 130, 134, 135–6, 139, 140–2, 153, 179, 185–7t, 187–9, 192
    warrior/swordsman grave 9, 132, 168, 181
    see also women 201, 215

age-at-death 2, 7, 10, 11, 12–13, 15, 19, 23t, 24, 37, 39, 41, 45, 48t, 49, 50, 51, 61–2, 62t, 64, 71–2, 80, 81t, 82, 85t, 96, *96*, 107, 110t, 116–17, 129t, 133, 133t, 134t, 137t, 159 n9, 170, 221, 227–47, *228*, 230–1t, *232*, 233–4t, 236t, *237*, 238t, 240–1t, *241*, 244t, 257–8; *see also* mortality profile/rate
age category 10, 61–2, 71, 80, 108, 110t, 116–7, 127–9, *128*, 154, 159 n2, 165, *168*, 181–4, *182*, 183t, 185–6t, 190, 219t, 228, 229–30, 230t, 231–2, 233–4t, 235, 236t, 238t, 254, 255, 260 n2
    cultural 108, 116–7
agency
    of children 2, 180, 193, 246
    of the living 1, 15, 180, 227
agriculture 47, 78, 80, 125, 214
Alakul' culture, 7, 13, 125, 126–9, *126*, 127t, *128*, 129t, 130, 133, 136–9, *137–8*, 137t, 141, 143, 144
    Alakul' 137
    grave goods 137–9, 137t, *138*, 141
    Stepnoye VII 127t, 133, *134*, 136, *137–8*, 138
amulet/amuletic 7, 113, 130, 136, 139, 141, 184, 190, 191, 206, 207, 208
Anglo-Saxon sites 2, 14
    Berinsfield, Wallingford 180
    Great Chesterford 6, 9, 14, 179–93, *182*, 183t, 185–8t
    Mill Hill, Deal, Kent 180
animals
    disturbance to burials/graves 37, 64, 139, 181
    dog burial 130, 181, 188
    domestic 47, 127t, 130
    gnawing 37
    horse burial 181, 144 n4
    livestock breeding 125, 126, 143
    sacrifice 127t, 130, 133, 135, 136, 138–9, 143, 144 n4
animal bone 27, *27*, 53, 66, 68, 102, 130, 171, 172, 175

object/grave good 62t, 66, 68, 102, 130, 131, 133–4t, 135, 136, 137t, 140, 141, 142, 167, 171, 172, 173
   antler 171
   arrowhead 131, 140
   astragali (gaming piece) 130, 133–4t, *135*, 135t, 136, 137–8, 137t, 139, 140–1, 143
   awl 66, 68
   tooth pendant/amulet (drilled fang) 130, 133–4t, 135, *135*, 136, 137, 137t, 140, 141
anthropology/anthropologist 6, 21, 28, 44, 47, 61, 97, 103, 113–4, 128–9, 140, 144 n5, 149–50, 163, 165, 170, 171, 174, 175–6, 176 n3
archaeothanatology 6–7, 149–50, 230, 246
Argos, Greece 110t, 114, 118
arrowhead 9, 30, 130–2, 133t, 140, 144 n4, 170
   bone 131, 140
   bronze 9, 131, 132, 133t, 140, 170
   iron 9, 170
   stone 9, 30, 131, 140
artificial cranial deformation 83
Ayios Konstantinos Methana sanctuary, Greece 110t, 113, *114*
Ayios Stephanos, Greece 109, 110t, 113, 117

Ballyhanna, Co. Donegal 14, 227–47, *228*, 230–1t, *232*, 233–4t, 236t, *237*, 238t, 240–1t, *241*, 244t
   variant burial positions and orientations 227, 229–38, 231t, *232*, 233–4t, 236t, *237*, 238t, 242–5, 244t
   grave goods and furnishings 238–43, 240–1t, *241*, 244t
barrow/tumulus/barrow cemetery 9, 126, 127t, 129, 131, 132, 136, 140, 166, 170, 173
bead 68, 72, 130, *131*, 133t, 134t, 135, 137t, *138*, 184, 185–7t, 187, 188t, 239, 240–1t
   amber 9, 170, 185–6t, 187, 188t
   amethyst 114
   bone 66
   bronze 135, 136, 140, 141
   glass 185–6t, 187, 188t, 239, 243
   glass-paste/faience 114, *135*, 141
   necklace 114, 187
   rosary/paternoster 239, 254
   shell 8, 38, 53, 66
   stone 8, 66, 112, *135*
   waistband/belt 43, 53, 68
   workshop 53
bioarchaeology/bioarchaeological analysis 2–3, 4, 10, 11, 35, 36–7, 157, 158, 222
body orientation 15, 52, 58, 60, 62t, 63, 64, 66, 72, 173, 180, 184, 185–6t, 189, 203, 208, 227, 229, 230–1, 231t, 242–3, 244
body position 6, 7–8, 10, 11, 12, 13, 14, 15, 21, 26–7, *27*, 37, 38–9, *38*, 43, 45, 48, 50–1, *51*–2, 58, 60, 62t, 64–6, *65*, 72, 81t, 83, *84*, 85, 86, 87, 98, 129–30, 133, 134, 137, 141, 143, 163, 171, 172, 173, 175, 180, 181, 184, 189, 190, 200, 201, 203, *204*, 205, 208 n2, 215, 227, 228, 229–38, 231t, *232*, 233–4t, 236t, *237*, 238t, 242–3, 244t, 245, 246, 247 n2, n3, n5
   adult 51, *51*, 85, 86, 87, 129–30, 141, 143, 170, 175, 184, 203, 205, 215, 229
   'face to face'/'embracing' 136, 141, 143, 170
   hocker 10, 203
   neonate/perinate 37, 38–9, 50–1, *51*, 81t, *204*, 232–4, 233–4t, 236t, 238t, 243
   seated 27
body processing and treatment 9, 10, 12, 57, 91–104, *93*–4, 94t, *95*, 98t, *99*, 100t, 176
   curation 10
   decomposition before burial 7, 10, 12, 39, 91, 100, 102, 103, 173
   dismemberment 9, 10, 12, 91, 100–1, 100t, *101*, 102, 103, 171, 176
   exposure 171, 173, 174, 176
bone/tooth object 62t, 66, 68, 102, 130, 131, 133–4t, 135, 136, 137t, 140, 141, 142, 167, 171, 172, 173
bone/skeletal element preservation 37, 51, *52*, 63–4, 66, 68, 80, 95–6, 103, 128, 148, 150, 152, 154, 157, 181, 200, 203, 230, 230t, 233, 235, 237, 243, 247 n3, 258
   missing bones/elements 64, 91, 96, 97–8, 98t, 100, 101, 102–3, 170, 175
bronze 130, 188t
   adze 131, *131*, 133t, 140, 142
   amulet 136
   anklet 188t
   arrowhead 9, 131, 132, 133t, 140, 170
   awl 133t, 140, 142
   axe 133t, 140
   battle-axe 9, 131, 132
   bead 135, 136, 140, 141

bracelet 134t, 135, *135*, 136, 137t, *138*, 140, 141, 188t
brooch 187, 188t, 189, 191
foil 9, 170
girdle hanger 187
hair decoration 134t, 137t, 140, 141
headdress 136, 239
jewellery 181
knife 9, 132, 133t, 134t, 135, *135*, 136, 142
metallurgy 125, 130
ornaments *135*, 136, 140
pendant *135*, *138*, 140
pin 239, 240t, 242
ring/finger ring 9, 113, 137t, 140, 141, 188, 188t
tools 142
spearhead 9, 131, 132, 133t, 140
weaponry 127t, 130–1
Bronze Age 5, 7, 109, 141
  barrows/tumuli 9, 126, 127t, 129, 131, 132, 136, 140, 170
  Early BA Bulgarian Thrace 7, 9, 10, 12–13, 91–104, *93–4*, 94t, *95*, 98t, *99*, 100t, *101*
  Late Helladic Mycenae 107–20, 109–10t, *111–12*, *114*, *116*
  Pre- and protopalatial Minoan Crete 13, 147–58, *148*, 149t, *152*, *156*
  southern Trans-Urals 7, 13, 125–45, *126*, 127t, *128*, 129t, *130–2*, 133–4t, *134–5*, *137–8*, 137t
burial containers
  box 7, 112, 113
  coffin 14, 201, 203, 205, 215, 216, 220, *220*, 222, 254, 255, 257, 258
  jar/pot/amphora/pithos 7, 10, 12, 53, 77, 80, 81t, 83–7, *84*, 85t, 91–104, *93*, 94t, *94*, 112, 117, 119, 152–3, 155, *156*, 157, 159 n5, n8, 174; *see also* jar burial
  shroud 201, 208, 229, 254
  textile/woven 38
burial environment 10, 11, 35, 37, 150
burial practices *see* funerary/mortuary/burial practices
burial repository/grave
  brick built 200–1
  cave/rock shelter 21, *22*, 23, 25–6, 149t, 150, 151, 155–6
  cist 85t, 86, 112, 113, *114*, 150, 237

  on wooden plank 216, 222
  pit 9, 12, 13, *20*, 23, 26, 28, 60, 62–3, 62t, 66, 71, 83, 86–7, 91, 93, 94t, 95, 96, 100, 103, 112, 130, 132, 133, 134, 140, 163, 165, *165*, 170, 171, 173, 174, 175, *175*, 181, 208, 234, 249, 250, 254, 255, *256–7*, 259
    pit field 163, *165*, 170, 171–2, 173, 175, *175*
    with wooden ceiling 9, 132–3, 134, 137
  tomb 27t, 129–30, 132–3, 136, 150
    barrow/tumulus 9, 126, 127t, 129, 131, 132, 136, 140, 166, 170, 173
    chamber 114, 115, 118
    'house-tomb'/burial building 149t, 151, *152*, 153, 154, 156–7, *156*
    megalithic tomb/dolmen *20*, 22–3, 25
    rock-cut 150
    tholos 150, 151–2, 153, 155, 157, 158

Çamlıbel Tarlası, central Anatolia 12, 77–87, 78t, *79*, 81t, *84*, 85t
Canterbury, St Gregory 14, 211, 214–223, *212*, *215*, 216–9t, *218*, 221t
Çatalhöyük, Central Anatolia 6, 7, 9, 11, 35–41, *36*, *38*, *40*
cave/rock shelter 21, *22*, 23, 25–6, 149t, 150, 151, 155–6
cemetery/burial ground/funerary area 3, 4–5, 6, 7, 13
  Neolithic 19–31, *22*, 23–4t, *25*, 47, 159 n4, n5
    Kadruka, Sudan 9, 11–12, 23, 43–54, *46*, 48t, *49*, *51–2*
    Sultana-Malu Roşu, Romania 6, 8–9, 12, 57–73, *59–60*, 62t, *63*, *65*, *67*, *69–70*
  Chalcolithic
    Çamlıbel Tarlası, central Anatolia 12, 77–87, 78t, *79*, 81t, *84*, 85t
  Bronze Age 7
    Early Bronze Age Bulgarian Thrace 7, 9, 10, 12–13, 91–104, *93–4*, 94t, *95*, 98t, *99*, 100t, *101*, 125–45, *126*, 127t, *128*, 129t, *130–2*, 133–4t, *134–5*, *137–8*, 137t
    Pre- and protopalatial Crete 147–58, *148*, 149t, *152*, *156*
    Late Helladic Mycenae 107–20, 109–10t, *111–12*, *114*, *116*
  Anglo-Saxon 2, 6

Great Chesterford 6, 9, 14, 179–93, *182*, 183t, 185–8t
medieval 7, 9, 14
  Ballyhanna, Co. Donegal 14, 227–47, *228*, 230–1t, *232*, 233–4t, 236t, *237*, 238t, 240–1t, *241*, 244t
  Canterbury, St Gregory 14, 211, 214–23, *212*, *215*, 216–9t, *218*, 221t
  Copenhagen, St Clemens 197–208, *198*, *202*, *204–5*, *207*
  Gloucester, St Oswald 14, 211, *212*, 214–23, *215*, 216–9t, *218*, 221t
  Taunton, St Peter and St Paul 14, 211, *212*, 214–23, *215*, 216–9t, *220*, 221t, 234, 245
19th century
  Kilkenny Union Workhouse 8, 15, 249–60, *251*, *253*, *255–7*
ceramic
  burial container
    amphora 174
    jar 7, 10, 12, 81t, 85t, 85–7, 91–104, *93*, 94t, *94*, 112, 117, 119, 152, 153
    pithos 92, 149t, 157
  grave goods 9, 27, *27*, 50, 53, 66, 68, 72, 84–5, 113, 114, 127t, 130, *131*, 132, 133, 134, *135*, 136, 137, *138*, 139, 168, 170, 171, 172, 173, 174, 185t, 187–8t, 191, 239, 240–1t, *241*; *see also* jar burials
Chalcolithic 12, 92, 95
  Çamlıbel Tarlası, central Anatolia 12, 77–87, 78t, *79*, 81t, *84*, 85t
charcoal, inside skull 201, 207, *207*
'cherished child' 8, 9
childhood, concept of 2, 3, 4, 14, 108, 179, 192–3
child-rearing activities/child-care 6, 29, 121, 157
Christian/Christianity 14, 15 181, 204, 252
  baptism 204, 239–40
    chrisom cloth 239–40
  unbaptised 14, 204–5
  burials, *see* Ballyhanna *and* Kilkenny Union Workhouse
  church regulations 7
  grave goods, *see* Ballyhanna
  priest/erenagh 245, 246
  shrine 171, 239
cist 85t, 86, 112, 113, *114*, 150, 237
clothing 13, 141, 143
  accessories/fastening 13, 68, 115, 125, 130, 136, 139, 141, 143, 168, 170, 172, 191, 201
  gender distinctive 13, 125, 141
  hobnail 188t, 191
coffin 14, 201, 203, 205, 215, 216, 220, *220*, 222, 254, 255, 257, 258
  nails 216, *220*, 254
collective burial *see* double *and* multiple burial
commemoration 4, 6, 108, 119
comparison between child and adult burial 7, 8, 9, 12, 13–14, 15, 21, 28, 29, 38, 39, 43, 50–6, *52*, 64–6, 71–3, 77, 85–7, 102, 107, 109, 115, *128*, 129–30, 129t, *130*, 132–3, *134*, 136–8, *137*, 142, 143, 147, 150–2, 155–7, 163, 179, 183–4, 187–9, 203, 211, 216–9, 229, 235–46, 254–7, *257*
Copenhagen, St Clemens, Denmark 14, 197–208, *198*, *202*, *204–5*, *207*
  chronology 199–201, 208 n2
  grave goods and stones 197, 201, 205, *205*, 206–8
  grave markings 197, 200, 201
  grave preparation/inclusions 200, 201
  historical sources 197–9
  re-use of graves 200, 205
cosmology 118, 119
cremation 93, 163, 164–8, *164*, *167–8*, 170, 174, 175, 176, 181
Crete, Pre- and protopalatial Minoan Bronze Age 13, 147–58, *148*, 149t, *152*, *156*
  Agia Photia *148*, 151
  Archanes, Phourni *148*, 149t, 153, 154, 155, 157
  chronology 149t, 153, 155, 159 n4
  grave goods 153–4
  Hagios Charalambos cave *148*, 149t, 151, 155
  'house-tomb'/burial building 149t, 151, *152*, 153, 154, 156–7, *156*
  Kamilari *148*, 149t, 158
  Kephala Petras rock shelter 149t, 151, 155–6
  Knossos 151
  Lebena *148*, 149t, 157
  Livari Skaidi *148*, 149t, 150, 151, 154, 155
  Mallia, Pièrres Meulières *148*, 149t, 157
  Moni Odigitria *148*, 149t, 150, 152, 154, 155, 157
  Nopigeia *148*, 149t, 151, 155

Petras *149*, 150, 151
Pseira *148*, 149t, 150, 151, 155, 158
secondary burials 149t, 151-2, 153, 154, 155, 157
Sissi *148*, 149-50, 149t, 151, 153, 154, 155, 156-7, *156*, 158
tomb/*tholos* 150, 151-2, 153, 155, 157, 158
Zakros, Pezoules Kephala *148*, 149t, 157
cribra orbitalia 14, 211, 212, *212*, 213, 217, 217t, 219-21, 219t, 221t, 223
cult site 13, 163, 171-2, 175
'cultural child' 108

dairy products/milk 83, 250-1
churn 83
fatty acid residue 83
Dayaks, Indonesia 6, 28, 30
decomposition 98-100, 103, 236, 237
before burial 7, 10, 12, 39, 91, 100, 102, 173
demography /representation by age category 6, 11, 12, 13, 21, 23t, 28, 29, 30, 36, 39, 47, 49, 53, 58, 64, 78, 87, *128*, 133, 154, 167, 168, 174, 176, 181-4, *182*, 183t, 189, *202*, 203, 229, 230t, 257-8
missing/under-represented age categories 6, 12, 13, 21, 23, *23*, 28, 107-20, 110t, 147, 148, 151-2, 154, 155, 157-8, 181, 188-9, 258
see also age category *and* mortality profile/rate
depiction of children in ancient art 147-8, 159 n3
depth of burial 63, 133, 134, 136, 140, 172, 180, 181, 184, 189-90, 203, 229, 245
diet *see* nutrition
dentition 50, 61, 77, 82, 83, 95, 116-17, 132
dental attrition and pathology 82-3, 129
stress markers/indicators (inc. enamel hypoplasia) 14, 81t, 129, 211-23, *212*, *215*, 216-9t, *218*, *220*, 221t
disarticulated bone 80, 94, 98-9, *99*, 100, 102, 103, 115, 166, 173, 174, 199, 201
disease 2, 97, 117, 129, 206-7, 208, 212, 213-4, 222, 250-2, 259
anaemia 213
cholera 253-4
dental 14, 81t, 82-3, 129, 211-23, *212*, *215*, 216-9t, *218*, *220*, 221t
'famine fevers' 253
respiratory 250, 251
rickets 251
scurvy 221
Scheuerman's 235
tuberculosis 213, 222
see also infection
dismemberment 9, 10, 12, 91, 100-1, 100t, *101*, 102, 103, 171, 176
disturbance to burials 66, 81t, 149, 190, 201, 230
by ploughing 181
by animals 37, 64, 139, 181
by excavation 148, 149, 151, 153, 155, 157, 190
by later/secondary graves/burials 37, 149t, 151-2, 153, 154, 155, 157, 174, 190, 199-201, 203, 229, 234, 255-7
looting 134, 139, 148, 149
smashing/damaging of burial pots 85, 92
domestic/settlement context for burials 6, 9-10, 13, 19, 21, 24, 24t, *25*, 26, 28-30, 72-3, 77, 80, 85t, 86-7, 92, 95, 155, 163, 165-6, *165*, 170, 171, 174, *175*, 181
beneath house/floor 85t, 86, 87, 95, 104, 108, 112, 171, 181
extramural 12, 57-73, *59-60*, 62t, *63*, *65*, *67*, *69-70*, 82, 86, 87, 108, 113, 114-15, 117, 217, 220
foundation deposits 6, 29, 38, 174, 181, 189
intramural 10, 15, 36, 38, 39, 73, 86, 87, 107-20, 109-10t, *111-12*, *114*, *116*, 127t, 143, 159 n5, 217, 220; *see also* Kilkenny Union Workhouse
double burials 20, 26, 80, 82, 94t, 97, 98, 102, 103, 136, 143, 168, 170, 190
dye 112

economic activity of children 139, 142-4
enamel hypoplasia 14, 81t, 129, 211, 212-3, *212*, 216, 216t, 219-21, 219t, 221t, 223; *see also* dentition
environmental stress 37, 39
ethnicity 83, 107
ethnography 21, 44, 117, 142
extramural burials 12, 57-73, *59-60*, 62t, *63*, *65*, *67*, *69-70*, 82, 86, 87, 108, 113, 114-15, 117, 217, 220

fertility 6
foetus 102, 104, 110t, 113, 118, 133, 141, 184, 189 *see also* neonate/perinate *and* stillborn

forgetfulness/forgetting 179, 191–2, 258–9
fortress 13, 163, 172, 175
fostering 10
foundation deposit/burial 6, 29, 38, 174, 181, 189; *see also* domestic/settlement context for burial *and* house
funerary mound *see* barrow
    Kadruka, Sudan 9, 11–12, 43–54, *46*, 48t, 49, *51–2*
    stelae 117
funerary/mortuary/burial practices (general) 2, 3, 4, 6, 7, 11, 12, 13, 14, 15, 19–20, 21, 26–7, 28, 29–30, 36, 37, 38, 39, 41, 43, 44, 45, 50, 52, 54, 57, 58, 73, 77–8, 82, 83–4, 85–6, 87, 91, 92, 101, 102, 103, 104, 107, 108, 109, 111, 117, 118, 119, 140, 147, 150, 151, 152, 153, 155, 157, 159 n5, n6, 164, 173, 174, 175, 176, 180, 184, 185–6t, 189–90, 200, 208, 227, 228–9, 235, 239, 245, 246, 254
furness 78, 80

gender 2, 7, 8, 13, 30, 45, 72, 80, 104, 107, 139–42, 180, 184, 192–3, 200
    'gender identification' 2, 72, 139, 141
    gendered clothing 13, 125, 141
    gendered grave goods 7, 13, 30, 72, 118, 125, 136, 139–41, 143, 180, 184, 185–6t, 187–9, 191, 192–3
    marker 139–40
    socialisation 13, 125, 139, 143, 144
geophysics 61
Getae 163–4, 172
Geto-Dacian, Iron Age burials 9, 13, 163–76, *164–9*
    chronology 163–4, 166, 167, 168, 170
    Dulceanca *164–5*, 171
    Grădiștea *164–5*, 170–1
    Hunedoara, Grădina Castelului 9, *166*, 167, 168, *169*, 170, 172–3, 175, 176, 176 n2
    Mologa II *164–5*, 171–2
    Stolniceni *164*, *165*, 173, 176 n2
    Șuvița Hotarului, Căscioarele *165*, 170, 172
    Telița Celic Dere 9, *164*, *166*, 170, 175
GIS 11, 58, 61, 68–9, *69–70*, 71
Gloucester, St Oswald 14, 211, *212*, 214–23, *215*, 216–9t, *218*, 221t
gold 58, 181
    death mask 115

    diadem 115
    foil wrapping to body 8, 115, *116*
    model of scale 115
    pendant 114
grave *see* burial repository
    mass 8, 254, 255–7, *256*, 257t
grave goods 1, 6–9, 11–13, 14, 15, 21, 26–7, *27*, 37, 38, *40*, 41, 43, 45, 49, 52–3, 57, 58, *67*, *70*, 71, 72, 112, *112*, 113, 125, 131, *131*, 133, 133t, 140, 144 n4, 181, 184–9, 185–8t, 191–2, 200, 201, 227, 228, 230, 238–43, 240–1t, *241*, 244t, 246, 247 n4
    adult 8–9, 53, 72, 115, 134, 139, 140, 142, 153, 185t, 187, 192
    amulet/amuletic 7, 113, 130, 136, 139, 141, 184, 190, 191, 206, 207, 208
    apotropaic 8, 14
    bead 8, 9, 38, 43, 53, 66, 68, 72, 114, 130, *131*, 133–4t, 135, *135*, 136, 137t, *138*, 140, 141, 170, 184, 185–7t, 187, 188t, 239, 240–1t, 243, 254
    bone/tooth 62t, 66, 68, 102, 130, 131, 133–4t, 135, 136, 137t, 140, 141, 142, 167, 171, 172, 173
    bronze 9, 113, *131*, 132, 133t, 134t, *135*, 136, 137t, *138*, 140, 141, 142, 170, 181, 187, 188–9, 188t, 239, 240t, 242
    ceramics 9, 27, *27*, 50, 53, 66, 68, 72, 84–5, 113, 114, 127t, 130, *131*, 132, 133, 134, *135*, 136, 137, *138*, 139, 168, 170, 171, 172, 173, 174, 185t, 187–8t, 191, 239, 240–1t *241*
    chariot 127t, 144 n4
    clothes/clothing attachments/accessories 13, 115, *116*, 125, 136, 139, 141, 143, 168, 170, 172, 191, 201
    coin 188t, 201, 222
    comb 114, 201
    composite bow 131, 133t
    container 185t, 187, 188, 187–8t, 192
    figurine 113
    gaming piece (astragali) 130, 133–4t, *135*, 135t, 136, 137–8, 137t, 139, 140–1, 143
    gendered 7, 13, 30, 72, 118, 125, 136, 139–41, 143, 180, 184, 185–6t, 187–9, 191, 192–3
    gold 8, 114, 115, *116*, 181
    horn 131, 133t, 140, 144 n4
    horse harness 8, 127, 130

inappropriate to age 9, 30, 141, 168, 170, 175, 188-9, 191
iron 9, 170, 187-9, 188t, 191, 192, 239, 240-1, 240-1t, *241*, 242
ivory 112, 114
jewellery 9, 27, 113, 114, 134t, 137t, *138*, 141, 170, 181, 184, 185-6t, 187, 188, 188t, 191, 201, 254
medallion/Miraculous Medal 254, *255*
metal slag 78t, 239, 240-1t, *241*, 242
metal weapons 9, 127t, 130-1, 132, 133t, 135, 140, 141, 142, 144 n4, 167, 168, 170, 175, 180, 181, 184, 185-6t, 187-9, 188t, 191
necklace 68, 72, 113, 114, 153, 187, 191
neonate/perinate 37, *40*, 41, 131, 133t, 140, 144 n4, 188t, 239-40
ochre 27, *27*
ornament/adornment 114t, 127t, 130, 133t, 134t, *135*, 136, 137, 137t, 139, 140, 141, 143, 144 n5, 168, 172
pendant 66, 68, 72, 114, 130, 133t, 134t, 135, *135*, 137t, *138*, 140, 141
shell 8, 27, *27*, 38, 43, 53, 66, 113, 114, 134t, 135, 137t, 139, 141, 153
silver 9, 170, 181, 222
stone 8, 9, 14, 27, *27*, 30, 53, 66, 78, 92, 131, *131*, 132, 133, *135*, 140, 142, 170, 171, 200, 201, 203, 205, *204-5*, 206-8, 240t, 242, 243
tools 9, 127t, 130, 140, 142
wood and bone 53, 113
Great Chesterford, Anglo-Saxon burial ground 6, 9, 14, 6, 9, 14, 179-93, *182*, 183t, 185-8t
composition of cemetery 181, *182*
demography 181-4, *182*, 183t, 189
Grave #99 188-9, 191
grave goods 181, 184-9, 185-8t
non-normative burials 184-7, 185-6t, 190-1
Great Irish Famine 9, 15 *see* Kilkenny Workhouse
Gurgy, Les Noisats, France *20*, *22*, *28*,30

Halvay III, Thrace 9, *126*, 127t, 131, 132
horn
cheek-piece 131, 133t, 140, 144 n4
composite bow 131, 133t, 140
house 9, 29, 78t, 80, 85t, 86, 87, 95, 104, 118

burials associated with 85t, 86, 87, 95, 104, 108, 112, 171, 181
foundation deposit 6, 29, 38, 174, 181, 189
household 6, 57, 72, 118, 119, 172, 190
hunting/hunter 9, 30, 135, 214

identity 27, 43, 54
'artificial' 57, 73
embodied 8
gender 2, 7, 8, 13, 30, 45, 72, 80, 104, 107, 139-42, 180, 184, 192-3, 200
social/cultural/collective 2, 11, 12, 27, 35, 40-1/ 57, 72, 73, 107, 117-8, 119, 148, 151, 157, 159 n3, 246, 260
ideology 1, 142
infanticide 104, 190
infection/infectious disease 97, 212, 213-4, 219, 221
interment 8, 44, 87, 108, 118, 119, 132, 134, 137, 143, 191, 200, 228, 229, 235, 245-6, 255, *256*
after exposure 171, 173, 176
after storage 37, 38
rapid after death 7, 38, 39
intramural burial 10, 15, 36, 38, 39, 73, 86, 87, 107-20, 109-10t, *111-12*, *114*, *116*, 127t, 143, 159 n5, 217, *220*; see also Kilkenny Union Workhouse
Ireland 7
Ballyhanna, Co. Donegal 14
Great Irish Famine 9, 15
Kilkenny Union Workhouse 8, 15, 249-60, *251*, *253*, *255-7*
Iron 9, 170, 187-9, 188t, 192, 239, 240-1, 240-1t, *241*, 242
arrowhead 9, 170
bar 240t, 243
buckle 9, 170, 188, 188t
clamp 9, 170
coffin nails 216, *220*, 254
dagger/scabbard 9, 170
hobnail 188t, 191
knife 9, 170, 188, 188t, 191, 240-1, 240t, *241*, 242
nail 188, 188t, 191, 192, 239, 240-1t, *241*, 242
ore 78
shield 188
spearhead 188-9, 188t, 191

sword 119
Iron Age 5, 9, 141
    Getae 163-4, 172
    Geto-Dacian burials 9, 13, 163-76, *164-9*
    Greece 174
    Southern Thrace 172, 173-4
iron deficiency 213, 221
ivory
    comb 114
    figurine 147
    handle 112
    hairpin 114

jar burial 7, 10, 12, 81t, 85t, 85-7, 91-104, *93*, 94t, *94*, 112, 117, 119, 152, 153
jewellery 9, 27, 113, 114, 134t, 137t, *138*, 141, 170, 181, 184, 185-6t, 187, 188, 188t, 191, 201, 254

Kadruka, Sudan 9, 11-12, 23, 43-54, *46*, 48t, 49, *51-2*
Kamennyi Ambr-5, Thrace *126*, 127t, 129, 130, *130-2*
Kefkandi, Greece 109, 119t, 113, 117
Kilkenny Union Workhouse 8, 15, 249-60, *251*, *253*, *255-7*
    fever hospital 258
    poverty in 19th-century 249, 250-3
    social context of burials 257-9
    workhouse conditions 252-4
kinship/family 6, 8, 10, 12, 13, 31, 40, 41, 57, 72, 73, 118-9, 132, 137, 144, 155, 156-7, 158, 159 n8, 189, 190, 192, 205, 222, 235, 245-6, 258; *see also* family
    sibling 2, 10, 156-7, 242
Krivoe Ozero, Thrace *126*, 127t, 129, *130*, 133-4, *134*

Lerna, Greece 110t, 112-3
life stage 3, 4, 7, 14, 43, 44, 107, 129, 139, 157, 158, 179, 181-3, 183t, 189, 193, 229 *see also* age category *and* threshold/liminal age
Lisakovskiy, Thrace *126*, 127t, 129, 136, *137*
lithics *see* stone
location of burials/arrangement of burial ground 6, 9, 11, 12, 13, 19, 20, 21, 24-6, *25*, 28, 29, 37-9, 41, 43, 50-2, *52*, 58, 60, 71, 73, 82, 87, 95, 103, 107, 108, 109, 111-12, 119, 125, 129-30, 136, 137, 140, 143, 150-1, 170, 171-3, 175, 201, 214, 215-9, 216-8t, 255-6

magic/magic charms 6, 118, 242
manipulation of body 1, 2, 6, 7, 12 *see also* body processing
marriage 44, 141
massacre 28
medieval 7, 9, 14
    Ballyhanna, Co. Donegal 14, 227-47, *228*, 230-1t, *232*, 233-4t, 236t, *237*, 238t, 240-1t, *241*, 244t
    Canterbury, St Gregory 14, 211, 214-23, *212*, *215*, 216-9t, *218*, 221t
    Copenhagen, St Clemens 197-208, *198*, *202*, *204-5*, *207*
    Gloucester, St Oswald 14, 211, *212*, 214-23, *215*, 216-9t, *218*, 221t
    Taunton, St Peter and St Paul 14, 211, *212*, 214-23, *215*, 216-9t, *220*, 221t, 234, 245
megalithic tomb/dolmen 20, 22-3, 25
memory 15, 39, 41, 107, 119, 259; *see also* forgetfulness/forgetting
metallurgy 78, 80, 125, 130
    slag 78t, 239, 240-1t, *241*, 242
methods/methodology 2, 4, 7, 10, 11, 15, 21, 31, 36, 45, 48, 50, 61-2, 80, 94-5, 149, 150, 174, 181-2, 198, 199-200, 203, 208 n2, 229-30, 246
micro-excavation 10, 95, 98, 103
micro-stratigraphy 61
migration 3, 100
Minoan 13, *see* Crete
missing
    children/age categories 6, 12, 13, 21, 23, *23*, 28, 107-20, 110t,147, 148, 151-2, 154, 155, 157-8, 181, 188-9, 258
    bones 64, 91, 96, 97-8, 98t, 100, 101, 102-3, 170, 175
Modi, Troizinia, Greece 110t, 113, 115
mortality profile/rate 11, 21, 28, *49*, *63*, 72, 104, 109, 113, 129t, 133, 136, 183t, 189, 254-5, 258, 259; *see also* age-at-death *and* age category
mortuary practices, *see* funerary/mortuary/ burial practices
Mougins, Aven de Bréguières, France 21, *22*, 24
mourning/mourners 119, 140, 192
multiple burial 14, 20, *20*, 21, 23, 24-7, 31 n3, 80, 127t, 130, 133, 136, 139, 143, 150, 151-2, *152*, 153, 157, 170, 184, 185-6t, 190, 200-1, 203, 204, *204*, 205, *205*, *206-7*, *207*
    double *20*, 26, 80, 82, 94t, 97, 98, 102, 103, 136, 143, 168, 170, 190

mass grave 8, 254, 255–7, *256*, 257t
Mycenae 8, 115
    Grave Circle A 8 115
        Grave III 8, 115
            gold foil body adornment and gold objects 8, 115, *116*
    Grave Circle B 115
Mycenaean 7, 12–13, 107–20, 109–10t, *111–12*, *114*, *116*
    chronology 109, 109t, 113, 116
    sanctuary 113, *114*, 118

necropolis 9, 11–12, 21, 23, 167–70, 163, 165, 166, *166*, 168–9, 171, 172–3, 174, 175–6, *175*
    Hunedoara, Grădina Castelului 9, *166*, 167, 168, *169*, 170, 172–3, 175, 176, 176 n2
    Kadruka, Sudan 9, 11–12, 23, 43–54, *46*, 48t, 49, *51*–2
    Telița Celic Dere 9, *164*, *166*, 170, 175
Neolithic 4, 11, 95, 159 n4, n5
    Çatalhöyük, Central Anatolia 6, 7, 9, 11, 35–41, *36*, *38*
    Early 21, *22*, 23–4t, 24, *25*, 28, 29
    Eneolithic 58–9
        Sultana-Malu Roşu, Romania 6, 8–9, 12, 57–73, *59*–60, 62t, *63*, *65*, *67*, *69*–70
    France 9, 19–31, *22*, 23–4t, *25*
        Cerny Culture 30
    Kadruka, Sudan 9, 11–12, 43–54, *46*, 48t, 49, *51*–2
    Late 21, *22*, 23, 23–4t, *25*, 26–7, *27*, 28
    Lepenski Vir, Serbia 29
    Middle 21, *22*, 22, 23–4t, 24–6, *25*, 26–7, *27*, 29, 30
    Talheim, Germany 28
    Varna I, Bulgaria 58
    Vlasac, Serbia 29
neonate/perinate 7, 9, 11, 24, 25, 28, 29, 30, 35–41, *36*, *38*, *40*, 48t, 50, 61, 62–4t, 68, 69, 71, 80, 81t, 82, 83, *84*, 96, *96*, 97, 101, 102, 104, 108–11, 110t, 113, 115, 116, 117, 133, 143, 151, *152*, 153, 154, 155, 157, 183–4, 190, 204, 212, 229, 230–1, 230–1t, 233, 234–5, 235t, 236t, 238t, 239, 240–1t, 243, 244t
    binding/strapping 37
    body position 37, 38–9, 50–1, *51*, 81t, *204*, 232–4, 233–4t, 236t, 238t, 243
    grave goods 37, *40*, 41, 131, 133t, 140, 144 n4, 188t, 239–40
    stillborn 11, 49, 96, 117, 190, 204, 207

nutrition/nutritional disorder/malnutrition 83, 97, 212–3, 214, 221, 222, 229, 250–2, 259
    starvation/poverty 249, 250–2
'osteological child' 108
Osteological Paradox 11
osteoarchaeology/osteology/osteologist 2, 11, 12, 13, 21, 45, 50, 61, 78, 80, 87, 95, 102, 103, 147, 150, 184, 192, 200, 201, 203, 204, 206, 227, 229
oven, clay 80

palaeoecological study 61
Papuans, New Guinea 6, 28
parasite 212, 213, 221
pathology/palaeopathology 12, 36, 64, 77, 80, 81t, 82–3, 97, 117, 118–9, 128–9, 170, 172, 211–23, *212*, *215*, 216–9t, *218*, *220*, 221t
    cribra orbitalia 14, 211, 212, *212*, 213, 217, 217t, 219–21, 219t, 221t, 223
    enamel hypoplasia 211, 212–3, 216, 216t, 219–21, 219t, 221t, 223; *see also* dentition
    periostitis 14, 211, 212, 213–4, 217–9, 218t, *218*, 219–21, 221t, 219t, *220*, 222, 223
pendant 66, 68, 72, 114, 130, 133t, 134t, 135, *135*, 137t, *138*, 140, 141
    tooth pendant/amulet (drilled fang) 130, 133–4t, 135, *135*, 136, 137, 137t, 140, 141
    bronze *135*, *138*, 140
    gold 114
    stone 66
periostitis 14, 211, 212, 213–4, 219–21, 219t, *220*, 221t, 222, 223
Petrovka culture 13, 125, 126–9, *126*, 127t, *128*, 129t, 130, 133–6, 134–5, 134t, 136, 137, *138*, 143, 144
    Krivoe Ozero *126*, 127t, 129, *130*, 133–4, *134*
    Stepnoye VII 127t, 133, *134*, 136, 137–8, *138*
physiological stress 2, 11, 14, 208
phytoliths 38
pit, burial 9, 12, 13, *20*, 23, 26, 28, 60, 62–3, 62t, 66, 71, 83, 86–7, 91, 93, 94t, 95, 96, 100, 103, 112, 130, 132, 133, 134, 140, 163, 165, *165*, 170, 171, 173, 174, *175*, *175*, 181, 208, 234, 249, 250, 254, 255, *256*–7, 259
    pit field 163, *165*, 170, 171–2, 173, 175, *175*
    with wooden ceiling 9, 132–3, 134, 137
plant remains 101–2
'political child' 9–10
'potential child' 8, 9
puberty 44, 54

radiocarbon dates 60, 71, 228
rebirth/regeneration 7, 113, 118–9
religion/religious belief 1, 29, 44, 118, 119, 143, 204, 206, 227, 239, 245–6
   sanctuary 9, 113, *114*, 118
   shrine 171, 239
   see also Christian/Christianity
responses to child death 2, 8, 11, 14, 15, 31, 35, 36, 37–41, 43–4, 57, 73, 140, 148, 150, 158, 175, 179, 180, 191, 193, 197, 200, 205, 207–8, 227–8, 229, 245–6, 254, 258–9
   cultural 11, 12
   emotional 2, 3, 11, 14, 31, 35, 36, 37–41, 57, 148, 150, 158, 191, 197, 207–8, 227–8, 245–6, 258–9
rite of passage 118, 151
ritual (general) 7, 8–9, 11, 12, 13, 14, 29, 44, 47, 57, 58, 71, 72, 73, 92, 102–4, 117, 118, 133, 143, 144 n4, 148, 159 n2, 163, *164*, 167, 171, 172, 173, 180, 191–2, 193, 197, 207–8, 227, 243, 246
   'ritual child' 9
   taboo/exclusion rite 117
Roman/Romano-British 104, 117, 176, 181
   coin 188t

sacrifice
   animal 127t, 130, 133, 135, 136, 138–9, 143, 144 n4
   human 29, 44, 54 n5, 174, 175
settlement site 6, 13, 35, 71, 78, 127, 127t, 163, 165–6, *165*, 173, 175, *175*
   fortified 58, 127, 127t, 163, 172, 173, 175
   tell 10, 58, 59, 60, 72, 91–104, *93–4*, 94t, *95*, 98t, *99*, 100t, *101*, 155
   see also domestic/settlement context for burials
sex 2, 19, 39, 45, 48, 80, 104, 118, 120–1, 180, 183t, 184, 188, 192–3, 237; see also gender
shell 27, *27*, 38, 43, 53, 66, 113, 114, 134t, 135, 137t, 139, 141, 153
   bead 8, 38, 53, 66
   conch 114
   Dentalium 66, 68
   Lithoglyphus naticoides 66
   necklace 153
   ostrich egg 53
   snail 8
   Spondylus 8, 66, 68

spoon 43
shroud 201, 208, 229, 254
siblings/biological relations see kin/kinship and family
silver 181
   coin 222
   earring 9, 170,
Sintashta Culture 7, 13, 125, 126–9, *126*, 127t, *128*, 129–33, *130-2*, 129t, 133t, 136, 138, 139–42
   grave goods 127t, 130, 140–1
   Halvay III 9, *126*, 127t, 131, 132
   Kamennyi Ambr-5 *126*, 127t, 129, 130, *130-2*
   Krivoe Ozero *126*, 127t, 129, *130*, 133–4, *134*
   Stepnoye-1 *126*, 127t, 130–1, *130*
social/cultural change and development 29, 58, 61, 92, 118, 214
social organisation/structure 1, 8, 11, 13, 28, 41, 47, 117, 118–9, 125, 143–4, 148, 150, 154–5, 157, 158, 180
social role of children 3, 19, 28, 158
southern Trans-Urals 7, 13, 125–45, *126*, 127t, *128*, 129t, *130-2*, 133–4t, *134-5*, *137-8*, 137t
   Alakul' culture 7, 13, 125, 126–9, *126*, 127t, *128*, 129t, 130, 133, 136–9, *137-8*, 137t, 141, 143, 144
   barrow cemeteries 127t, 136, 140
   chronology 126
   fortified settlements 127t
   Petrovka culture 13, 125, 126–9, *126*, 127t, *128*, 129t, 130, 133–6, *134-5*, 134t, 136, 137, 138, 143, 144
   Sintashta culture 7, 13, 125, 126–9, *126*, 127t, *128*, 129–33, *130-2*, 129t, 133t, 136, 138, 139–42
   Srubnaya culture 126, 127, 130, 144 n3
   Yamnaya cultures 126, 130
spatial analysis 1, 6, 68–71, *69-70*
spoon
   ceramic 43, 53
   shell 43, 53
stable isotopes 82–3
status/social/economic status 2, 4, 8–9, 11, 13, 14, 19–20, 21, 30–1, 35, 39–40, 47, 58, 72, 107, 112, 113, 117–9, 132, 143, 159 n8, 170, 173, 175, 181, 189, 192, 197, 200, 201, 211, 213, 214, 216, 221–3, 245–6
stelae 117
Stepnoye-1, Thrace *126*, 127t, 130–1, *130*

Stepnoye VII, Thrace 127t, 133, *134*, 136, *137-8*, 138
stillborn 11, 49, 96, 117, 190, 204, 207
stone 14, 140, *152*, 171
   arrowhead 9, 30, 131
   bead 8, 66, 112, *135*
   carnelian 112
   debitage 80
   grinding stone 9, 170
   malachite 66
   marble 66
   non-artefact in grave 14, 78, 200, 201, *204*, 205, *205*, 206-8, 239
   quartz 239, 240-1t, 241, *241*
   pendant, 66
   'pillow' 8, 203, *242*, *243*
   polished axe head 53
   seal stone 113
   slab/anvil *131*
   tool 9, 132, 133t, 140, 142
   tool kit 30
stress 18, 39, 254
   environmental 37, 39
   markers/indicators 14, 211-23, *212*, *215*, 216-9t, *218*, *220*, 221t; see also dentition and pathology/palaeopathology
   psychological 2, 11
Sultana-Malu Roşu, Eneolithic cemetery, Romania 6, 8-9, 12, 57-73, *59-60*, 62t, *63*, *65*, *67*, *69-70*
superstition 197, 207-8
symbolism 12, 19, 28, 29, 30, 44, 57, 58, 72, 73, 82, 107, 108, 113, 115, 119, 139, 141, 172, 189, 191-2, 206, 207

taboo/exclusion rite 117
taphonomy 5, 10, 28, 37, 45, 64, 103, 148, 154
Taunton, St Peter and Paul 14, 211, *212*, 214-23, *215*, 216-9t, *220*, 221t, 234, 245
tell site 10, 58, 59, 60, 72, 91-104, *93-4*, 94t, *95*, 98t, *99*, 100t, *101*, 155
theoretical perspective/approaches 1-2, 4

Thrace, Bulgaria
   Early Bronze Age 7, 9, 10, 12-13, 91-104, *93-4*, 94t, *95*, 98t, *99*, 100t, *101*, 125-45, *126*, 127t, *128*, 129t, *130-2*, 133-4t, *134-5*, *137-8*, 137t
   Iron Age 172, 173-4
   Malko Tarnovo 172, 173
threshold/liminal age 21, 43-4, 53, 54, 114, 116-18, 157; see also age category
tomb 127t, 129-30, 132-3, 136, 150
   barrow/tumulus 127t, 131, 132, 136, 140, 166, 173
   chamber 114, 115, 118
   'house-tomb' 149t, 151, *152*, 153, 154, 156-7, *156*
   megalithic tomb/dolmen *20*, 22-3, 25
   rock-cut 150
   tholos 150, 151-2, 153, 155, 157, 158
toys 2, 189, 241-2
transition between life stages see threshold age
trauma 2, 37, 97, 102, 213
   birth 97, 214
   social 250, 257-9
tree trunk/tree burial 6, 28
trepanation 170
tumulus see barrow

violence 129, 171, 172, 174, 175

weapons
   arrowhead 9, 30, 130-2, 133t, 140, 144 n4, 170
   composite bow 131, 133t
   metal 9, 127t, 130-1, 132, 133t, 135, 140, 141, 142, 144 n4, 167, 168, 170, 175, 180, 181, 184, 185-6t, 187-9, 188t, 191
women 26, 28, 72, 170, 175, 180, 181, 189, 190, 200, 201, 215, 242

Yamnaya cultures 126, 130